T0137438

Springer Theses

Recognizing Outstanding Ph.D. Research

Aims and Scope

The series "Springer Theses" brings together a selection of the very best Ph.D. theses from around the world and across the physical sciences. Nominated and endorsed by two recognized specialists, each published volume has been selected for its scientific excellence and the high impact of its contents for the pertinent field of research. For greater accessibility to non-specialists, the published versions include an extended introduction, as well as a foreword by the student's supervisor explaining the special relevance of the work for the field. As a whole, the series will provide a valuable resource both for newcomers to the research fields described, and for other scientists seeking detailed background information on special questions. Finally, it provides an accredited documentation of the valuable contributions made by today's younger generation of scientists.

Theses are accepted into the series by invited nomination only and must fulfill all of the following criteria

- They must be written in good English.
- The topic should fall within the confines of Chemistry, Physics, Earth Sciences, Engineering and related interdisciplinary fields such as Materials, Nanoscience, Chemical Engineering, Complex Systems and Biophysics.
- The work reported in the thesis must represent a significant scientific advance.
- If the thesis includes previously published material, permission to reproduce this must be gained from the respective copyright holder.
- They must have been examined and passed during the 12 months prior to nomination.
- Each thesis should include a foreword by the supervisor outlining the significance of its content.
- The theses should have a clearly defined structure including an introduction accessible to scientists not expert in that particular field.

More information about this series at http://www.springer.com/series/8790

Julián Villamayor

Influence of the Sea Surface Temperature Decadal Variability on Tropical Precipitation: West African and South American Monsoon

Doctoral Thesis accepted by
Universidad Complutense de Madrid,
Madrid, Spain

Springer

Author
Dr. Julián Villamayor
Department of Earth Physics
and Astrophysics
Universidad Complutense de Madrid
Madrid, Spain

Supervisor
Prof. Elsa Mohino
Universidad Complutense
de Madrid
Madrid, Spain

ISSN 2190-5053 ISSN 2190-5061 (electronic)
Springer Theses
ISBN 978-3-030-20329-0 ISBN 978-3-030-20327-6 (eBook)
https://doi.org/10.1007/978-3-030-20327-6

This Springer imprint is published by the registered company Springer Nature Switzerland AG
The registered company address is: Gewerbestrasse 11, 6330 Cham, Switzerland

To my parents and Jessi, for their unconditional support.

Supervisor's Foreword

The Earth's climate is a fascinating system made of multiple and complex interactions among its constituents the atmosphere, hydrosphere, cryosphere, lithosphere, and the biosphere. The awareness of climate change has put forward to society that the climate system is far from static and that it rather presents variability at very different timescales. In particular, climate can vary from one decade to the next, which can have important societal implications. Being able to foresee these changes would undoubtedly be of great value, as it would have applications in long-term planning. A necessary first step in this direction is enhancing our understanding of climate variability at decadal-to-multidecadal timescales, which is the general objective of this Ph.D. Thesis.

The continental regions in the Tropical Atlantic are among the most affected by changes in their rainfall regime. Water availability in the semiarid region of Northeast Brazil in South America affects the economy and shapes the population in the region. Also in South Africa, the Amazon's economic and ecological resources highly depend on rainfall. These regions are affected by the seasonal occurrence of the South American Monsoon and by its climate variability. In turn, to the east, the West African Monsoon modulates rainfall over the Sahel. This region is prone to drought occurrence, which has devastating humanitarian and economic impacts. The long-lasting Sahel drought in the 1980s stirred a great scientific effort to better understand its causes. Through its impact on the economy, rainfall variability at decadal timescales in these regions can also lead to stress in local population and migrations, affecting indirectly populations in other regions.

One of the main sources of tropical rainfall variability at decadal-to-multidecadal timescales is the change in the sea surface temperatures (SST). Previous research suggested links among changes in the rainfall regimes of the Sahel, Northeast Brazil, and Amazon regions and patterns of decadal-to-multidecadal variability in SSTs. However, the mechanism underlying such links and their simulation by state-of-the-art general circulation models had not been addressed in a systematic way, which constitutes the specific objective of this Ph.D. Thesis.

In this Thesis, Julián analyses the decadal-to-multidecadal patterns of SST variability simulated by 17 state-of-the-art general circulation models and evaluates their impact on rainfall associated with the South American and the West African Monsoons and the mechanisms underlying such impact. The analysis allows Julián to delve into issues like the effect of the external forcing in shaping this impact, the changes of the impacts under a climate change scenario with a strong increase of these forcings, and the relative contribution of the different SST variability patterns to the decadal variability of rainfall. Finally, as a case study, Julián analyses the late nineteenth century anomalously humid period in the Sahel, which has been little documented due to the scarcity of observations. By performing simulations with an atmospheric general circulation model driven by observed sea surface temperatures as boundary conditions, Julián is able to pinpoint the oceanic basin responsible for this humid period and explain the mechanism causing it.

Throughout his Thesis, Julián has been supported and advised by different internationally renowned researchers. For the analysis of the South American Monsoon, he enjoyed a 3-month visit at the University of São Paulo (Brazil) under the supervision of Tércio Ambrizzi, expert on climate variability in the region. For the West African case study, Julián collaborated with Myriam Khodri, Juliette Mignot, and Serge Janicot, experts in climate modeling, decadal variability, and the West African Monsoon, at the Institute Pierre Simon Laplace (IPSL) in Paris (France).

This Ph.D. Thesis constitutes a step forward in our understanding of changes in rainfall regimes in the regions studied and is also of great use to improve decadal prediction systems and to foresee the societal impacts of such changes. I, therefore, invite you to immerse yourself in this fascinating reading.

Madrid, Spain Elsa Mohino
March 2019

Abstract

Introduction

The Sahel is the semiarid West African region between the Sahara desert and the wet tropical savanna. The Sahel rainfall depends on the West African Monsoon (WAM) system and peaks between July and September. The rainfall regimes of the Amazonia and Northeast regions, located in northern Brazil, depend on the South American Monsoon system. The Amazonia is the region covered by the Amazon River basin, where heavy rains occur throughout the year but with a rainier season extending from December to May. The Northeast is a semiarid region with a short rainy season between March and May.

Precipitation regimes in these three regions have undergone changes over time with important humanitarian, environmental and economic consequences and have been a major topic of study (e.g., Rodríguez-Fonseca et al. 2015; Zhou and Lau 2001; Marengo et al. 2016). At decadal-to-multidecadal timescales, these changes have been mainly associated with the global sea surface temperature (SST) variability. Particularly, the Sahel precipitation has been associated with the global warming (GW), the Atlantic Multidecadal Variability (AMV), and the Interdecadal Pacific Oscillation (IPO) modes of decadal-to-multidecadal SST variability (e.g., Mohino et al. 2011a). The Amazonia and Northeast rainfall changes have been related to the Pacific and the Atlantic SST variability at decadal timescales (e.g., Grimm and Saboia 2015), which is led by the AMV and IPO.

Climate study through Global Circulation Models (GCMs) is crucial to understand climate changes and assessing its effects. So, in the first part of this Thesis, a multi-model analysis is done addressing the influence of the main decadal-to-multidecadal modes of SST variability on precipitation in the Sahel, Amazonia, and Northeast using different GCMs simulations from the 5th phase of the Coupled Model Intercomparison Project (CMIP5) (Taylor et al. 2012).

Particularly, in the Sahel, the decadal-to-multidecadal precipitation variability along the recent past and even for the future has been extensively studied, but barely prior to the twentieth century. Only a few studies suggest that the Sahel experienced a long wet period throughout the late nineteenth century. This motivates the second part of this Thesis, which seeks to reproduce this period with an atmospheric GCM (AGCM) forced with observed SST since 1854.

Objectives

The objective of this Thesis is to achieve a better understanding of the SST decadal-to-multidecadal variability on rainfall in the Sahel, Amazon, and Northeast regions. For that purpose, a multi-model analysis is done aiming to characterize the main modes of SST variability (GW, AMV, and IPO) in observations and CMIP5 simulations, assess their impacts on precipitation in the three regions and the causes of such links. Other goals are to seek whether these links are expected to change in the future, discuss an eventual role of the radiative forcing on the AMV and IPO, and assess the contribution of the SST modes to the total decadal-to-multidecadal rainfall variance in the regions of interest. A final objective is to find out whether the long rainy period of the late nineteenth century can be reproduced with an AGCM forced with observed SST and the factors that caused it.

Data and Methodology

Monthly data from different simulations of 17 CMIP5 models are used. The simulations analyzed are historical (simulates the recent past with imposed observed external radiative forcing), piControl (radiative forcing is fixed to pre-industrial values), RCP8.5 (future projections with a representative concentration pathway of high concentrations of greenhouse gases), and historicalGHG (similar to the historical simulation but with greenhouse gas forcing only) (Taylor et al. 2012). For the sake of robustness in the observational results, different SST and precipitation databases and reanalyses are analyzed.

A set of simulations is performed using the fifth version of the *Laboratoire de Météorologie Dynamique* (LMDZ) AGCM (Hourdin et al. 2013). In the first set of simulations, the LMDZ is run with imposed observed boundary conditions over 1854–2000. Second, a set of sensitivity experiments has been done for 1854–1910 imposing full variability of the SST only in the Atlantic or in the Indo-Pacific while the rest is fixed to the climatological seasonal cycle.

The methodology used is based on mathematical tools commonly used in climate studies, such as EOF, linear regression, and correlation analysis and filtering of time series, among others.

Results and Conclusions

The results are presented in two parts:

1. The first part shows the results of the multi-model analysis. CMIP5 models, on average, can reproduce the main observed features of the GW, AMV, and IPO and their impacts. The main results and conclusions obtained are:

   - The GW has been prone to aerosol changes in the recent past. This induces inter-model differences but does not affect the way CMIP5 models, on average, reproduce the rainfall response: a drying in the Sahel and more precipitation in Amazonia and Northeast of Brazil. The GW reduces the WAM low-level circulation in response to a tropical SST warming. It also enhances convection over northern Brazil through anomalous Walker circulation related with the tropical Pacific SST anomalies (SSTA) in observations. But CMIP5 models fail in reproducing the tropical Pacific SSTA in the GW pattern, affecting the reliability of the simulated precipitation response in northern South America.
   - During positive AMV phases, the Sahel and Amazonia precipitation are enhanced and reduced in the Northeast (the opposite during negative phases). Positive (negative) AMV induces interhemispheric pressure gradient promoting anomalous northward (southward) shifts of the Intertropical Convergence Zone.
   - The IPO has negative impact on rainfall in the three regions. Positive (negative) IPO produces Walker circulation anomalies from the tropical Pacific with anomalous subsidence (rise) over West Africa and northern South America.

The results also show that the aerosol radiative forcing effects induce inter-model uncertainties as to the simulated AMV, which shows slight differences between historical and piControl. The IPO signal, instead, shows no noticeable differences. This suggests that the AMV may have a component of external forcing, while the IPO is dominated by internal variability.

The RCP8.5 future projections reveal a different GW pattern and impacts on rainfall to the historical simulations. However, they show similar AMV and IPO behaviors. This suggests that changes in the precipitation response to the GW in the three regions studied are expected under the future scenario described by the RCP8.5 projections, but not in the case of the AMV and the IPO.

A multi-linear regression analysis between the GW, AMV, and IPO indices and the precipitation index of each region show that CMIP5 models, in general, do not reproduce the observed contribution of each mode of SST to the total decadal-to-multidecadal rainfall variability.

The proper simulation of the decadal-to-multidecadal rainfall variability in the regions studied is related to the correct SSTA distribution in the simulated patterns and with the monsoon atmospheric circulation sensitivity to the SST changes.

2. The second part of results shows that the LMDZ model reproduces a long Sahel rainy period in the late nineteenth century in response to observed SST forcing since 1854. The sensitivity experiments show that the Atlantic SST plays a dominant role inducing such a precipitation enhancement through enhanced convection over the Sahel and more moisture supply from the tropical Atlantic.

References

Rodrguez-Fonseca, B., Mohino, E., Mechoso, C. R., Caminade, C., Biasutti, M., Gaetani, M., Garcia-Serrano, J., Vizy, E. K., Cook, K., Xue, Y., et al.: Variability and Predictability of West African Droughts: A Review on the Role of Sea Surface Temperature Anomalies, J. Clim., **28**, 4034–4060, (2015)

Zhou, J., Lau, K. M.: Principal Modes of Interannual and Decadal Variability of Summer Rainfall Over South America, Int. J. Climatol., **21**, 1623–1644, 10.1002/joc.700, (2001)

Marengo, J. A., Torres, R. R., and Alves, L. M.: Drought in Northeast Brazil–Past, Present, and Future, Theoret. Appl. Climatol., pp. 1–12, 10.1007/s00704-016-1840-8, (2016)

Mohino, E., Janicot, S., and Bader, J.: Sahel Rainfall and Decadal to Multi-Decadal Sea Surface Temperature Variability, Clim. Dyn., **37**, 419–440, 10.1007/s00382-010-0867-2, (2011a)

Grimm, A. M., Saboia, J. P.: Interdecadal Variability of the South American Precipitation in the Monsoon Season, J. Clim., **28**, 755–775, (2015)

Taylor, K. E., Stouffer, R. J., and Meehl, G. a.: An Overview of CMIP5 and the Experiment Design, Bulletin of the American Meteorological Society, **93**, 485–498, (2012)

Hourdin, F., Foujols, M.-A., Codron, F., Guemas, V., Dufresne, J.-L., Bony, S., Denvil, S., Guez, L., Lott, F., Ghattas, J., et al.: Impact of the LMDZ Atmospheric Grid Configuration on the Climate and Sensitivity of the IPSL-CM5A Coupled Model, Clim. Dyn., **40**, 2167–2192, (2013)

Parts of this thesis have been published in the following journal articles:

- López-Parages, J., **Villamayor, J.**, Gómara, I., Losada, T., Martín-Rey, M., Mohino, E., … & Suárez, R. (2013): Nonstationary interannual teleconnections modulated by multidecadal variability. *Física de la Tierra*, **25**, 11–39. http://dx.doi.org/10.5209/rev_FITE.2013.v25.43433
- **Villamayor, J.** & Mohino, E. (2015): Robust Sahel drought due to the Interdecadal Pacific Oscillation in CMIP5 simulations. *Geophys. Res. Lett.*, **42**, 1214–1222.
http://dx.doi.org/10.1002/2014GL062473
- Rodríguez-Fonseca, B., Suárez-Moreno, R., Ayarzagüena, B., López-Parages, J., Gómara, I., **Villamayor, J.**, … & Castaño-Tierno, A. (2016): A Review of ENSO Influence on the North Atlantic. A Non-Stationary Signal. *Atmosphere*, **7**, 87.
https://doi.org/10.3390/atmos7070087
- **Villamayor, J.**, Ambrizzi, T. & Mohino, E. (2018a): Influence of decadal sea surface temperature variability on northern Brazil rainfall in CMIP5 simulations. *Clim. Dyn.*
https://doi.org/10.1007/s00382-017-3941-1
- **Villamayor, J.**, Mohino, E., Khodri M., Mignot, J. & Janicot, S. (2018b): Atlantic Control of the Late Nineteenth-Century Sahel Humid Period. *Journal of Climate*, **31**, 8225–8240.
https://doi.org/10.1175/JCLI-D-18-0148.1

Acknowledgements

It's funny to remember the day that Belén invited Rober and I to her office to offer to join the Tropa group, where I met Elsa and heard the concept of *decadal variability* for the first time. Rober and I left there jumping for joy because we were going to be paid for keep studying and asking "dude, what is *cadal*?". That's how it all started.

That's why I want to heartily thank Belén, for giving us such a great opportunity, and Elsa, for being keen to guide my work and for teaching me everything I know about research. I have always thought that teaching is one of the most beautiful and worthwhile things that can be done in life. That's why I want to especially recognize in these acknowledgements the enormous work of Elsa that is behind this Thesis.

Thanks to all the Tropa Family for all these fun years. Martuki, you are the best colleague that one can have, as well as a great pastry and *bichito* caretaker. Thank you because your perseverance and passion for what you do have inspired me more than once. Jorge and Íñigo (Altintop!), I have'd really great moments with you. Antonio, good luck with your Thesis and thanks for your willingness to always lend a hand. Tere, I'm very happy for your recent and well-deserved successes. Irene, although we have not coincided much, I also remember you. Belén, you're a fantastic group leader. Your vitality is contagious and it can be felt in the environment of Tropa. Elsa, thanks again not only for being a great boss and director of my Thesis but also for being a fantastic colleague. To my dearest friend Coumba. You are one of the most interesting people I know. I miss our conversations mixing Spanish, French, and English. Ibrahima (P.A.), my best *Senegañol* friend. It is good luck to meet people like you. I hope to see you both again soon and often. To all the colleagues of Tropa in general, thank you very much because I have learned a lot from you and enjoyed your company. Long live to Tropa!

Rober, my *compi* par excellence. Official *tropaellero*. You went from being a nice guy from the faculty and my notes dealer during the university years to being my partner and friend during the master to end up being inseparable during the Ph. D. Great friend and better (if possible) composer of music hits. I don't know what the future holds. I think a post-doc together would be too much to ask. Or not… ? Who knows? It has been a pleasure, Dr. Suárez.

I don't forget the rest of good friends and colleagues who don't belong to the Tropa group but do to the Family: Jesús, Ade, Mariano, Jon, Luís, Ana, Javi, Marta, Álvaro, and Blanca. Pare, with you I have also got along very well from the beginning, as if we were already friends from before and now you are one of the important ones. By the way, we have many outstanding family plans.

Many thanks to Tercio and Iracema for your help during my stay at the USP and to Myriam, Juliette, and Serge for your warm welcome at LOCEAN. I also appreciate the important work of the external reviewers and members of the tribunal of this Thesis.

Also thanks to my lifelong friends, los Alamierdos, without whom I could have done the Thesis exactly the same, although these years would have been much more boring. Seriously, thanks for always supporting me and being my escape route.

Finally, I want to thank my family for the support they have always given me. In particular, thank you Mom and Dad for instilling in me the importance of the studies. Both, together with Lucía, are the best reference I could have had. Jessi, I don't know what I would have done without you these months. I can't say enough. Thank you. Much of the effort put into this book is your merit. You are my most powerful engine. Miguel, thank you for being so respectful with my times before being born and for making being your father so easy. You are a good son.

Financial Support and Others

This Ph.D. has been supported by the scholarship BES-2013-063821 (*Ayudas para contratos predoctorales para la formación de doctores*) from the Spanish Ministry of Economy and Competitiveness (MINECO), within the national project MULCLIVAR (CGL2012-38923-C02-01-MINECO). The research leading to the results of this thesis has also received funding from the European project PREFACE (EUFP7/2007–2013 Grant Agreement 603521). Part of the results of this thesis has been obtained as a result of two 3-month stays at *Universidade de São Paulo* (Brazil) and *Institut Pierre Simon Laplace* (IPSL) in Paris (France), both funded by the grants for short stays EEBB-I-15-09241 and EEBB-I-16-10979 of the MINECO, respectively.

Thanks to the World Climate Research Programme's Working Group on Coupled Modelling, which is responsible for CMIP, and the climate modeling groups for producing and making available their model output. For CMIP, the U.S. Department of Energy's Program for Climate Model Diagnosis and Intercomparison provides coordinating support and led development of software infrastructure in partnership with the Global Organization for Earth System Science Portals.

Contents

Acronyms

20CRV2c	NOAA-CIRES 20th Century Reanalysis version 2c
AEJ	African Easterly Jet
AGCM	Atmospheric Global Circulation Model
AMO	Atlantic Multidecadal Oscillation
AMOC	Atlantic Meridional Overturning Circulation
AMV	Atlantic Multidecadal Variability
ANOVA	Analysis of variance
AOGCM	Atmosphere-Ocean Global Circulation Model
ASWI	African Southwesterly Index
ATLVAR	Refers to the sensitivity experiment performed with the LMDZ model in which the Atlantic SST variability imposed
CMIP5	Coupled Model Intercomparison Project phase 5
CRU TS3.24.01	Climatic Research Unit time series version 3.24.01
DJFMAM	Refers to the season from December to May
ENSO	El Niño/Southern Oscillation
EOF	Empirical Orthogonal Function
ERA-20C	European Center for Medium-Range Weather Forecasts reanalysis of the 20th Century
ERSST.v4	Extended Reconstructed Sea Surface Temperature version 4
GCM	Global Circulation Model
GHGs	Greenhouse gases
GPCC v7	Global Precipitation Climatology Centre Version-7
GW	Global Warming
HadISST1	Hadley Center sea ice and sea surface temperature version 1
HLVP	High-level velocity potential
ICOADS	International Comprehensive Ocean-Atmosphere Dataset
INPVAR	Refers to the sensitivity experiment performed with the LMDZ model in which the Indo-Pacific SST variability is imposed
IPCC	Intergovernmental Panel on Climate Change
IPO	Interdecadal Pacific Oscillation

IPSL	*Institut Pierre Simon Laplace*
ITCZ	Intertropical Convergence Zone
JAS	Refers to the season from July to September
LMDZ	*Laboratoire de Météorologie Dynamique* "Zoom"
LLW	Low-level westerlies
NAO	North Atlantic Oscillation
OGCM	Ocean Global Circulation Model
PC	Principal Component
PDO	Pacific Decadal Oscillation
RCP8.5	Representative Concentration Pathway 8.5
REF	Refers to the simulations of reference performed with the LMDZ model in which all the SST variability is imposed
SACZ	South Atlantic Convergence Zone
SAH	Sahel precipitation
SALLJ	South American Low Level Jet
SAM	South American Monsoon
SPG	Surface pressure gradient
SST	Sea surface temperature
SSTA	Sea surface temperature anomalies
TEJ	Tropical Easterly Jet
THC	Thermohaline circulation
UDEL v4.01	University of Delaware Air Temperature and Precipitation version 4.01
WAM	West African Monsoon
WAWJ	West African Westerly Jet

Part I
Introduction to the Thesis

Chapter 1
Introduction

Abstract This Ph.D. Thesis addresses the influence that the sea surface temperature (SST) has on the tropical precipitation changes at time scales from one to several decades, which is typically referred to as decadal-to-multidecadal variability. Specifically, the focus of this work is on the effect on rainfall variability of the Sahel in West Africa and the Amazon and the Northeast of Brazil in South America. This Chapter begins with a brief explanation of the motivation of this study. Thereafter, some basic concepts are introduced to enter upon the subject, such as the global climate system, the main mechanisms that govern it and its variability. Then, the main modes of low-frequency SST variability and the mechanisms leading the tropical rainfall in the regions of interest are explained in detail. The last section of this Chapter is dedicated to review the research done on the relationship between the SST and the Sahel, the Amazon and the Northeast of Brazil precipitation low-frequency variability.

1.1 Motivation

The tropical precipitation is tightly related to the Intertropical Convergence Zone (ITCZ), which is the near-equatorial band of maximum surface wind convergence that is associated with strong convective activity. The ITCZ seasonally migrates meridionaly from south to north and back along the year determining the rainy seasons in tropical regions (Fig. 1.1). Over the Atlantic sector, the ITCZ latitudinal shifts vary year to year resulting in precipitation variability in the surrounding continental regions, especially in West Africa and northern South America (Nobre and Shukla 1996). The precipitation variability is particularly relevant for the West African region of the Sahel and the Amazonia and the Northeast of Brazil in South America, which are very sensitive to changes in their rainfall regimes.

The Sahel is the part of West Africa extending zonally between the Sahara desert to the north and the tropical rainy savanna to the south (roughly between 10°and 18°N) (Fig. 1.2). Climatologically, it is a dry region with a mean annual rainfall rate of 200 mm to the north and 600 mm to the south (Nicholson 2013). This region is

J. Villamayor, *Influence of the Sea Surface Temperature Decadal Variability on Tropical Precipitation: West African and South American Monsoon*, Springer Theses, https://doi.org/10.1007/978-3-030-20327-6_1

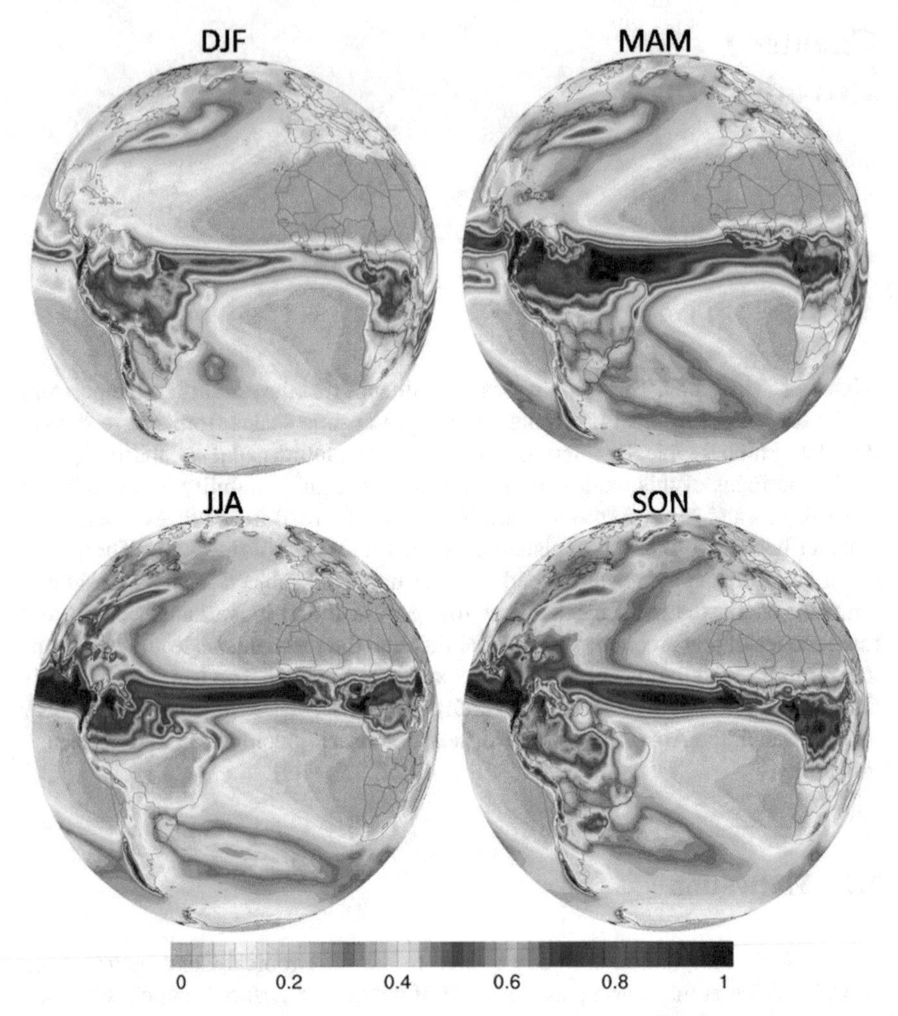

Fig. 1.1 Total seasonal precipitation (mm day^{-1}) from December to February (DJF), March to May (MAM), Jun to August (JJA) and September to November (SON) averaged over 1979–2012 using ERA-Interim reanalysis data. Image from Climate Reanalyzer (https://ClimateReanalyzer.org), Climate Change Institute, University of Maine, USA

extremely sensitive to rainfall changes during its rainy season, going from July to September (JAS), when the ITCZ reaches its northernmost position over West Africa. Indeed, the precipitation amounts in the Sahel have undergone important changes over time, with strong variability at different time scales (Fig. 1.3). In particular, the Sahel is one of the world's regions with the most marked rainfall variability at decadal time scales, which refers to changes in climate from some decades to others. Considering the precariousness of the countries in the Sahel and the climatic characteristics of the region itself, it is not surprising that such a change in the rainfall

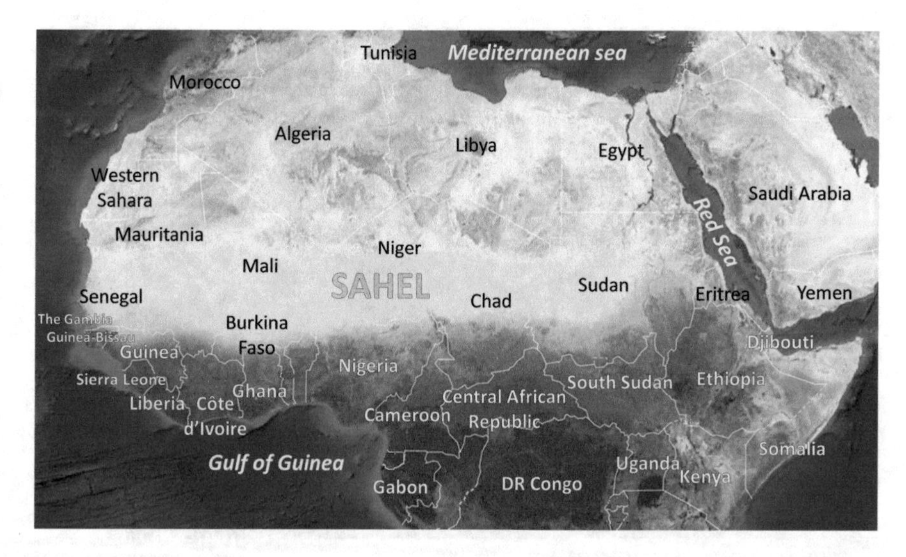

Fig. 1.2 The Sahel region in West Africa. Modified picture from Google Earth; data provided by SIO, NOAA, U.S. Navy, NGA, GEBCO Landsat/Copernicus Mapa GISrael ORION-ME

regime and the long persistence of drought conditions had dramatic economic and humanitarian consequences (Cook and Vizy 2006; Giannini et al. 2013). The decadal variability of the Sahel rainfall along the 20th century has therefore been the focus of numerous research works (e.g. Folland et al. 1986; Giannini et al. 2003; Caminade and Terray 2010; Mohino et al. 2011a; Rodríguez-Fonseca et al. 2015).

The Northeast[1] of Brazil is the northeastern tropical region of the country (Fig. 1.4). It is mostly a plateau area with a semiarid precipitation regime, in which typically no more than 400 mm of precipitation per year are recorded (Kousky 1979; da Silva 2004). This region has a short rainy season between Mach and May, when the ITCZ shifts to the south spanning this region. The Northeast is particularly prone and sensitive to changes on precipitation, especially to droughts. The local economy

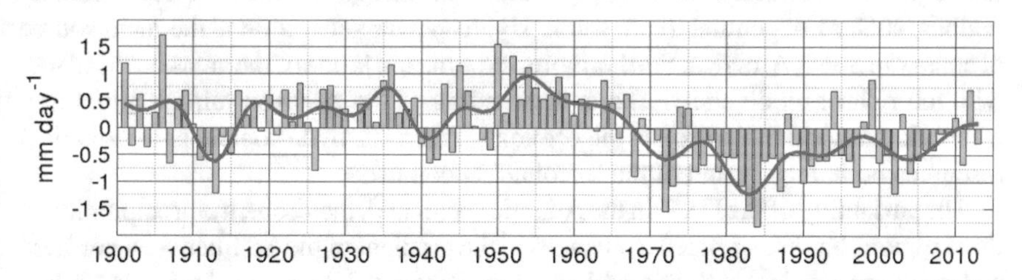

Fig. 1.3 JAS seasonal precipitation anomalies with respect to the 1901–2013 mean averaged over the Sahel region (between 17.5°W–10°E and 10°–17.5°N). Bars represent the inter-annual values and the curve is the 8-year low-pass filtered index. Data from GPCC.v7

[1] Also referred to as *Nordeste* in the literature.

Fig. 1.4 The Amazon region and the Northeast of Brazil in the north of South America. Modified picture from Google Earth; data provided by: SIO, NOAA, U.S. Navy, NGA, GEBCO INEGI Landsat/Copernicus

of the region is mostly based on crops and livestock. Hence, the lack of water supply has severe socioeconomic impacts. During drought years, massive migratory waves of climate refugees to other regions of Brazil have occurred, leading to problems of overpopulation in the major cities (Marengo 2008; Marengo et al. 2016).

The Amazon, in turn, is one of the world's wettest areas, totaling 2000 mm of mean annual precipitation. It has a long rainy season that goes from December to May (DJFMAM), during the annual migration of the ITCZ over this region. The Amazon rainfall regime has alternated between wet and dry periods with marked decadal frequency and impacts on hydrological and environmental resources (Robertson and Mechoso 1998; Dettinger et al. 2001; Marengo 2004, 2009). The Amazon is the cradle of multiple rivers that supply water and energy resources to more remote regions, such as Argentina to the south. Hydro-power generation is the main source of energy in South America. Furthermore, the Amazon is one of the areas on the planet with the richest biodiversity. The understanding of the Amazon rainfall variability is therefore important to protect the economic interests, hydrological and ecological resources both, locally and of the surrounding countries.

The variability of the ITCZ in the Atlantic sector and, therefore, the precipitation in these regions has been related to changes in the gradient of the SST between northern and southern tropical Atlantic and to anomalous tropical atmospheric circulation associated with variations in the tropical Pacific SST (Chiang et al. 2002; Ruiz-Barradas et al. 2000).

Throughout the 20th century, the Sahel has experienced abrupt changes of its precipitation regime. It went from a sequence of mostly rainy years in the mid-20th century, over the 1950s and 1960s, to a long period, between the 1970s and 1980s, in which severe drought years predominated. In turn, there has been a recovery of the Sahel precipitation during the last decade of the 20th century (Nicholson 2005; Lebel and Ali 2009). It is broadly agreed that these changes in the Sahel rainfall rate at decadal time scales are induced by the SST variability in the Atlantic (Knight et al. 2006; Zhang and Delworth 2006; Ting et al. 2009; Martin and Thorncroft 2014), the Pacific and the Indian Oceans (Caminade and Terray 2010; Bader and Latif 2003), and that are amplified by land surface processes (Folland et al. 1986; Zeng and Neelin 1999; Giannini et al. 2003; Kucharski et al. 2013).

Rainfall variability in the Amazonia and Northeast regions has attracted the attention of several studies due to its relevant impacts (e.g., Rao and Hada 1990; Wainer and Soares 1997; Zhou and Lau 2001; Yoon and Zeng 2010; de Albuquerque Cavalcanti 2015). In both regions, precipitation shows strong interannual variability (Hastenrath and Heller 1977; Moura and Shukla 1981; de Albuquerque Cavalcanti 2015). There is broad consensus on the dominant role of the tropical Pacific SST in driving rainfall variability at these time scales (Zhou and Lau 2001; Souza and Ambrizzi 2002; Ambrizzi et al. 2004; Kayano and Andreoli 2006; Rodrigues et al. 2011). The lack of continuous and long term records of precipitation both in the Amazon and the Northeast hinder the analysis of decadal variations (Marengo 2004; Nobre et al. 2006). However, some works have identified variations at decadal timescales in streamflow and rainfall records (Wainer and Soares 1997; Marengo et al. 1998; Robertson and Mechoso 1998; Houghton et al. 2001; Grimm and Saboia 2015; Marengo 2004). This long-term precipitation variability has been related to SST variations at decadal-to-multidecadal timescales in both the Atlantic and the Pacific Oceans through induced changes in the atmospheric circulation associated with the tropical rainfall in the north of South America (Nobre and Shukla 1996; Wainer and Soares 1997; Robertson and Mechoso 1998; Zhou and Lau 2001; Marengo 2004; Andreoli and Kayano 2005; da Silva 2004).

The climate variability and the associated dynamic mechanisms that induce it are usually studied through the use of general circulation climate models. They also can be used for decadal climate prediction. Such predictions could be highly valuable for planification in different sectors (agriculture, fisheries, energy production and consumption, etc.) and to establish prevention protocols to mitigate the negative impacts of climate variability in vulnerable areas, such as the Sahel, the Amazon and the Northeast of Brazil (Latif et al. 2004; Keenlyside et al. 2008; Meehl et al. 2009; Latif and Keenlyside 2011; Doblas-Reyes et al. 2013; Gaetani and Mohino 2013; García-Serrano et al. 2015; Mohino et al. 2016). Nevertheless, these models have unresolved errors in the simulated mean state of climate, called biases, which have important implications in different aspects of the simulated climate (e.g. Lin 2007; Wahl et al. 2011; Wang et al. 2014; Oueslati and Bellon 2015; Richter 2015). Thus, it is important to assess the ability that models have to reproduce the link between the main patterns of long-term climate variability and tropical rainfall. Also having a good knowledge of the evolution of the decadal variability in the past is highly relevant for understanding and assessing future changes.

As previously stated, the decadal-to-multidecadal variability of the SST in the principal ocean basins is the main modulator of the long-term rainfall changes in the tropical regions of the Sahel, the Amazon and the Northeast. These changes in precipitation have strong impacts with relevant consequences in these three regions. Hence, it is important to understand the relationship between the SST and precipitation variability at decadal time scales. This Ph.D. Thesis researches into decadal-to-multidecadal variability through a multi-model analysis of climate models simulations. This analysis allows to identify the atmospheric mechanisms and to assess the ability of the models to reproduce them. In addition, the occurrence of a decadal wet period in the Sahel during the late-19th century is studied for the first time by means of model simulations.

In the following section, some basic concepts about the climate system are introduced in order to understand how it varies. We also present what are the main tools used for its study. Then, the main modes of the decadal-to-multidecadal SST variability and the monsoon systems of interest of this Thesis are presented, followed by a review of the works addressing the study of the decadal precipitation variability in the Sahel, Amazonia and Northeast. The following Chapter introduces the data and methods used in this Thesis, prior to the presentation of the results obtained and the conclusions drawn.

1.2 The Climate System

The Earth's atmosphere, oceans, cryosphere, land surface and biosphere constitute the full climate system. The climate system is mainly regulated by the balance between the energy received from the Sun and its loss to space. The different climate components provide feedbacks that regulate the energy balance of the entire system. These components can also interact with each other through energy or matter exchanges (Fig. 1.5). Therefore, any variation of the components of the climate system may result in changes in the global climate.

The next subsection presents some aspects of the climate system that can alter its energy balance. Then, as this Thesis focuses on the tropical climate variability for which the most influential components are the atmosphere and the ocean, only these two components are introduced in more detail in this chapter.

1.2.1 Thermodynamics of the Climate System

The climate system is a thermodynamic one fueled by the energy of the incoming solar radiation (solar irradiance). Almost 40% of this radiation is reflected back to space by the atmosphere or the Earth's surface reflectivity. This is known as the albedo effect. But the rest of the incoming short-wave radiation of the Sun is absorbed by the climate system and heats the Earth's surface. The Earth's surface, in turn, radiates

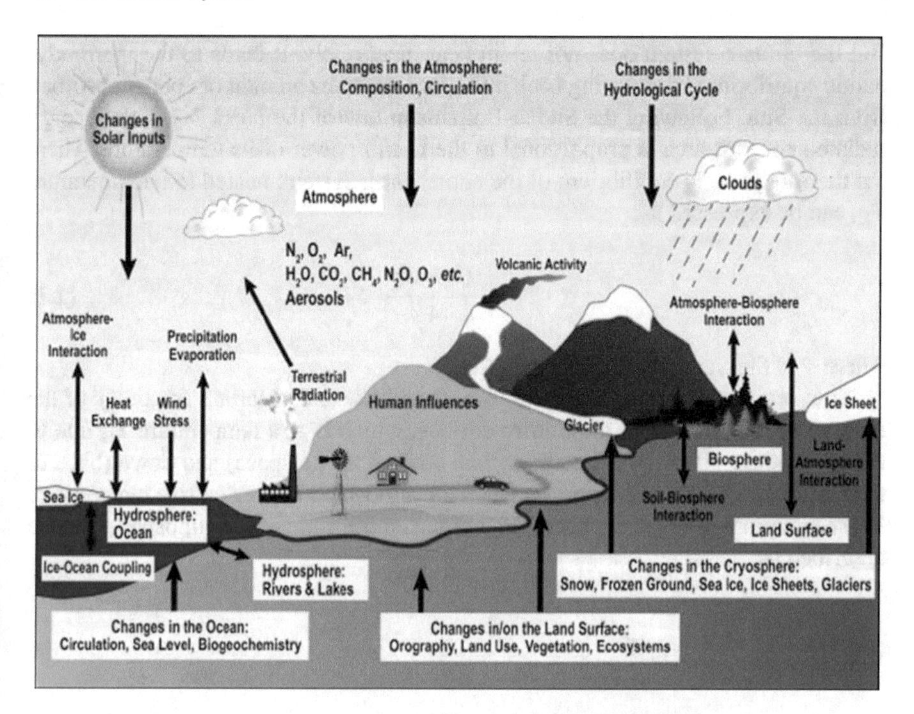

Fig. 1.5 Schematic representation of the components of the climate system, their processes and interactions (thin arrows) and some aspects that may change (bold arrows). Figure from IPCC (2007). Figure 1 from Le Treut, H., R. Somerville, U. Cubash, Y. Ding, C. Mauritzen, A. Mokssit, T. Peterson and M. Prather, 2007: Historical Overview of Climate Change: The Physical Science Basis. Contribution of Working Group I to the Fourth Assessment Report of the Intergovernmental Panel on Climate Change [Solomon, S., D. Qin, M. Manning, Z. Chen, M. Marquis, K.B. Averyt, M. Tignor and H.L. Miller (eds.)]. Cambridge University Press, Cambridge, United Kingdom and New York, NY, USA

the stored energy as long-wave radiation back to the atmosphere, which absorbs part of it maintaining the temperature on Earth. This process is known as the greenhouse effect and is crucial to regulate the energy balance of the entire climate system. The capacity of the climate system to accumulate energy depends on the composition of the atmosphere, in particular of the concentration of the so-called greenhouse gases (GHGs). The GHGs are gases that interact with the long-wave radiation emitted by the surface absorbing and releasing heat in the atmosphere. So, the GHGs concentration in the atmosphere can regulate the global temperature. These gases primarily are the water vapor, carbon dioxide, methane, nitrous oxide and ozone.

Being S_0 the radiative solar energy per unit area perpendicular to the solar radiation direction, the Earth receives a total radiation that is proportional to its circumferential section and is distributed throughout its spherical surface. Hence, the average incoming solar radiation per unit area (at the top of the atmosphere) is $S_0/4$. Considering the albedo as the fraction α of the total incoming solar radiation that is reflected by the Earth as a whole, the energy that the climate system absorbs is $(1-\alpha)S_0/4$ (Fig. 1.6).

But the climate system does not retain heat indefinitely. It tends to the thermodynamic equilibrium by radiating back into space the same amount of energy absorbed from the Sun. Following the Stefan-Boltzmann law of the *black body*, the energy radiated per unit area is proportional to the fourth power of its temperature. Then, the thermodynamic equilibrium of the entire Earth system, heated to a temperature T_E, can be expressed as:

$$\sigma \cdot T_E^4 = \frac{(1-\alpha)}{4} \cdot S_0, \tag{1.1}$$

where σ is the Stefan-Boltzmann constant.

Now, lets consider the atmosphere as a *gray body* that absorbs a fraction f of the long-wave radiation from the Earth's surface, which is at a temperature T_s, due to the greenhouse effect. This energy is then emitted up (to space) and down (back to the surface) in the same proportion when the atmosphere is heated to a temperature T_a (see diagram in Fig. 1.6). Then, the thermal equilibrium of the atmosphere can be described as:

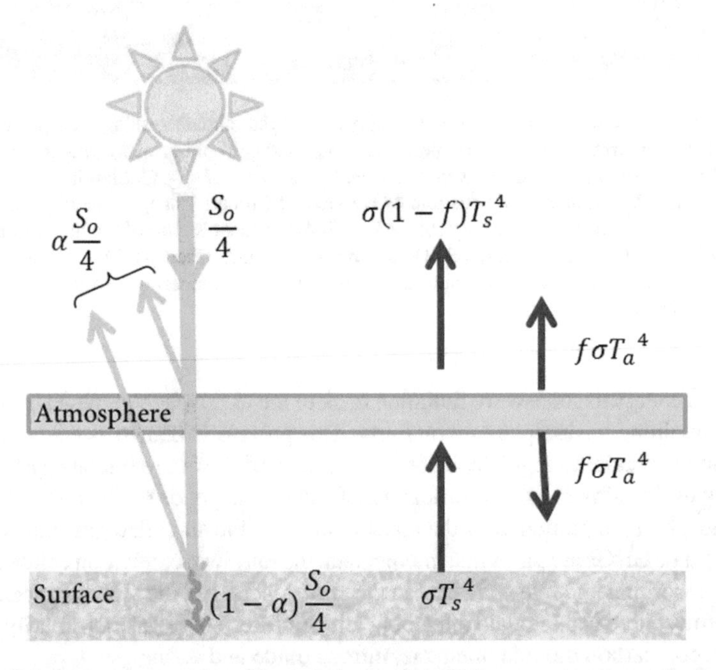

Fig. 1.6 Schematic representation of the greenhouse effect. Yellow arrows represent the short-wave solar radiation, with the total solar energy per unit area that the climate system reflects and absorbs expressed. Red arrows represent the outgoing long-wave radiation of the climate system per unit area emitted by the Earth's surface to the atmosphere and through it and the one emitted by the atmosphere upward and downward. The meaning of the terms of the expressions are explained in the text

$$f \cdot \sigma \cdot T_s^4 = 2 \cdot f \cdot \sigma \cdot T_a^4 \tag{1.2}$$

In turn, the thermodynamic equilibrium of the Earth's climate system (differentiating between the atmosphere and the surface) at the top of the atmosphere can be expressed as:

$$\frac{(1-\alpha)}{4} \cdot S_0 = (1-f) \cdot \sigma \cdot T_s^4 + f \cdot \sigma \cdot T_a^4, \tag{1.3}$$

where the first term of the right hand side of the equation is the outgoing long-wave radiation of the Earth surface that is not absorbed by the atmosphere and the second term is the radiation emitted by the atmosphere (Fig. 1.6).

In summary, considering both Eqs. 1.2 and 1.3, the Earth's surface equilibrium temperature can be estimated though the following expression:

$$T_s = \left(\frac{(1-\alpha) \cdot S_0}{2 \cdot (2-f) \cdot \sigma} \right)^{1/4} \tag{1.4}$$

Regarding Eq. 1.4, the Earth's surface temperature depends on the albedo (α) and the greenhouse effect of the atmosphere (f), which is proportional to the atmospheric GHGs concentration. This dependence occurs in such a way that the smaller the albedo and the larger the greenhouse effect, the higher the temperature and vice-versa. (For further details, see Liou 2002).

1.2.2 Radiative Forcing

The energy balance of the climate system can be altered if a disturbance on the agents that regulate it is imposed. This forced energy imbalance is called the radiative or external forcing, which is defined as the difference between the incoming solar irradiance and the outgoing net radiation from the Earth (Stocker 2014). Radiative forcing is, therefore, an important factor that induces changes on climate, especially related to the global mean temperature. Regarding the origin of the causes that induce the radiative forcing, it is classified into anthropogenic and natural forcings.

Human activity can modify some properties of the climate system with important implications in the radiative forcing. For example, the combustion of fossil fuels in the industrial activity, which has been rapidly developed since the 19th century, generates large amounts of GHGs, especially carbon dioxide. Then, the GHGs emissions produce anthropogenic forcing on the climate system through alterations of the greenhouse effect of the atmosphere (f in Eq. 1.4). Also the different land uses imply changes in the albedo (α in Eq. 1.4) of the Earth's surface.

Natural forcing is principally produced by the solar irradiance variability and the volcanic activity. The solar irradiance (S_0 in Eq. 1.4) depends on the solar activity and astronomical cycles that vary at different time scales, including centennial and millennial ones. These variations affect the solar radiative forcing and, hence, induce changes in the climate system. Volcanic eruptions are sporadic but have long lasting and strong radiative forcing effects. They can inject mineral particles, known as aerosols, in the stratosphere where they have long lifetimes.

Aerosols are tiny liquid or solid particles suspended in the air. Some of them have natural origin, like the aforementioned volcanic particles, sand dust, sea salt, etc. But others are emitted by human sources, such as industrial emissions or smoke. Aerosols have different and complex effects on the radiative forcing (Stocker 2014). On the one hand, they have a direct effect on the solar irradiance that is principally to scatter the short-wave radiation (i.e., enhancing α in Eq. 1.4) and therefore cool the Earth's surface. But depending on the composition and color, some aerosols can also absorb long-wave radiation (i.e., increasing f in Eq. 1.4), thus warming the atmosphere. On the other hand, aerosols have an indirect effect in which they interact with clouds. In the troposphere, aerosols enhance the concentration of cloud condensation nuclei, leading to changes in the clouds properties such as the albedo. Therefore, aerosols are a source of natural or anthropogenic radiative forcing, depending on its origin.

Summarizing, the climate system receives energy from the Sun and maintains a thermodynamic equilibrium. In turn, the incoming solar irradiance can vary depending on the radiative forcing inducing changes in the climate system. Then, due to the Earth's curvature and since the reflectivity of the Sun's radiation is not globally homogeneous (e.g., the ice and snow cover or the sand of the Sahara desert have very high albedo), the surface is not equally heated across the globe. These regional differences of heat are key to generate the dynamics of the climate system, which distributes energy to approximate to the global thermodynamic equilibrium. Below, some basic concepts of the atmospheric and ocean systems are introduced (for more detailed information, see Hartmann 1994).

1.2.3 Atmospheric Circulation

The general circulation of the atmosphere is induced by the combined effect of the incoming solar energy and the Earth's rotation. The solar irradiance is not uniformly received throughout the Earth's surface. It is more intense around the equator and less at the poles (Fig. 1.7). The atmosphere reaction is to distribute this energy producing poleward circulation of heat from the tropical band (Fig. 1.8). Near the equator, the air near the heated surface warms and rises. Then it flows meridionally in the upper troposphere toward colder latitudes of both hemispheres and sinks down to the subtropical high-pressure belts. The air at the surface flows from subtropical latitudes equatorward and converges near the equator carrying colder air to warm regions. These meridional cells of circulation on both sides of the equator are called the Hadley cells (Hadley 1735).

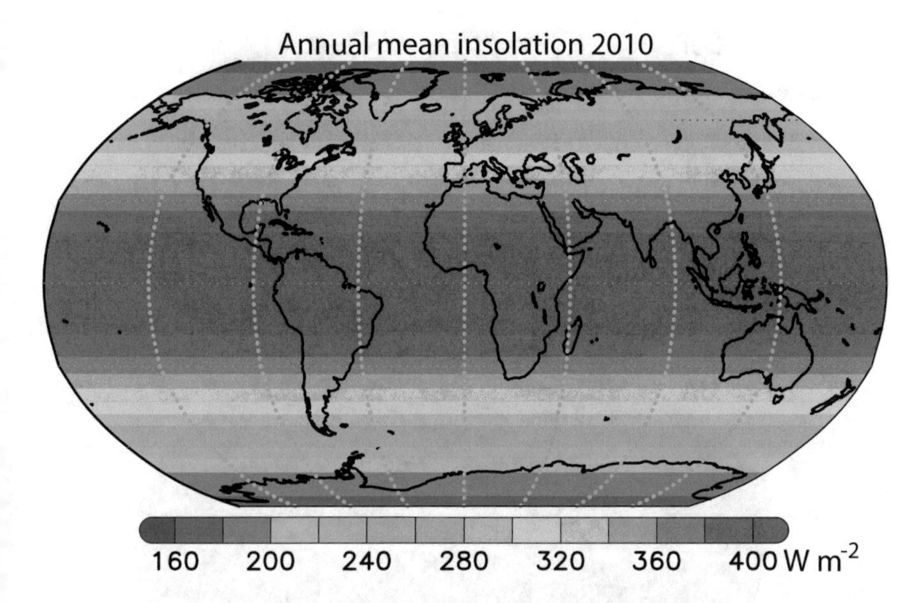

Fig. 1.7 Annual mean insolation before interacting with the atmosphere in 2010. Data from ERA Interim Reanalysis (Dee et al. 2011)

As a consequence of the rotation, any trajectory described on Earth is subject to the Coriolis effect (Coriolis 1835). This effect consists of a deviation of any trajectory on a rotatory system of reference, such as the Earth. In case of the wind, the Coriolis

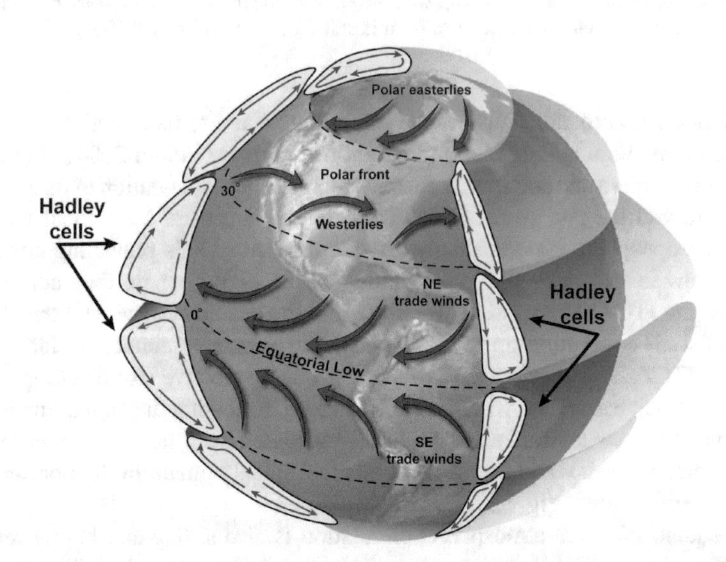

Fig. 1.8 Schematic representation of the main large-scale structures of the atmospheric circulation. Figure from NASA

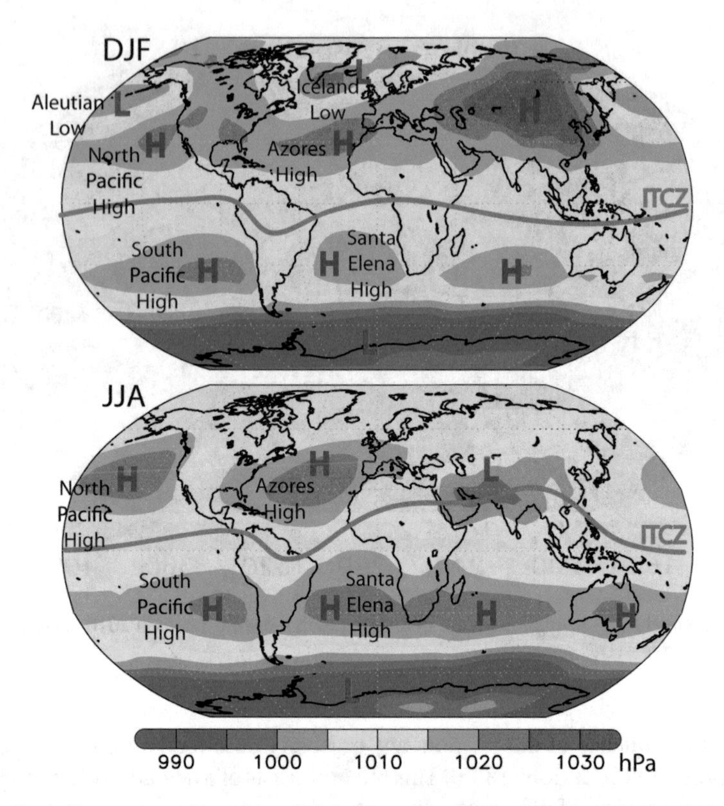

Fig. 1.9 Prevailing centers of low (L) and high (H) sea-level pressure, averaged over 1901–2000 in boreal winter (top) and summer (bottom), with the name of the most important ones and schematic representation of the ITCZ location. Data from HadSLP2 (Allan and Ansell 2006)

effect deflects it perpendicularly to the rotation axis and to its direction, being this deflection maximum at the poles and null at the equator (Holton 2004). So that the wind in the northern and southern hemisphere experiences a deviation of its trajectory to the right and the left, respectively. Thus, the equatorward meridional winds of the lower troposphere experience a westward deviation. This prevailing equatorial easterly flow is known as the trade winds and the region where they converge is known as the ITCZ. The wind convergence and the heated surface near the equator force moist air convection up to the upper troposphere that condenses into clouds along the ITCZ. This mechanism produces strong convective storms comprising a tropical rain belt associated with the ITCZ. In turn, following the annual meridional transition of the solar irradiance, the ITCZ oscillates from north to south leading the characteristic rainy and dry seasons of the tropical climate in the northern and southern hemispheres (Fig. 1.9).

The equatorial zonal atmospheric circulation is also strong and highly relevant and is known as the Walker circulation. This circulation comprises six cells, with branches of rising moist air over eastern Africa, western tropical Pacific and northern

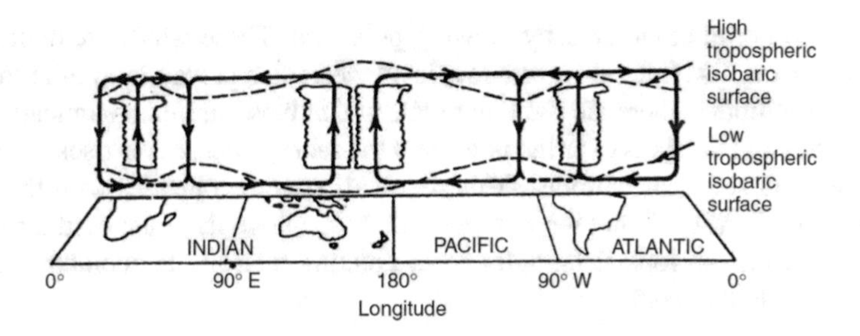

Fig. 1.10 Schematic representation of the vertical and longitudinal equatorial circulation, particularly referred to as the Walker Circulation. From Lau and Yang (2003). © Elsevier. Used with permission

South America, as well as subsidence of dry air over western Indian Ocean and eastern tropical Pacific and Atlantic (Lau and Yang 2003) (Fig. 1.10). The biggest and most persistent Walker circulation is developed over the wide equatorial Pacific. In the Pacific Walker cell, the trade winds force the surface water to drift toward the west side of the basin, where it is strongly heated by the solar irradiance, while the colder deep water rises near the South American coast. This water distribution induces low surface pressure above the warmer west part of the equatorial Pacific and high ones to the east. As a consequence, warm and moist air masses rise above the Maritime Continent forming clouds and producing heavy rains over Southeastern Asia. In the upper troposphere, dry air diverges and flows eastward sinking as it cools over the eastern equatorial Pacific basin, close to the western coast of South America. Finally, the surface branch that closes the Pacific Walker cell flows westward reinforcing the trade winds.

At higher latitudes the circulation is more complex. Thermodynamically only two Hadley-like cells would be expected, with meridional circulation from the equator up to both poles. However, due to the deflection of winds caused by the Coriolis effect, the meridional circulation cells are twisted and divided into three cells in each hemisphere.

In mid-latitudes, the mean air flow describes the so-called Ferrel cells. The meridional circulation is weaker than the one of the Hadley cells and has opposite direction. It is a thermally indirect cell in which cold air rises in subpolar latitudes, in the low-pressure belts, and flows in the high troposphere to the subtropics, where it sinks. The aforementioned low- and high-pressure belts coincide with the rising and sinking branches of the meridional circulation cells in subpolar and subtropical latitudes, respectively. These belts are not as such but are distributed in semi-permanent and stable centers of low and high pressure with associated cyclonic and anticyclonic gyre, respectively, as consequence of the Coriolis effect. Some of them are the well-known, such as the Aleutian Low of the North Pacific, the Iceland Low and the Azores High in the North Atlantic and the Santa Helena High in the South Atlantic (Fig. 1.9). In between the subpolar and subtropical latitudes, the low-level winds continue the Ferrel circulation by blowing poleward. These winds are deflected to the east by Coriolis effect, therefore low-level westerlies prevail in mid-latitudes.

In polar latitudes, there are the Polar cells which have thermodynamically direct circulation. Cold air sinks over the poles and the more tempered air rises to the high troposphere in subpolar latitudes. This rising air stream coincides with the ascent branch of the Ferrel cell and the low-pressure belt, where the polar cold air and the warm one from subtropical latitudes meet creating a strong temperature gradient known as the Polar Front.

1.2.4 Ocean Circulation

The oceans cover two-thirds of the Earth's surface and, like the atmosphere, they constitute an important component of the climate system because of its ability to store and distribute radiative energy from the sun. But the ocean water has much higher heat capacity than the air of the atmosphere. The heat capacity of a system is the amount of energy needed to raise one degree of its temperature. Therefore, the higher the heat capacity, the slower a body warms and, in turn, the longer it retains this energy. This is why it is said that the oceans have thermal inertia. Any changes in temperature persist much more in the oceans than in the atmosphere or on the continental surface, since the latter have lower heat capacity. Then the oceans gradually transmit this heat to the atmosphere. That is why the oceans are considered as important drivers of the global climate and the SST a potential predictor (Rowell 1998; Suárez-Moreno and Rodríguez-Fonseca 2015).

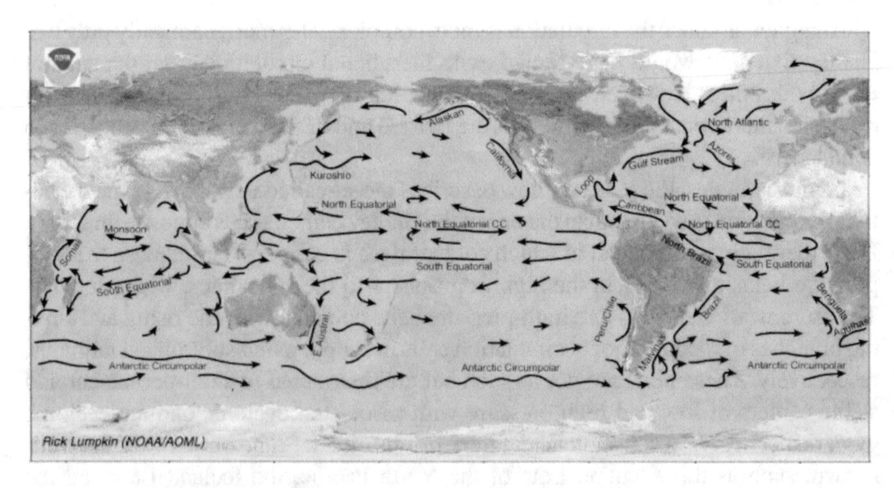

Fig. 1.11 Schematic representation of the wind-driven surface ocean currents. Figure from the U.S. National Oceanic and Atmospheric Administration (copyright free)

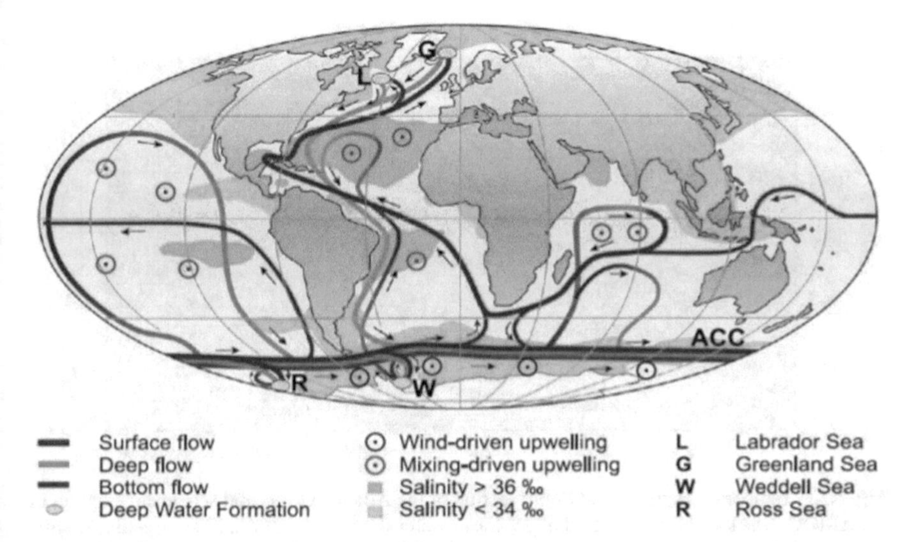

Fig. 1.12 Schematic global map of the thermohaline circulation. From Rahmstorf (2006). © Elsevier. Used with permission

In the surface, where the oceans have the strongest interaction with the atmosphere, the ocean circulation is strongly influenced by the atmospheric one. For example, the equatorial surface currents follow the trade winds direction. Due to the Coriolis effect, in the northern hemisphere the northward currents are deflected eastward and the southwards ones westward, and vice versa in the southern hemisphere. As a consequence, two clockwise ocean gyres of surface circulation prevail in the northern Atlantic and Pacific and three of opposite sense in the Indian Ocean and the southern Atlantic and Pacific. This coincides with the anticyclonic circulation of the extratropical low level winds around the surface pressure maximums induced by the subsiding branches of Hadley and Ferrel cells. Therefore, the warm tropical waters flow poleward in the western edges of the oceans while equatorward currents transport cold waters from high latitudes throughout the eastern side of the basins (Fig. 1.11).

The deep oceanic circulation is a consequence of the different buoyancy of water masses, which depends on their density. The relation between temperature and salinity regulates the water density. The colder and saltier, the denser a water mass is. This is why the deep ocean circulation is known as the thermohaline circulation (THC). Thereby, the denser water masses tend to sink to the deepest layers of the oceans, receiving the name of deep water, and the less dense ones remain in upper levels.

The formation of deep water is key to the THC operation. The main source of deep water formation is the North Atlantic. This basin is particularly salty mainly because of the continuous supply of the Mediterranean Sea water, which has elevated salt concentration due to the high evaporation rate (Bozec et al. 2011). When this salty water reaches the subpolar North Atlantic, it cools resulting in very dense water

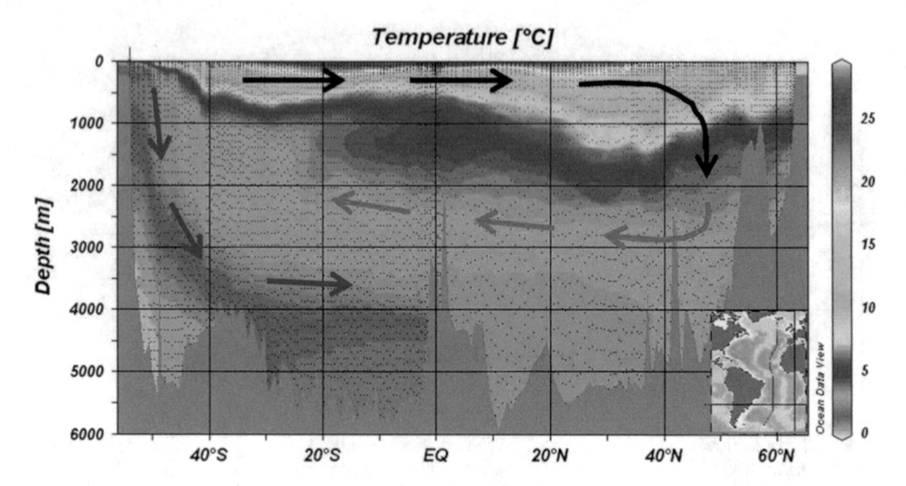

Fig. 1.13 Temperatures in a North South section of the Atlantic Ocean and schematic diagram of the AMOC. Black arrows indicate shallow warm water flow. To the north, blue ones the Atlantic deep water circulation and purple ones the Antarctic deep water. Adapted figure obtained from the Ocean Data View, odv.awi.de

masses that sink to the deep ocean, underneath the warmer surface water. The deep water then flows southward into the Southern Hemisphere (Fig. 1.12).

Part of the deep water in the South Atlantic joins the Antarctic Circumpolar Current. It is a clockwise current of dense water at the bottom of the Southern Ocean that flows around the Antarctic. At the same time, the Antarctic Circumpolar Current distributes deep water to the Indian and Pacific basins from the south. In both oceans, the deep water gradually mixes with upper masses and warms as it passes though the equator. Then progressively rises up to near the surface along its way to the north of the basins. The Indian Ocean is so warm and the Pacific has so low salinity that the surface waters are prevented to sink (Broecker 1997). Instead, the warm surface water is returned to the Atlantic from the south (Broecker 1992; Rahmstorf 2006).

Another part of the Atlantic deep water upwells to the south of the basin, close to the Antarctica. Then it flows back northward close to the surface to occupy the gap left in the subpolar North Atlantic by the sinking deep water masses (Fig. 1.13). In this way, the Atlantic branch of the THC closes its cycle, which is referred to as the Atlantic Meridional Overturning Circulation (AMOC). The AMOC then constitutes a mechanism of warm water transport from the south to the north of the Atlantic Ocean, were it transmits heat to the atmosphere. This inter-hemispherical imbalance of heat and the changes in the AMOC determines the global climate system and its variability (Srokosz and Bryden 2015).

1.2.5 Climate Variability and Teleconnections

Climate variability refers to the departures of the climate system from a mean state of it averaged over a reference long period. Such departures are known as anomalies and are defined as the difference between the state of climate at a given time and the mean value of the state over a long period called the climatology. Climate varies on several temporal scales beyond weather events, which refer to day-to-day changes of the atmosphere state. Hence, climate variability describes the variations on the state of the atmosphere or the ocean over time periods of a month, season, years and even one or more decades. Variations over periods longer than one century are defined as climate change. This Thesis addresses the longest term climate variability, from one to several decades, which is known as decadal-to-multidecadal variability (also referred to as long or large time scale variability in this Thesis).

Some recurrent modes of variability can be identified in certain climate variables as patterns of anomalies that are regularly repeated with a specific frequency. There are modes at all time scales and they may be characteristic of a localized region, of a very broad one or even global. Regarding to its origin, the climate variability can be classified as internal or external. Changes produced within the climate system itself are referred to as internal climate variability, such as anomalies produced by natural processes in the aforementioned climate circulation mechanisms. In turn, when the climate system responds to variations in the external radiative forcing, whether due to natural or anthropogenic causes, such changes are referred to as external climate variability. As explained before, the GHGs and aerosols have an important role in the climate system as they can alter the energy balance through the radiative forcing (Taylor and Penner 1994). The former are globally distributed almost equally. They regulate the aforementioned greenhouse effect and are, for the most part, of natural origin. However, industrial activity over the last century has increased the concentration of carbon dioxide. So that the variations in climate associated with these changes in the carbon dioxide concentration are considered as anthropogenic external forcing. The aerosols, in turn, produce scattering of the solar radiation, increasing the atmospheric albedo and therefore dampening the Earth heating. In contrast to GHGs, the aerosols effects have strong regional character. The aerosols are microparticles that are suspended in the atmosphere and which may come from industrial activity, fossil fuel and biomass burning, like the sulphate aerosols, or come from natural origin, such as from volcanic eruptions. Thus, the external radiative forcing produced by the aerosols can then be considered anthropogenic or natural, depending on the origin of the aerosols that cause it.

As seen before, the climate system is a complex one in which different subsystems are interconnected and in constant interaction among one another. A consequence of this interaction is the connection between different remote regions of the globe by means of the propagation of climate anomalies through the atmosphere (Gill 1980). These climate linkages are known as teleconnections.

To better illustrate the concepts of climate variability and teleconnections, the El Niño/Southern Oscillation (ENSO) (Philander 1990) is used as example. The ENSO is an inter-annual mode of climate variability, which means that produces year-to-year variations. It is the result of ocean-atmosphere interactions, broadly consisting of the oscillation of the SST in the center and east of the tropical Pacific accompanied by changes in the winds and pressure along the equator. The positive phase of ENSO is known as the El Niño phenomenon and is characterized by a tropical Pacific warming above normal associated with a rainfall increase in Peru and Ecuador and a decrease in Southeast Asia. The negative phase is known as La Niña and has broadly opposite effects. Typically, the duration of each phase is between 9 and 12 months, reaching its maximum amplitude during the boreal winter. The ENSO completes its cycles in periods ranging from 2 to 7 years (Jin et al. 2006).

The oceanic component of the ENSO is well located in the tropical Pacific. However, its atmospheric component is more widespread and has nearly global implications. This is because of the teleconnections between climate variations in the tropical Pacific region with other remote areas (Trenberth et al. 1998). The SST pattern of ENSO is related with anomalies in the surface pressure gradient between the east and west sides of the tropical Pacific and in the trade winds. During El Niño years, the eastern tropical Pacific is anomalously warmer than the western side coinciding with low and high surface pressure anomalies at each side, respectively, and weakened trade winds, while the opposite occurs during La Niña events. Such changes can modify the global circulation by means of alterations in the Hadley cells and the Walker circulation (Souza and Ambrizzi 2002; Ambrizzi et al. 2004). This way, climate variations in the tropical Pacific sector propagate through *atmospheric bridges* (Klein et al. 1999) to remote extratropical and tropical regions. For example, it is known that the ENSO is related to changes in the North Atlantic Oscillation (NAO) (Brönnimann 2007; García-Serrano et al. 2011), which is a mode of variability of the surface pressure difference between the subpolar and subtropical parts of the North Atlantic. The NAO, in turn, plays an important role in regulating the winter climate in Europe (Rodwell et al. 1999; Hurrell et al. 2003). The ENSO is also teleconnected with subsidence and precipitation anomalies in West Africa during the northern summer (Janicot et al. 1996, 1998, 2001; Mohino et al. 2011b) and northern South America in the austral summer through Walker circulation anomalies (Rao and Hada 1990; da Silva 2004; Andreoli and Kayano 2005).

There are many other modes of variability at different time scales that are highly relevant to the climate, but the subject of study of this Thesis is the low-frequency variability. So in Fig. 1.3, the climate variability at decadal-to-multidecadal time scales, which refers to changes in climate from several decades to others, will be address.

1.2.6 The General Circulation Models

The Global Circulation Models (GCMs) are computational tools that reproduce the climate system processes. They solve the fundamental equations of the laws and principles of physics governing the processes in each component of the global cli-

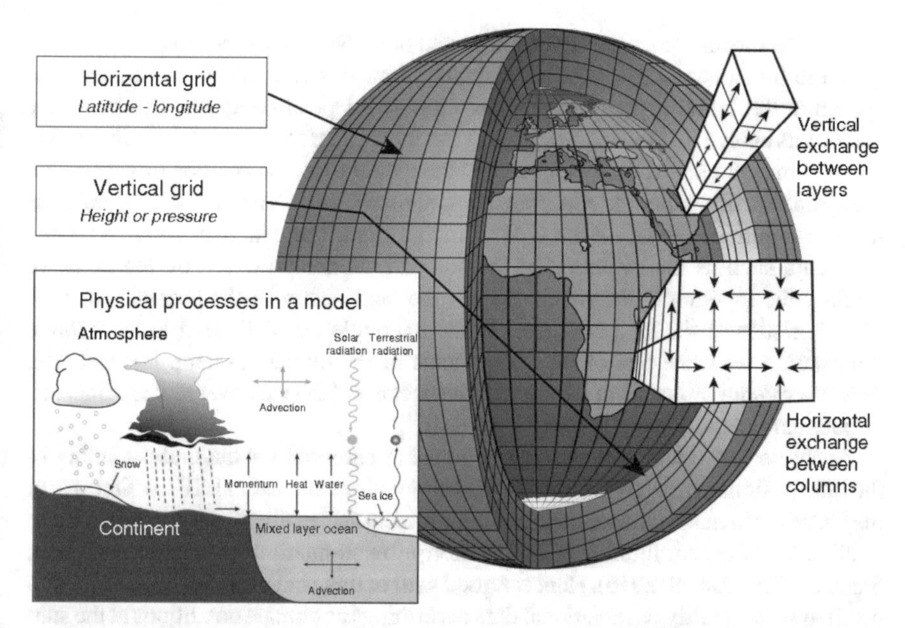

Fig. 1.14 Schematic representation of a GCM. From Edwards (2011). © John Wiley & Sons Ltd. Used with permission

mate system, such as energy and mass exchanges, by means of numerical methods (Hartmann 1994; Edwards 2011). To do that, GCMs consider a discretized three-dimensional space in a grid, as well as the temporal evolution (Fig. 1.14). At each grid point, the values of the model variables are represented and the fundamental equations are solved, obtaining values of the meteorological variables for a certain instant in the future. GCMs are the most advanced tool currently available for understanding climate variability and change, assessing their effects and predicting weather and climate.

A shortcoming of the spatial discretization is that in the real climate system there are some processes that occur on smaller scales, such as some involved in cloud formation. To tackle this problem, some of these small-scale processes are parameterized. But parameterizing accurately these processes is one of the most challenging aspects of climate modeling, being an important source of uncertainty of the GCMs.

The GCMs are composed of different modules, each of which reproduces the processes of each component of the climate system. The main components are the atmospheric GCMs (AGCMs) and ocean GCMs (OGCMs). These models are coupled to form a fully coupled GCM.[2]

[2]The coupled GCMs are also usually referred to as CGCMs or AOGCMs, because of the coupling between an AGCM and an OGCM. In this Thesis the acronym GCM is used to refer by default to the coupled GCMs.

Climate models require a set of initial and boundary conditions to properly integrate the fundamental equations. Initial conditions describe the state of the climate system at the initial instant of the simulation and depend on space. The boundary conditions evolve in time and provide the model with information of the climate system that may not be explicitly included in the model (e.g., distribution of vegetation, topography, carbon dioxide concentrations, aerosols, etc.). The boundary conditions that are required depend on the type of model used. For example, coupled GCMs need data such as the evolution of the solar irradiance produced by the radiative forcing, the topography of the continents, the bathymetry of the oceans and some characteristics of the land coverage. While uncoupled models need to incorporate information from the rest of the components of the climate system. For instance, AGCMs require inputs from the state of the ocean and the ice cover that are imposed as boundary conditions.

Accurate initial and boundary conditions are essential for the understanding of the future climate through the use of GCM simulations (Flato et al. 2013). One use of the GCMs is to make climate predictions or forecasts to estimate the actual evolution of climate in the near future y means of short-term simulations (Kirtman et al. 2013; Sanchez-Gomez et al. 2016). Hence, a good skill of this kind of simulations requires a detailed set of quality observational data describing the climate conditions at the start of the modeling experiment. In turn, long-term simulations performed with GCMs are used to make future climate projections at long time scales, such as decadal-to-multidecadal. These climate projections aim at understanding the long-term changes of the climate system, which depend on the climate response to the evolution of the external forcings in time. Therefore, such projections are essentially a boundary condition problem, requiring good information on all factors driving climate changes over time (Meehl et al. 2009).

GCMs have biases that promote large uncertainties in their simulations. Several different biases have been reported in numerous studies. One is the so called double ITCZ, which results in excessive precipitation and is associated with simulated equatorial SST that is colder than in observations, specially in the Pacific basin (Mechoso et al. 1995; Lin 2007; Li and Xie 2014; Oueslati and Bellon 2015). In the Atlantic basin, the simulated tropical SST is warmer than the observed one (Richter and Xie 2008). This bias is associated with another one in the ITCZ over this sector, whose simulated annual mean position is slightly shifted to the south with respect to observations (Richter and Xie 2008; Richter et al. 2014; Richter 2015; Wahl et al. 2011). Also biases in the strength of the AMOC have been associated with errors in the mean state of the global SST (Wang et al. 2014). The origin of these biases is complex and is still a major subject of research aiming to improve models performance (e.g. Luo et al. 2005; Li and Xie 2014; Menary et al. 2015; Richter 2015).

1.3 Low-Frequency Variability of SST

As previously stated, the oceans transmit heat to the atmosphere much more persistently than the way in which the latter can induce changes in the former as a consequence of the high heat capacity of water. So, at low-frequency time scales, the main modes that drive climate variability have an important oceanic component which is typically imprinted in the SST. Previous studies have identified the main sectors of the global SST that drive the rainfall variability at decadal-to-multidecadal time scales over the Sahel, the Amazonia and the Northeast of Brazil.

The low-frequency variability of Sahel precipitation has been linked principally to decadal-to-multidecadal variations of the Atlantic SST (Folland et al. 1986; Rowell et al. 1995; Zhang and Delworth 2006; Knight et al. 2006; Ting et al. 2009; Martin and Thorncroft 2014; Martin et al. 2014), also to the Indian Ocean warming (Bader and Latif 2003; Giannini et al. 2003; Lu and Delworth 2005; Lu 2009) and partially to the Pacific SST variability at long time scales (Caminade and Terray 2010; Mohino et al. 2011a). The warming trend of the global SST observed over the last century has also been associated with a decrease in Sahel rainfall at longer than multidecadal time scales (Biasutti and Giannini 2006; Ackerley et al. 2011; Biasutti 2013).

The Amazonia and Northeast of Brazil rainfall changes have been principally related to the Pacific and the Atlantic SST variability at decadal timescales (Wainer and Soares 1997; Robertson and Mechoso 1998; Zhou and Lau 2001; Marengo 2004; da Silva 2004; Andreoli and Kayano 2005; Knight et al. 2006; Nobre et al. 2006; Fernandes et al. 2015; Grimm and Saboia 2015). At longer time scales, the observed long-term precipitation trends in both South American regions have been found to be uncertain and poorly significant compared to decadal changes, which may be partially attributed to the short sample of observations available (Marengo 2004, 2009; da Silva 2004; Zhou and Lau 2001).

The internal decadal-to-multidecadal variability of the Indian Ocean SST has been found to be tightly associated to that of the Atlantic and the Pacific basins, both in GCMs simulations (Martin et al. 2014) and in observations (Mohino et al. 2011a), suggesting that it does not have an independent mode of variability at these time scales. Hence, a detailed description of the main features of the global SST change at longer than multidecadal time scales and of the leading modes of the Atlantic and Pacific decadal-to-multidecadal internal SST variability is presented below.

1.3.1 Low-Frequency Variability of the Global SST

On a global scale, the main observed long term variation of SST over the last century is a tendency of warming almost generalized across the globe. This SST rise is part of the global warming (GW), which usually refers to the global mean surface air temperature rise. According to the fifth report of the IPCC (Stocker 2014), the global mean surface temperature has experienced a linear increase of 0.85 °C in the 1880–

2012 period, reaching in the first decade of the 21st century the highest temperatures of the instrumental record. There is broad consensus that the observed GW along the 20th century and until the present day has been induced by the steady increase of the GHGs concentration, especially the carbon dioxide, since the Industrial Revolution (Xie et al. 2010). This modifies the natural greenhouse effect that modulates the incoming solar irradiance, so the GW is considered as an externally forced mode of climate variability.

In addition, there are important climate feedbacks that can intensify or attenuate the GW induced by the GHGs. Even though the GHGs warming effect is rather uniformly distributed throughout the globe, the climate feedbacks produce different regional patterns of temperature changes. These regional differences are determinant to the specific GW impacts on the different areas of the globe. The most effective climate feedback is the one produced by the increase of the atmospheric content of water vapor. Since the air dew point is higher when it is warmer, the atmosphere heated by the effect of increased GHGs can contain more water vapor, which is the most efficient GHG, producing a positive feedback to GW. In turn, clouds, depending on the type, have both cooling and warming effects on climate: they enhance the albedo by reflecting part of the solar radiance from the top and also retain heat from

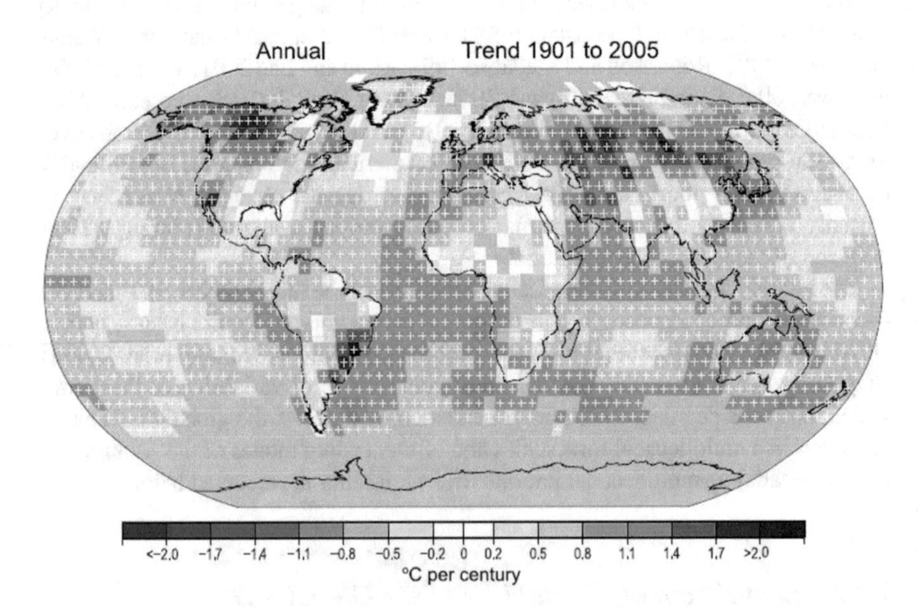

Fig. 1.15 Linear trend from 1901 to 2005 of annual temperature expressed in °C per century. From Trenberth et al. (2007). Figure 3.9 (left panel) from Trenberth, K.E., P.D. Jones, P. Ambenje, R. Bojariu, D. Easterling, A. Klein Tank, D. Parker, F. Rahimzadeh, J.A. Renwick, M. Rusticucci, B. Soden and P. Zhai, 2007: Observations: Surface and Atmospheric Climate Change. In: Climate Change 2007: The Physical Science Basis. Contribution of Working Group I to the Fourth Assessment Report of the Intergovernmental Panel on Climate Change [Solomon, S., D. Qin, M. Manning, Z. Chen, M. Marquis, K.B. Averyt, M. Tignor and H.L. Miller (eds.)]. Cambridge University Press, Cambridge, United Kingdom and New York, NY, USA

the surface (Karlsson et al. 2008). Nevertheless, the net effect of clouds is to produce a positive climate feedback (Soden and Held 2006). The main negative climate feedback comes from the tropospheric warming. The warmer the high troposphere is, the more low-frequency radiation is emitted to space, dampening the greenhouse effect (Bony et al. 2006). However, these feedbacks are important sources of uncertainty that difficult the understanding of the regional effects of global warming and the assessment of model future projections (Held et al. 2005; Biasutti et al. 2008; Xie et al. 2010). Both the clouds and the high troposphere climate feedbacks are unevenly reproduced by the climate models, with which the assessment of the GW effects is done (Bony et al. 2006; Soden and Held 2006). This is a problem considering that in the foreseeable future the GHGs concentration and the associated GW effects are expected to accelerate (Meehl et al. 2009).

The regional trends of the surface temperature along the last century depict, in general, a worldwide warming although not uniform (Fig. 1.15). Regarding the SST, the South Atlantic and the tropical and southern Indian Ocean show the highest tem-

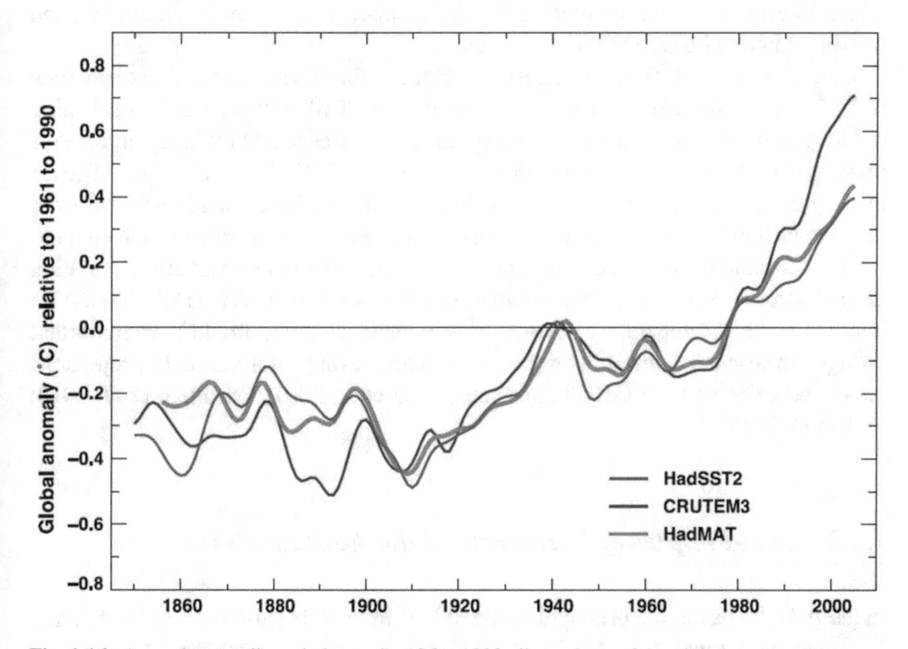

Fig. 1.16 Annual anomalies relative to the 1961–1990 climatology of the global mean temperature of sea surface (blue), land surface air (red) and night marine air (green) data bases. From Trenberth et al. (2007). Figure 3.8 from Trenberth, K.E., P.D. Jones, P. Ambenje, R. Bojariu, D. Easterling, A. Klein Tank, D. Parker, F. Rahimzadeh, J.A. Renwick, M. Rusticucci, B. Soden and P. Zhai, 2007: Observations: Surface and Atmospheric Climate Change. In: Climate Change 2007: The Physical Science Basis. Contribution of Working Group I to the Fourth Assessment Report of the Intergovernmental Panel on Climate Change [Solomon, S., D. Qin, M. Manning, Z. Chen, M. Marquis, K.B. Averyt, M. Tignor and H.L. Miller (eds.)]. Cambridge University Press, Cambridge, United Kingdom and New York, NY, USA

perature increase of around 1°C throughout the last century. In other regions, such as in the subtropics of the Pacific, the South Indian Ocean and the North Atlantic, the temperature raise is smaller or there has even been a cooling like in the extratropical North Atlantic, south of Greenland. But the most intense heating occurs in the continental surface, especially in northern North America, southeastern South America and the interior of Asia. While the Mississippi river basin, in southeastern North America, has experienced a cooling.

Therefore, it is evident that the GHGs external forcing has faster heating effect on continental surface and the surface air than in the oceans (Sutton et al. 2007) (Fig. 1.16). However, the bulk of the thermal energy produced by GW is stored in the oceans, especially in surface waters, due to its high heat capacity. The broad climate response to GW can be mainly determined by the moist and heat supplied by the superficial ocean to the global troposphere at low levels (Compo and Sardeshmukh 2009). Therefore the anthropogenic global component of the low-frequency SST variability is a good indicator of the GW and its associated atmospheric teleconnection can explain the GW effects on global climate (Bichet et al. 2015). Indeed, the GW is often referred uniquely to as its SST component (Mohino et al. 2011a), and so will be in this Ph.D. Thesis hereinafter.

One effect of the GW is the sea level rise due to the thermal expansion of shallow water in the oceans (Meehl et al. 2005; Lyman et al. 2010). Others are the weakening of the AMOC, the increase of tropical cyclones occurrence and intensity and the ice melting in the poles (Meehl et al. 2005; Stroeve et al. 2012; Xie et al. 2010). The ice melting, in turn, acts as a positive climate feedback by altering the land surface albedo, hence contributing to warm the Earth. An increase of the atmospheric capacity to hold water vapor during the last century, related to alterations in atmospheric circulation and a drying in tropical regions, has also been attributed to the GW (Dai 2011). Also interhemispheric temperature asymmetries in the GW pattern might have important changes in tropical rainfall, as reported by works using future climate projections under scenarios of high GHGs emission (Chou et al. 2007; Friedman et al. 2013; Park et al. 2015).

1.3.2 Low-Frequency Variability of the Atlantic SST

In the Atlantic basin, the principal mode of low-frequency variability is the Atlantic Multidecadal Variability (AMV) of the SST, also known as the Atlantic Multidecadal Oscillation (AMO) (e.g. Kerr 2000; Sutton and Hodson 2005; Knight 2005). The AMV is defined as the observed oscillation of the basinwide North Atlantic SST anomalies (SSTA), independent of the warming trend effect of the GHGs. It oscillates with a periodicity of approximately 60–80 years within a 0.4 °C range (e.g. Kerr 2000; Knight 2005) that is also manifest in the global mean surface temperature records (Schlesinger and Ramankutty 1994). Over the instrumental record, positive phases of the AMV occurred approximately during the 1870–1880s and the 1930–1950s and negative ones during the 1900–1920s and the 1960–1990s (e.g. Enfield

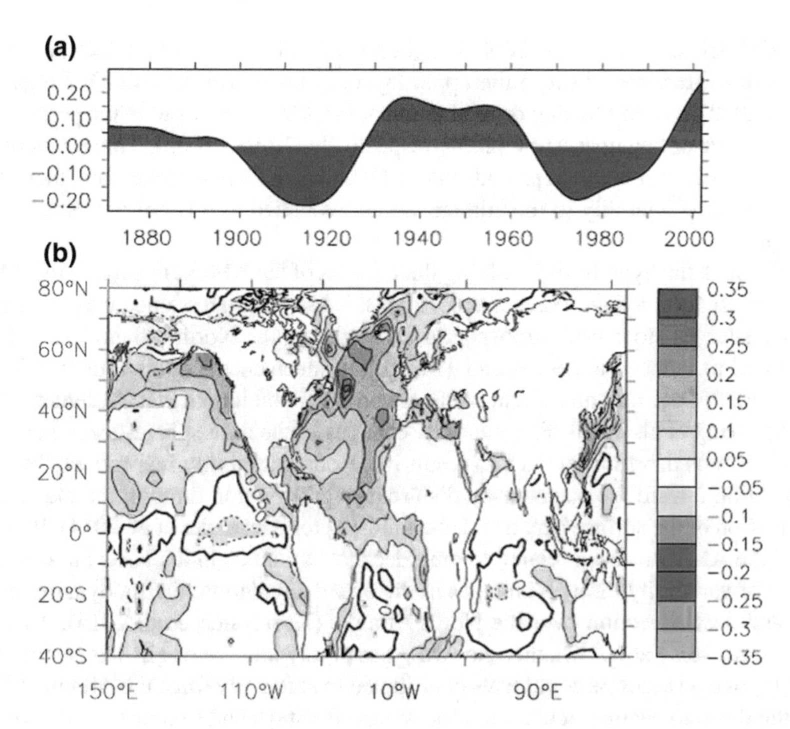

Fig. 1.17 **a** AMV index defined as the annual SSTA from 1871 to 2003 averaged between 0°–60°N and 75°–7,5°W, low-pass filtered and detrended (expressed in °C). **b** AMV spatial pattern defined as the regression of SSTA on the standardized AMV index (expressed in °C per standard deviation). From Sutton and Hodson (2005). © American Association for the Advancement of Science. Used with permission

et al. 2001; Sutton and Hodson 2005; Alexander et al. 2014) (Fig. 1.17a). Studies on reconstructed SST estimates from multiple types of proxies have shown that the AMV persisted hundreds (Delworth and Greatbatch 2000; Gray 2004; Mann et al. 2009; Svendsen et al. 2014) or even thousands (Knudsen et al. 2011) of years prior to the instrumental record. During its positive phase, the AMV spatial pattern of SST presents warm anomalies all across the northern half of the Atlantic basin and insignificant or even of opposite sign anomalies in the southern part. This forms a characteristic interhemispheric SSTA gradient in the broad Atlantic basin. In the North Atlantic, the SSTA depict a comma-shape pattern. Anomalies are more intense in the northernmost part of the North Atlantic (close to Newfoundland and south of Greenland) and extend southward along the eastern part of the basin to the northern half of the tropical Atlantic (Fig. 1.17b).

Regarding its origin, the AMV is usually defined as an internal mode of variability related to decadal variation of the internal ocean mechanisms involved in the AMOC. The subsurface ocean data records are short and scarce, so it is only possible to determine the AMOC decadal variability and its link to the AMV from climate models (Delworth and Mann 2000; Knight 2005; Medhaug and Furevik 2011). Since the AMOC transports heat to the upper layers of the North Altantic (Vellinga and Wood 2002), a strengthening or weakening of the AMOC is linked to the persistence of a positive or negative AMV phase, respectively (Knight 2005). This relationship with the oceanic circulation provide the AMV with certain persistence and, therefore, long term predictability to the climate variations linked to it (Latif and Keenlyside 2011).

Although the hypothesis involving fluctuations of the AMOC to explain the AMV is the most accepted (Alexander et al. 2014), it has been questioned and other processes such as stochastic atmospheric forcing over the North Atlantic have been proposed to induce the SST decadal variability mechanisms (Delworth and Greatbatch 2000; Delworth and Mann 2000; Kwon and Frankignoul 2012; Clement et al. 2015; Zhang et al. 2016). For example, changes in the atmosphere-ocean heat flux or in currents driven by wind that result in decadal variability respectively by heat integration inward the ocean or by disturbance propagation through the mean state circulation of the surface ocean and the atmosphere (Alexander et al. 2014). Particularly, the AMV has been related to multidecadal variations in the NAO atmospheric mode of variability through changes in the ocean circulation, thermodynamic forcing and heat advection over the North Atlantic (Keenlyside et al. 2015). There is also controversy about whether the AMV has purely internal origin, suggesting that anthropogenic aerosols could have contributed to externally force the Atlantic SSTA during the 20th century at decadal time scales (Rotstayn and Lohmann 2002; Terray 2012). From the 1940s there was a rapid increase of the sulfate aerosol emission, especially in the northern hemisphere, which produced a cooling trend of the North Atlantic SST, until the mid-1970s when it began to reduce (Thompson et al. 2010). These works argue that this variation of the sulfate aerosols concentration may have contributed to force the phase shift of the AMV observed around 1970. There are even some controversial works based on model simulations that claim that the AMV in the 20th century was mainly induced by the aerosols external forcing (Booth et al. 2012). Thus, there is no clear consensus as to whether the AMV is dominated by internal variability of the climate system itself or by external forcing (Zhang et al. 2013b).

Due to the large spatial extent of the SST pattern, the AMV has been linked to variations in climate phenomena of diverse regions. One is the hurricane activity in the Atlantic, which has been observed to intensify during years of positive AMV (Goldenberg 2001; Trenberth and Shea 2006; Zhang and Delworth 2006). The greater hurricane activity results from some regional effects of the anomalous North Atlantic warming in the 10°–14°N band of the tropical Atlantic. These are the raise of the SST and the stratospheric moisture as well as the sea level pressure decrease in this region during positive AMV. Such anomalies favor deep convection and atmospheric instability by means of the weakening of the vertical shear of wind, which are the most favorable conditions for the hurricanes development. The AMV has also been related to extratropical atmospheric circulation changes that induce precipitation and temperature anomalies during summer in Europe and North America (Enfield et al. 2001; McCabe et al. 2004; Sutton and Hodson 2005). During the positive phase of the AMV, there is temperature rise in North America and Europe as well as anomalous low surface pressure and more precipitation both in southern North America and western Europe. Another important impact of the AMV is its influence on tropical rainfall in the tropical Atlantic sector (Knight et al. 2006). The characteristic interhemispheric SSTA gradient of the AMV pattern produces anomalous surface pressure difference between north and south Atlantic and an associated anomalous latitudinal shift of the ITCZ accompanied by the tropical rain belt (Zhang and Delworth 2006; Ting et al. 2009; Mohino et al. 2011a; Folland et al. 2001; Alexander et al. 2014). Furthermore, the AMV has been suggested to modulate the non-stationarity of interannual teleconnections between remote areas (López-Parages et al. 2013). During negative AMV phases, the link between the interannual variability in the tropical Atlantic and Pacific basins is favored, with implications in the West African and Euro-Mediterranean climate.

The skill of GCMs to predict climate at decadal time scales is mainly provided by the prescribed atmospheric composition changes (Stocker 2014). However, the initialization of decadal predictions robustly improve GCMs skill in predicting the North Atlantic SST, and therefore the AMV, with respect to other regions at these time scales (Doblas-Reyes et al. 2013). Particularly, the possible relationship between the AMOC and AMV could provide a good opportunity for climate prediction at decadal time scales (Medhaug and Furevik 2011; Latif and Keenlyside 2011). It has been reported that an accurate ocean data assimilation (salinity and temperature), together with SST, would significantly improve the GCMs skill in predicting the North Atlantic state at decadal time scales (Pohlmann et al. 2004; Keenlyside et al. 2008; Smith et al. 2010). Therefore, the GCMs that properly reproduce the AMV teleconnections, as well as its connection with the AMOC, might have a good long-term predictability of climate.

1.3.3 Low-Frequency Variability of the Pacific SST

Apart from the signal of the GW, the SST low-frequency variability in the Pacific Ocean is dominated by the Interdecadal Pacific Oscillation (IPO) mode (Power et al. 1999; Dai 2013). The IPO is the basin-wide manifestation of the Pacific Decadal Oscillation (PDO), which is the main mode of decadal variability of the seasonal extratropical North Pacific SST in winter (Mantua et al. 1997). The time series of both the IPO and PDO variability modes are broadly similar (Salinger et al. 2001), showing no well-defined unique frequency, but some periodicities that can be grouped in a decadal and a multidecadal range of 15–25 and 60–70 years, respectively (Minobe 1999; Chao et al. 2000; Tourre et al. 2001; Mantua and Hare 2002; MacDonald and Case 2005). During the last century, the IPO has shown two positive phases in 1922–1944 and 1978–1998, and a negative one in 1946–1977 (Salinger et al. 2001) (Fig. 1.18a). When the IPO is in its positive phase, there are positive SSTA in the tropical Pacific SSTA, more to the east than in the western side of the basin. While the SSTA in the North Pacific are cold as well as in the southern basin, close to

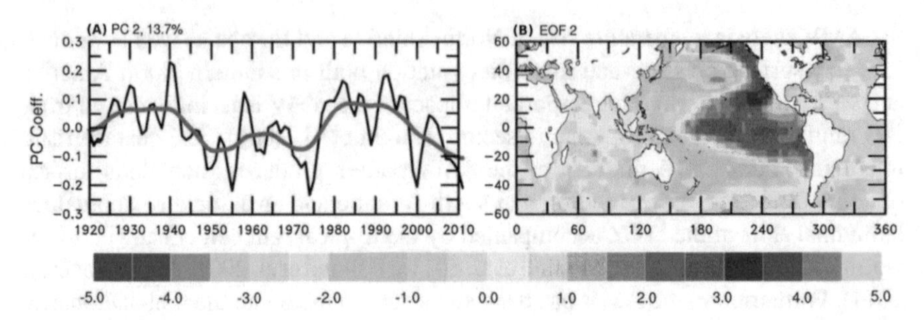

Fig. 1.18 IPO index (**a**) and pattern (**b**) defined, respectively, as the temporal and the spatial patterns of the second leading empirical orthogonal function (EOF2) of annual and global SSTA, 3-year moving averaged, between 1920 and 2011. The red line is the low-pass filtered IPO index using a 9-year running mean. Units of the pattern are K per PC coefficient. Adapted from Dai (2013). © Springer-Verlag. Used with permission

New Zealand (Trenberth and Hurrell 1994; Meehl et al. 2009) (Fig. 1.18b). During negative IPO phases, the effects are the opposite.

With regard to its origin, the IPO is typically related to an internal mode of the Pacific climate system (Meehl et al. 2009). In contrast to the Atlantic, the Pacific does not have a well-defined thermohaline circulation, so the IPO is not related to the internal variability of the ocean but to processes of the ocean-atmosphere interaction (Deser et al. 2010). Indeed this mode of variability is considered as the result of combined processes throughout the Pacific sector instead of a single phenomenon (Newman et al. 2016). But the mechanisms that generate the transient SSTA pattern of the IPO have led to extensive debate. Some works conclude that the IPO is stochastically determined by noise produced by different atmospheric and oceanic forcing processes, whose interaction gives persistence to the mode (Newman et al. 2003; Schneider and Cornuelle 2005; Shakun and Shaman 2009). While others have suggested a mechanism that connects the tropics and extratropics with atmosphere-ocean interactions involved that promotes the formation and persistence of the IPO pattern (Deser et al. 2004; Meehl and Hu 2006; Farneti et al. 2014; Newman et al. 2016). There is also no agreement as to whether the phase changes of the IPO are triggered by the tropical (principally by ENSO events) (Newman et al. 2003; Deser et al. 2004) or by the extratropical Pacific sector (such as perturbations in the Aleutian Low) (Miller and Schneider 2000).

Due to the broad extent of the pattern and all the climate processes involved in the IPO, it influences the decadal variability in different regions worldwide. The IPO has negative impacts on tropical rainfall over the Northern Hemisphere (Wang et al. 2013; Krishnan and Sugi 2003). In southwestern North America, anomalously rainy and drought years respectively coincide with positive and negative phases of the IPO (Meehl and Hu 2006; Dai 2013). Also changes in North Pacific marine ecosystems have been correlated with the IPO phases, showing positive and negative impacts on fishery productivity in the coasts of Alaska and the United States, respectively

(Mantua et al. 1997; Chavez et al. 2003). The IPO has been shown to modulate the intensity of ENSO and its impacts on remote regions. During positive phases of the IPO, the tropical Pacific is anomalously warmer and the relationship of El Niño with South American rainfall is more intense than during negative IPO years (Andreoli and Kayano 2005). It also modulates the ENSO teleconnections globally (Meehl and Hu 2006). For example, negative IPO phases favor a strong relationship between the Australian precipitation inter-annual variability and the ENSO, whereas during positive phases there is no such connection (Power et al. 1999; Arblaster et al. 2002). The IPO has also been related to offsets or intensifications of the GW trend related to its vast temperature signal in the tropical Pacific (Gu and Adler 2013; Meehl et al. 2013).

Current GCMs show very little skill in predicting the SST anomalies associated with the IPO (van Oldenborgh et al. 2012; Kim et al. 2012; Doblas-Reyes et al. 2013; Kirtman et al. 2013), which would compromise decadal predictability of phenomena related to it. Predictions based on a linear inverse model[3] show better skill to predict the IPO, on time scales of some years (Newman et al. 2003; Alexander et al. 2008).

1.4 The Monsoon System

The global monsoon system arises as the response of the climatic system to the annual cycle of the solar radiative forcing. It is characterized by a mode of the annual precipitation variations that mainly describes the difference of precipitation during summer and winter. It presents a contrast of rainfall between tropical and subtropical regions on both hemispheres. During the boreal summer there is more precipitation in the northern hemisphere and less in the south, while in winter the pattern is reversed (Fig. 1.19) (Wang et al. 2011). The latitudinal location where the majority of the

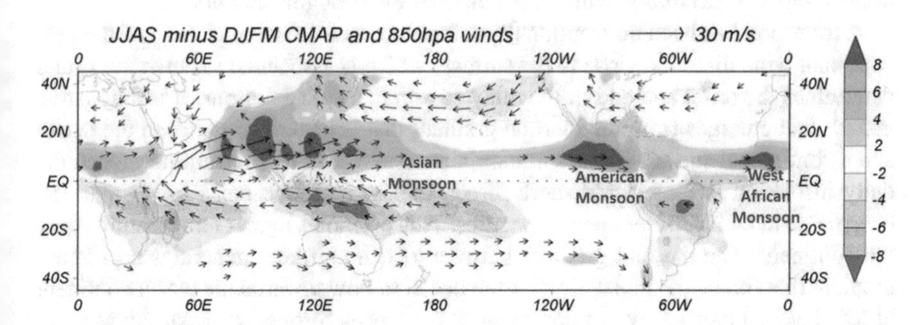

Fig. 1.19 Difference of precipitation (mm day^{-1}) and 850 hPa winds between June–September and December–March. Modified from Wang and Ding (2008). © Elsevier. Used with permission

[3]This method determines the multivariate noise from observed variations of the Pacific SST, assuming that the Pacific variability is determined by stochastic forcing.

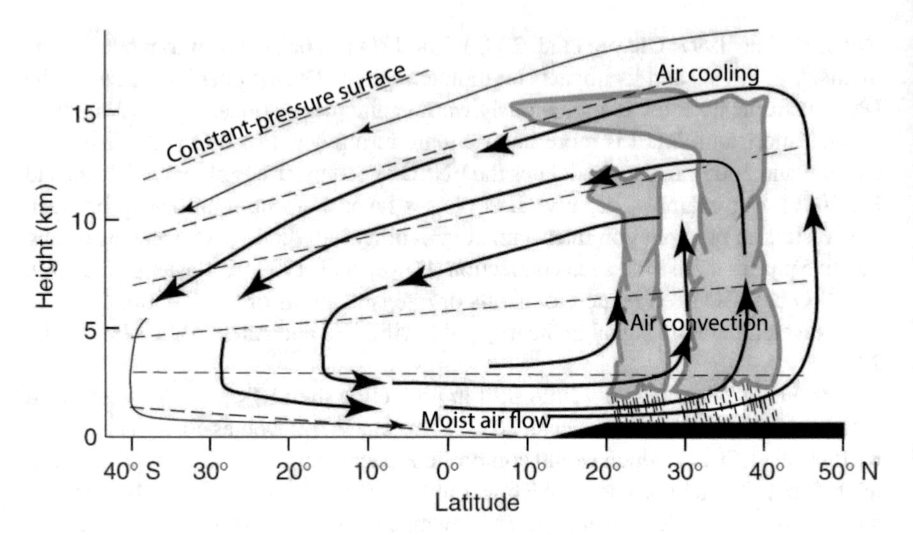

Fig. 1.20 Schematic representation of the monsoon circulation. Adapted from Webster and Fasullo (2003). © Elsevier. Used with permission

monsoonal precipitation falls is strongly related to the ITCZ position (Webster and Fasullo 2003). The band of maximum tropical rainfall is generally located poleward of the ITCZ latitudinal position over continental surface. While over oceanic regions the bulk of precipitation occurs in the same position as the ITCZ. However, the monsoon is a regional phenomenon characteristic of certain tropical and subtropical regions with unique local properties on which the monsoon system depends, such as topography and specific features of the ocean and continental surface. The regions with the most characteristic monsoon systems are South Asia, Australia, West Africa, North and South America (Fig. 1.19). These regions are in general very susceptible to the monsoon variability, which can cause severe droughts and floods.

A monsoon has been conventionally referred to as a seasonal change of the predominant wind direction close to the surface, which is associated to heavy rains and determines the onset and end of the rainy season in tropical regions. It is a dynamic system that emerges from the thermal gradient that is generated between the ocean and a continental surface in the summer hemisphere. During the summer, when the daily insolation is longer and more direct, the land surface heats faster than the ocean, due to their different heat capacities. This promotes higher temperatures over the continent. The thermal gradient between ocean and land generates a pressure gradient that forces the moist air over the ocean to flow towards the warm continent at low levels. Over the warm continent, the air rises through thermal convection. Such mechanism promotes a local decrease of the atmospheric pressure and forces moist air to condense in upper levels and precipitate. At high levels the air radiatively cools and descends again over the ocean, increasing the atmospheric pressure locally (Fig. 1.20).

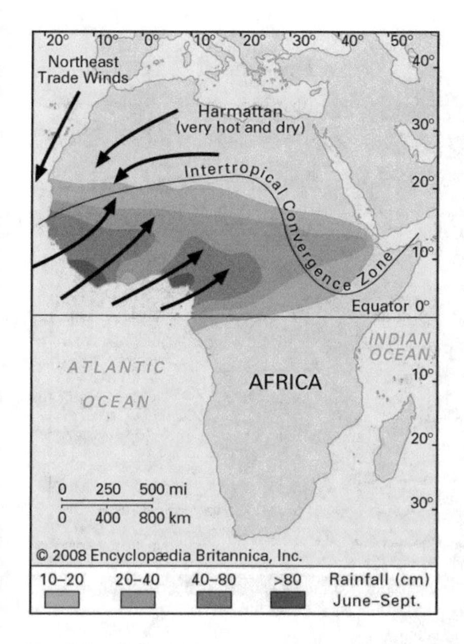

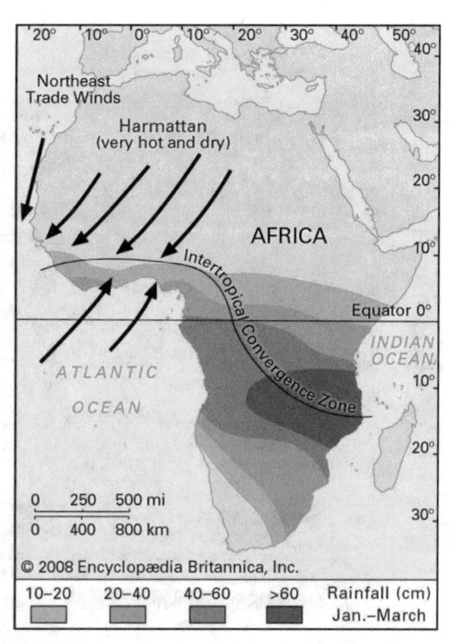

Fig. 1.21 Schematic representation of winds and rainfall of the West African monsoon in the boreal summer (left) and winter (right). By courtesy of Encyclopaedia Britannica, Inc., ©2008; used with permission

An overview of the mechanism of a monsoon system has been described so far. But this Thesis focuses on rainfall variability in tropical regions located in West Africa and north of South America. Therefore, the particular features of the West African Monsoon (WAM) and the South American Monsoon (SAM) systems are presented in further detail in the following sections.

1.4.1 The West African Monsoon

The WAM system is created by the ocean-land thermal contrast between the Gulf of Guinea in the tropical Atlantic and the west of the African continent, covering from the Gulf of Guinea coast to the Sahara desert to the north. Broadly, the rainfall band associated to the WAM follows the transient position of the ITCZ across the African continent throughout its annual cycle (Fig. 1.21). During the boreal summer the precipitation is zonally enhanced more intensely at around 10°N and gradually less to the north through the Sahel, up to the edge with the Sahara desert, at around 20°N. In the boreal winter, the monsoon rainfall occurs along central-south Africa.

However, the specific features of the WAM circulation in the boreal summer are somewhat more complex than described so far, as depicted in Fig. 1.22. At lower levels, the southwesterly monsoon flow advects moist air from the Tropical Atlantic

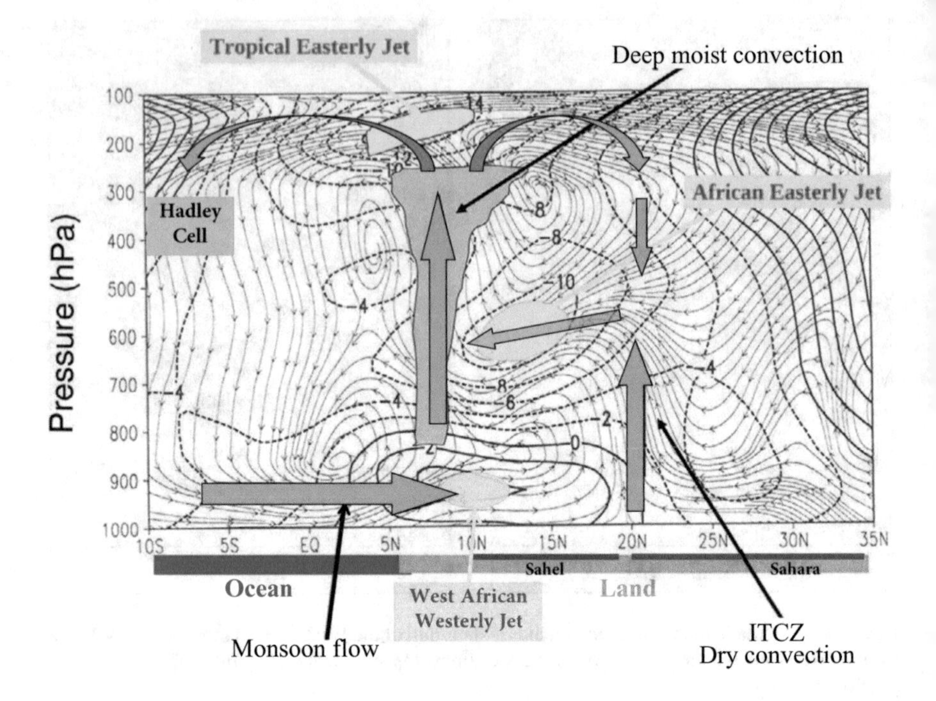

Fig. 1.22 Schematic representation of the main circulation structures in the vertical section associated with the WAM in the boreal summer. Mean July to September meridional circulation (stream lines) and associated mean zonal wind (m s^{-1}, contours). Green arrows indicate the main vertical and latitudinal air flows and yellow areas the zonal jets. Modified from Hourdin et al. (2010). © American Meteorological Society. Used with permission

northward to the continent. The majority of this moist flow sustains the cloud formation that produces heavy rains along a band of deep convection extending around 10°N. This convective band constitutes the rising branch of the Hadley cell. It forces deep air convection and, thus, its water vapor content is condensed to precipitate. The southward branch of wind of the Hadley cell in the high troposphere is deflected westward due to the Coriolis effect. These winds create a zonal belt at around 5°N of easterly winds which is referred to as the Tropical Easterly Jet (TEJ). The TEJ has its origin in the upper level Asian Monsoon circulation, which is very intense due to the strong thermal contrast between the Indian Ocean and the Tibetan plateau and to the steep orography. At low levels, there is the West African Westerly Jet (WAWJ) placed around 10°N that results from the convergence of the trade winds in the tropical Atlantic, close to the coast of Africa (Grodsky et al. 2003). The WAWJ introduces moisture inland from the west additional to the monsoon flow from the south.

North of the deep convection band, the Hadley cell of the northern hemisphere climatologically induces subsidence over the Sahara desert. The descending air is adiabatically heated and prevents moisture to penetrate, making the Sahara one of the warmest and driest regions in the world. During the boreal summer the northern Hadley circulation is weakened and the intense surface heating in the Sahara generates a heat low of surface pressure. Associated with the Saharan heat low, there are the northeasterly Harmattan winds that converge with the southwesterly monsoon flow at low levels (Nicholson 2009; Rodríguez-Fonseca et al. 2015). This creates a discontinuity of different air masses (dry from the north and wet from the south) at approximately 20°N. This band of wind convergence located in the border between the Sahel and the Sahara desert comprises the ITCZ over West Africa during the boreal summer. Associated with the wind convergence, there is dry air convection up to middle levels, where it meets the descending branch of the Hadley cell and diverges. This dry and warm air mass then flows southward above the cooler and wetter northward monsoon flow. At around 650 hPa, the air that flows southward is deflected by the Coriolis effect creating the mid-level African Easterly Jet (AEJ). Climatologically the core of the AEJ stands at around 15°N.

In the southern part of the Sahel, heavy rains are typically produced by rapid and small-scale squall lines forced by the deep convection. But to the north, where deep convection is inhibited, most precipitation occurs in mesoscale convective systems, which are larger-scale systems than the squalls and last for several hours, although they generate less rainfall (Nicholson 2013). In this part of the Sahel, the specific humidity provided by the monsoon flow is notably lower than to the south, specially close to the Sahara desert. Instead, the moisture supply in this region comes from recycled water, mostly from local evaporation, which is crucial for the Sahel rainfall (Trenberth 1999; Nicholson 2009).

1.4.2 The South American Monsoon

The SAM is part of a unique American monsoon system, together with the one of North America (Vera et al. 2006b; Liebmann and Mechoso 2011; Marengo et al. 2012). The intense precipitation associated with the SAM during the rainy season follows the annual migration of the ITCZ across South America. The SAM starts around September, after the boreal summer, with intensification of rainfall over the equatorial areas. The peak of the SAM occurs during the austral summer, from December to February, when the ITCZ presents its maximum southward displacement over the continent. It is then when the intense precipitation activity takes place in the south of the Amazon region. In the following months, the maximum rainfall shifts back north. Then, the annual precipitation of central and northernmost Amazonia peaks from March to May, respectively. The short rainy season of the semiarid Northeast region of Brazil also occurs between March and May (Moura and Shukla 1981; Rodrigues et al. 2011). It is then when the ITCZ reaches its southernmost position over this region, promoting deep convective cloud formation and occasional heavy rain events. Later, the rainy season associated to the SAM systems comes to its end.

The SAM mainly affects the north of the continent, which is mostly tropical, and as a consequence the seasonal contrast of temperature is softer than other monsoon

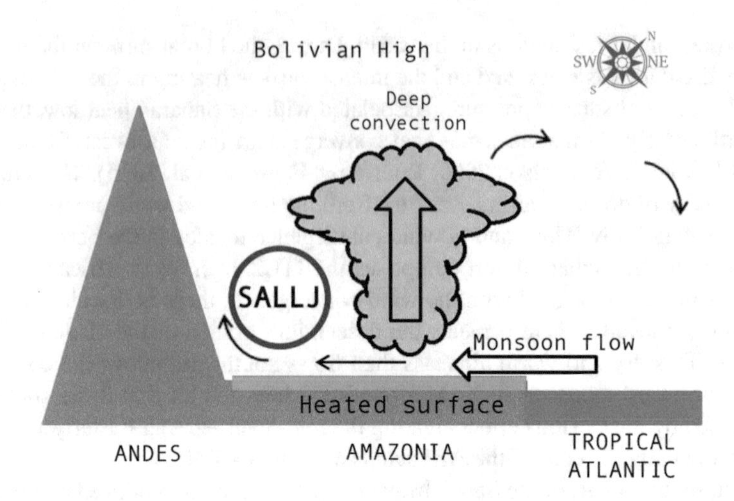

Fig. 1.23 Schematic representation of the northeast-southwest circulation of the SAM and the main associated large-scale atmospheric structures

systems involving subtropical regions, such as the WAM. However the SAM has a well-defined northeast-southwest monsoon circulation structure (Fig. 1.23). During the austral summer, the thermal contrast between land and ocean amplifies the trade winds that enhance the monsoon inflow of moisture from the tropical Atlantic. The humid air then rises over the heated continental surface through convention promoting rainfall. In addition, farther west, the Andean mountain range forces an abrupt air rise. At upper levels, over the rising branch there is a characteristic semi-stable anticyclonic structure with divergent wind called the Bolivian High (Lenters and Cook 1997). This characteristic structure only develops during the austral summer, when it can be found at 200 hPa and centered around 17°S–70°W. The Bolivian High is associated with enhanced deep convection and intense rainfall over Amazonia and northeastern Argentina. To the east and slightly north of the Bolivian High, an upper-level low with cyclonic circulation coexists. This structure produces air convergence with a branch of subsidence below, which closes the monsoon circulation.

The convection associated with the monsoon circulation of the SAM system forms the South Atlantic Convergence Zone (SACZ) (Grodsky and Carton 2003; Carvalho et al. 2004). The SACZ is a cloudy and rainy convective band that expands south-eastward from the Amazon to the southwestern tropical Atlantic, passing throughout the southeast of Brazil (Fig. 1.24). At low levels, the intense moisture flux entering inland from the northeast is forced by the steep orography of the Andes to flow along the eastern edge of the mountain range toward higher latitudes. This wind flow forms the South American Low Level Jet (SALLJ) (Marengo et al. 2004; Vera et al. 2006a). The SALLJ distributes large amounts of moisture toward Amazonia and central and southeast Brazil. The SALLJ advection of humid air also contributes to increase the SACZ.

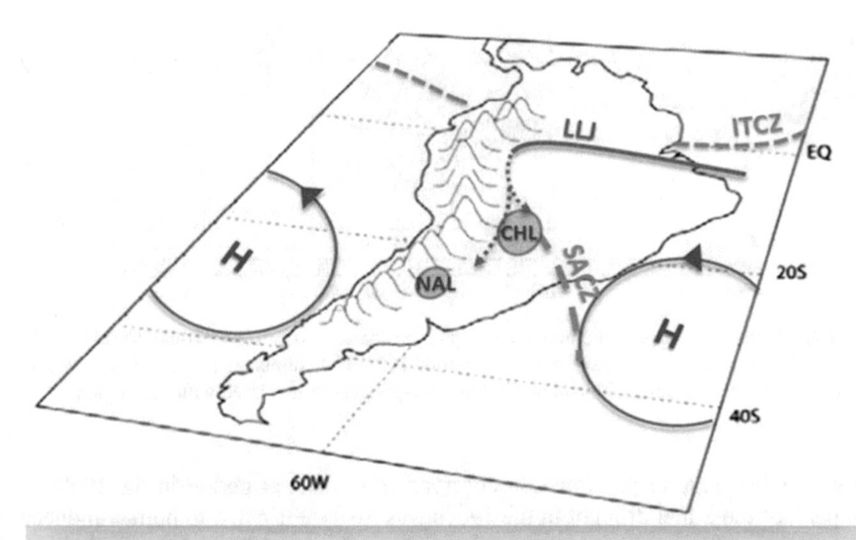

Fig. 1.24 Horizontal representation of the low-level circulation associated with the SAM system. *Source* eumetrain.org

1.5 Decadal Variability of Sahel and Northern Brazil Rainfall

This Thesis addresses the low-frequency variability of tropical rainfall in the regions of the Sahel, the Amazonia and the Northeast of Brazil and the influence of the SST modes of decadal-to-multidecadal variability. Before introducing the results, this section presents a review of other works that have also focused on this topic.

1.5.1 Decadal Variability of Sahel Rainfall

The variations of the WAM system that result in decadal-to-multidecadal changes of the Sahel rainfall have been related to the principal modes of SST variability at these time scales. Since the Sahel rainfall has shown marked decadal variability along the 20th century, it has focus the attention of many works (e.g. Folland et al. 1986; Giannini et al. 2003; Caminade and Terray 2010; Mohino et al. 2011a). The

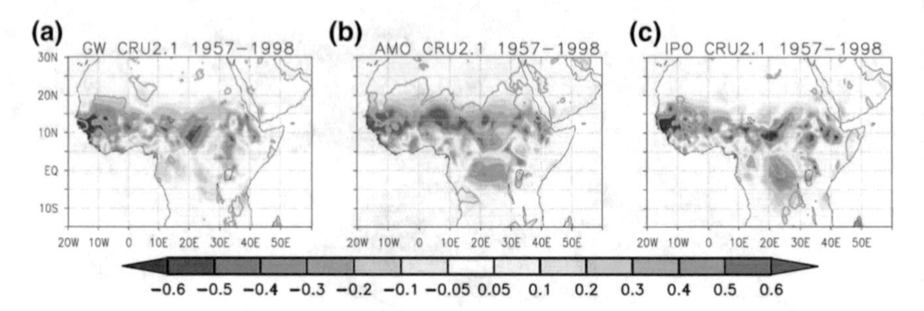

Fig. 1.25 Regression maps of the summer precipitation anomalies (mm day^{-1}) onto the (**a**) GW, (**b**) AMO and (**c**) IPO indices standardized for the period 1957–1998 (units are mm day^{-1} per standard deviation). Adapted from Mohino et al. (2011a). © Springer-Verlag. Used with permission

long term decrease of precipitation observed from the wet period in the 1950s to the peak of the Sahel drought in the 1980s was firstly attributed to human-induced changes in land use, which increased the surface albedo of the region affecting the WAM circulation (Charney 1975). But there is currently a broad agreement that the main driver of Sahel precipitation at decadal time scales is SST variability amplified by land surface processes (Zeng and Neelin 1999; Giannini et al. 2003; Kucharski et al. 2013). Therefore, the concern about the anthropogenic forcing of Sahel rainfall shifts now is related with the SST response to the gas emissions originating from human activity and the direct local effect of the carbon dioxide rather than to land uses (Dong and Sutton 2015; Biasutti 2016). Mohino et al. (2011a) suggested that the decadal oscillations of the Sahel precipitation is the result of the competition between the main internal and external modes of SST variability. Particularly, they showed a dominant role of the AMV in driving the Sahel precipitation at decadal timescales, a secondary one of the IPO and a minor but also important influence of the GW.

The GW has been linked to the observed decrease of Sahel rainfall along the second half of the 20th century (Fig. 1.25a). The quasi-linear trend of SST warming induced a tendency to enhance subsidence over West Africa resulting in the decay of Sahel rainfall in the second half of the 20th century, after the wet period of the 1960s (Giannini et al. 2003; Biasutti and Giannini 2006; Mohino et al. 2011a). This favored the Sahel drying while precipitation near the Gulf of Guinea coast was more abundant. These effects are mainly attributed to the tropical fringe of the GW pattern of SST. The tropical warming in the Indian and Pacific oceans is related to increased convergence over these areas and subsidence over West Africa that reduces Sahel rainfall. On the other hand, the tropical Atlantic warming reduces the thermal contrast between the Gulf of Guinea and West Africa weakening the WAM. As a result, the ITCZ presents an anomalous southward shift associated with the GW.

The future evolution of the Sahel rainfall regime has also been studied using future projections with a large sample of different climate models (Biasutti et al. 2008; Monerie et al. 2012; Biasutti 2013; Giannini et al. 2013). They show an inversion of the precipitation tendency as a response to increased GHGs concentration

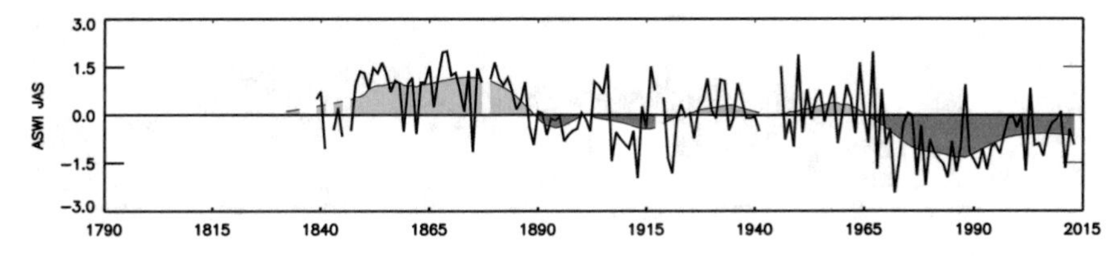

Fig. 1.26 Interannual (black index) and low-frequency (shaded curve) standardized ASWI for JAS. Adapted from Gallego et al. (2015). © John Wiley and Sons. Used with permission

with respect to the dry period of the 1980s and consistent with the late 20th century recovery. Such inversion consists of an increase of precipitation in the central and eastern part of the Sahel while the western part dries. The western Sahel drying is more pronounced during the beginning of the boreal summer and the precipitation increase occurs during the late monsoon season (Biasutti 2013). The drying results from anomalous subsidence over the region induced by the tropical SST warming, preventing deep convection. Whilst the rainfall increase is attributed to the land surface warming by the direct effect of the carbon dioxide, favoring moisture convergence and precipitation (Monerie et al. 2012; Gaetani et al. 2017). This recovery is related to an intensification of rainy events rather than to an increment of their occurrence (Panthou et al. 2014). However, these effects are difficult to assess due to the limitations of GCMs (Cook 2008). For example, the regional processes of convection, which are crucial for the Sahel rainfall, are in general poorly represented by the models (Panthou et al. 2014). Also because of the uneven response among the models to the direct effect of carbon dioxide or the distribution of increased SST due to the GHGs rise (Giannini et al. 2013; Biasutti 2013; Park et al. 2015; Gaetani et al. 2017).

Many works have highlighted the influence of the AMV on Sahel precipitation low- frequency variability (e.g. Zhang and Delworth 2006; Hoerling et al. 2006; Knight et al. 2006; Ting et al. 2009, 2011; Mohino et al. 2011a; Martin and Thorncroft 2014). The AMV has a positive relationship with the WAM (Fig. 1.25b). So during positive AMV phases the boreal summer precipitation is enhanced over the Sahel and reduced to the south, along the coast of the Gulf of Guinea, while the opposite occurs during negative phases. This impact is produced by the anomalous meridional shift of the ITCZ along the entire tropical Atlantic sector in response to the interhemispherical thermal gradient of the SST pattern of the AMV. When the North Atlantic basin is anomalously warm or cold, the ITCZ displaces to the north or to the south during the respective phases of the AMV.

The IPO has a negative impact on the Sahel rainfall (Fig. 1.25c), promoting dry or wet conditions during positive or negative phases, respectively. In contrast to the AMV, only a few works have shown the link between the IPO and the Sahel. Joly (2008) first showed a statistical link between them and Molion and Lucio (2013) suggest a possible modulation of the effect of ENSO on Sahel due to the IPO from observations. Mohino et al. (2011a) show the direct effect of the IPO on Sahel rainfall

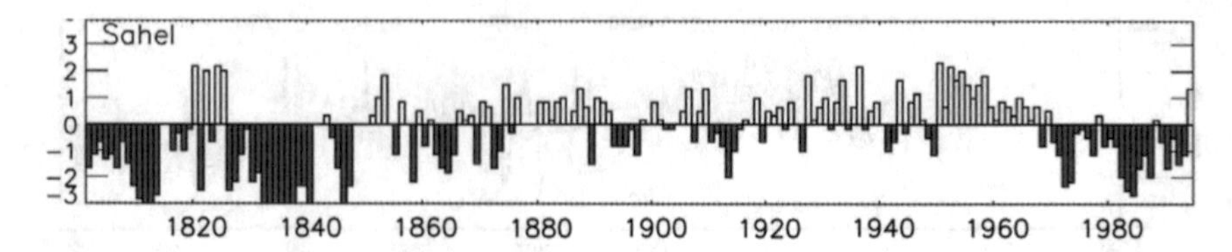

Fig. 1.27 "Wetness" index obtained from the semi-quantitative spatial reconstruction of the African precipitation averaging the regions spanned by the Sahel. Adapted from Nicholson et al. (2012). © University of Washington. Used with permission

decrease of the second half of the 20th century and the mechanisms involved in such connection by means of simulations with an atmosphere climate model. The tropical part of the IPO pattern is the most relevant to produce such connection. During the positive phase of the IPO, associated with the anomalous tropical SST warming there is enhanced upper-level convergence and subsidence over the Sahel that reduces rainfall (Mohino et al. 2011a). The opposite occurs during negative IPO phases.

Nevertheless, despite all the understanding of the WAM gathered for the observational period and even the future, there is very little documentation on the evolution of the Sahel precipitation during the 19th century. There are only a few sources that document this period. One of these works is the one of Gallego et al. (2015), in which they define an index based on the persistence of southwesterly low-level winds in a region over the Atlantic, close to West Africa (29°–17°W; 7°–13°N), and calculated from historical measurements of JAS wind direction, complete since 1839, which is called African Southwesterly Index (ASWI) (Fig. 1.26). The ASWI is strongly correlated with the observed Sahel precipitation since 1900 and is, therefore, presented as a good indicator of its variability. The other work addressing this early period is the one by Nicholson et al. (2012) who, in turn, made a semi-quantitative reconstruction of rain in Africa during the 19th century. It is based on the use of descriptive documentary data of rainfall throughout the continent to complete the scarce information of rain-gauge stations. To do this, the continent was divided into 90 regions with uniform precipitation regimes. Then, time series are defined taking positive/negative values from 1 to 3 depending on whether the year was moderately, abundantly or severely rainy/dry, according to the documentation compiled at each region and following a selection criterion. The Sahel spans 6 of these regions and its rainfall index is the average of their time series (Fig. 1.27). These two works show evidences of the existence of a humid period in the Sahel during some decades in the second half of the 19th century. Although the question as to the origin of this wet period has not yet been addressed and is unknown.

1.5.2 Decadal Variability of Northern Brazil Rainfall

The long-term trends in both the Amazon and the Northeast of Brazil regions have focus the attention of several studies (Robertson and Mechoso 1998; Marengo 2004, 2009; Marengo et al. 2009; da Silva 2004). But the observed data have typically short time span and are scarce over the regions. That is why the results regarding the trend of the entire period observed are somewhat confusing. Over the Northeast the observed tendency of precipitation is mostly to decrease, although not quite intensely nor statistically significant in all the stations studied of the area (da Silva 2004). In the Amazon regions the rainfall long-term trend is even more confusing, with different results among studies that depend on the data sets, the period, the season or the area of the Amazon that they analyze (Marengo et al. 1998; Marengo 2004; Espinoza Villar et al. 2009; Buarque et al. 2010). In general because of the short sample of the observed data available, the long-term rainfall trends in both the Amazon and the Northeast regions are suggested to be part of interdecadal changes rather than to a longer-than-decadal tendency (Zhou and Lau 2001; Marengo 2004, 2009). However, studies using atmospheric climate models forced with rising SST under increased carbon dioxide conditions show that the Amazon is expected to experience longer dry seasons but no significant changes during the SAM season (Harris et al. 2008). Others using future projections of climate models show an increase of drought events in the Northeast of Brazil during its rainy season associated with an increment of GHGs concentration (Marengo et al. 2016). Therefore, although the effects of the GW trend on Amazonia and the Northeast regions are still scarcely constrained in observations, climate models can help to assess the evolution of rainfall in both regions under intense GW conditions.

The AMV is related to the rainfall low-frequency variability in both, the Amazon and the Northeast regions, during their respective rainy seasons. This is associated with the anomalous ITCZ latitudinal shifts during the austral summer forced by the characteristic interhemispheric thermal gradient of the AMV pattern of SST

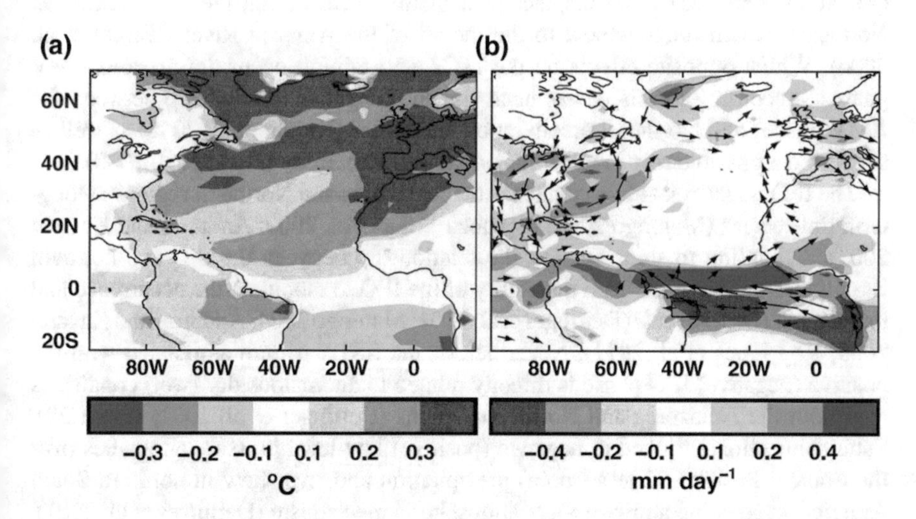

Fig. 1.28 Near surface annual temperature and March to May precipitation associated with positive AMV phase in a climate model simulation. Adapted from Knight et al. (2006). © John Wiley and Sons. Used with permission

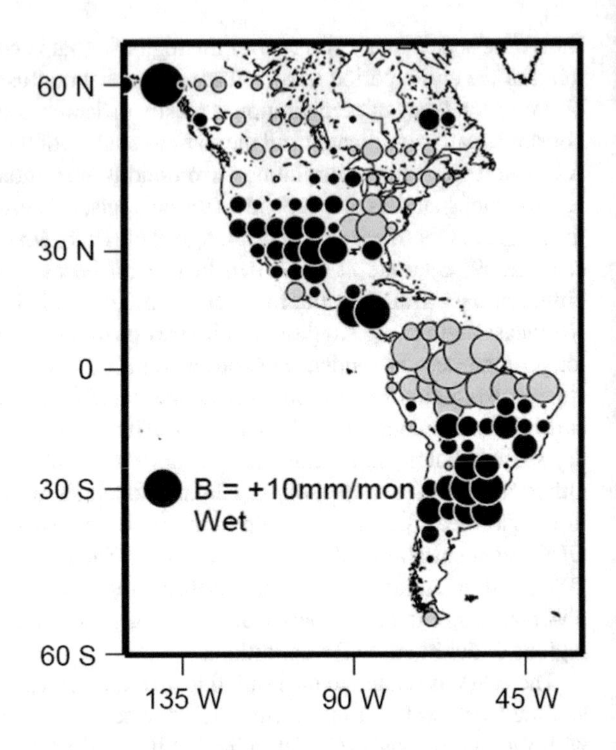

Fig. 1.29 Regression coefficients between the October-September mean precipitation and the IPO index defined as the first leading mode of an EOF analysis of the Pacific SST field from which the variability associated with the ENSO has been previously removed by linear regression. Dark and light circles indicate positive and negative values, respectively. Adapted from Dettinger et al. (2001). © Elsevier. Used with permission

(Fig. 1.28). During positive AMV phases the tropical SSTA gradient hinders the typical southern maximum displacement of the ITCZ during the rainy season in Northeast, which remains next to the mouth of the Amazon River (Knight et al. 2006). Whilst opposite effects on the ITCZ and rainfall occur during cold AMV phases. According to this, it has been shown a negative relationship between the AMV phases and Northeast precipitation anomalies (Knight et al. 2006) as well as suggested a possible role in the Amazon rainfall variability (Good et al. 2008).

The IPO is also related to changes in the Amazonia and Northeast regions at long-term time scales (Marengo 2009; Espinoza Villar et al. 2009; Andreoli and Kayano 2005). According to some studies, the relationship between the IPO and northern Brazilian rainfall comes from the ability of the IPO to modulate the occurrence and intensity of ENSO events (Dettinger et al. 2001; Marengo 2004; Andreoli and Kayano 2005; Rodrigues et al. 2011). Nevertheless, the SSTA pattern associated with the positive (negative) IPO phase is directly related to anomalous dry (wet) conditions over both the Amazonia and Northeast regions (Dettinger et al. 2001) (Fig. 1.29). Such connection is linked to negative (positive) low-level pressure anomalies over the tropical Pacific and less (more) precipitation and river flow in northern South America, suggesting an ENSO-like atmospheric mechanism (Dettinger et al. 2001).

1.6 Objectives

The main objective of this Thesis is to gain a better understanding of the influence of the decadal-to-multidecadal variability of SST on precipitation in three particular regions: The Sahel, the Amazonia and the Northeast of Brazil. In the first part of this Thesis, a multi-model analysis is done to determine whether the state-of-the-art CMIP5 models are able to reproduce the observed connection between the main modes of SST decadal-to-multidecadal variability and precipitation to understand the processes involved. The specific goals in this part are to:

- Characterize the main modes of SST variability (GW, AMV and IPO) in observations and CMIP5 simulations.
- Analyze the relationship between the SST modes and rainfall in the Sahel, Amazonia and Northeast that the models reproduce, on average, comparing with observations and shed some light on the reasons for models spread.
- Look into the associated atmospheric dynamics that lead to such links between the SST modes and precipitation in the models and its consistency with observations.
- Discuss the possible differences between externally forced and unforced simulations and seek whether in the future projections given by CMIP5 models these relationships are expected to change or not.
- Assess the contribution of the main modes of SST decadal-to-multidecadal variability to the total rainfall variance at these time scales in the regions of interest, both in observations and CMIP5 simulations.

In a second part of this Thesis, a particular case study is addressed. We seek to provide evidences that support the existence of an anomalous decadal rainy period in the Sahel over the late-19th century and find out the processes that caused it. The key questions that are addressed in this part are:

- Can the long rainy period of the late-19th century be reproduced with an AGCM forced with observed SSTs?
- What are the atmospheric mechanisms that explain such humid period and what is their relation with the SST?
- Which is the key basin controlling the Sahel decadal shifts in rainfall in the late-19th century?

References

Ackerley, D., Booth, B.B., Knight, S.H., Highwood, E.J., Frame, D.J., Allen, M.R., Rowell, D.P.: Sensitivity of twentieth-century Sahel rainfall to sulfate aerosol and CO_2 forcing. J. Clim. **24**, 4999–5014 (2011)

Alexander, M.A., Matrosova, L., Penland, C., Scott, J.D., Chang, P.: Forecasting Pacific SSTs: linear inverse model predictions of the PDO. J. Clim. **21**, 385–402 (2008). https://doi.org/10.1175/2007JCLI1849.1

Alexander, M.A., Halimeda Kilbourne, K., Nye, J.A.: Climate variability during warm and cold phases of the Atlantic Multidecadal Oscillation (AMO) 1871–2008. J. Mar. Syst. **133**, 14–26 (2014). https://doi.org/10.1016/j.jmarsys.2013.07.017

Allan, R., Ansell, T.: A new globally complete monthly historical gridded mean sea level pressure dataset (HadSLP2): 1850–2004. J. Clim. **19**, 5816–5842 (2006)

Ambrizzi, T., Souza, E.B., Pulwarty, R.S.: The Hadley and Walker regional circulations and associated ENSO impacts on South American seasonal rainfall. In: The Hadley Circulation: Present, Past and Future, pp. 203–235 (2004)

Andreoli, R.V., Kayano, M.T.: ENSO-related rainfall anomalies in South America and associated circulation features during warm and cold Pacific decadal oscillation regimes. Int. J. Climatol. **25**, 2017–2030 (2005). https://doi.org/10.1002/joc.1222

Arblaster, J., Meehl, G., Moore, A.: Interdecadal modulation of Australian rainfall. Clim. Dyn. **18**, 519–531 (2002)

Bader, J., Latif, M.: The impact of decadal-scale Indian Ocean sea surface temperature anomalies on Sahelian rainfall and the North Atlantic Oscillation. Geophys. Res. Lett. **30** (2003)

Biasutti, M., Giannini, A.: Robust Sahel drying in response to late 20th century forcings. Geophys. Res. Lett. **33**(L11), 706 (2006). https://doi.org/10.1029/2006GL026067

Biasutti, M.: Forced Sahel rainfall trends in the CMIP5 archive. J. Geophys. Res. Atmos. **118**, 1613–1623 (2013)

Biasutti, M.: Hydrology: what brings rain to the Sahel? Nat. Clim. Change **6**, 897–898 (2016)

Biasutti, M., Held, I.M., Sobel, A.H., Giannini, A.: SST forcings and Sahel rainfall variability in simulations of the twentieth and twenty-first centuries. J. Clim. **21**, 3471–3486 (2008). https://doi.org/10.1175/2007JCLI1896.1

Bichet, A., Kushner, P.J., Mudryk, L., Terray, L., Fyfe, J.C.: Estimating the anthropogenic sea surface temperature response using pattern scaling. J. Clim. (2015). https://doi.org/10.1175/JCLI-D-14-00604.1

Bony, S., Colman, R., Kattsov, V.M., Allan, R.P., Bretherton, C.S., Dufresne, J.L., Hall, A., Hallegatte, S., Holland, M.M., Ingram, W., Randall, D.A., Soden, B.J., Tselioudis, G., Webb, M.J.: How well do we understand and evaluate climate change feedback processes? J. Clim. **19**, 3445–3482 (2006). https://doi.org/10.1175/JCLI3819.1

Booth, B.B.B., Dunstone, N.J., Halloran, P.R., Andrews, T., Bellouin, N.: Aerosols implicated as a prime driver of twentieth-century North Atlantic climate variability. Nature **484**, 228–232 (2012). https://doi.org/10.1038/nature10946

Bozec, A., Lozier, M.S., Chassignet, E.P., Halliwell, G.R.: On the variability of the mediterranean outflow water in the North Atlantic from 1948 to 2006. J. Geophys. Res. Ocean. **116**, 1–18 (2011). https://doi.org/10.1029/2011JC007191

Broecker, W.S.: The great ocean conveyor. In: AIP Conference Proceedings, vol. 247, pp. 129–161, AIP (1992)

Broecker, W.S.: Thermohaline circulation, the Achilles heel of our climate system: Will man-made CO_2 upset the current balance? Science **278**, 1582–1588 (1997). https://doi.org/10.1126/science.278.5343.1582

Brönnimann, S.: Impact of El Niño-Southern Oscillation on European climate. Rev. Geophys. **45**, RG3003 (2007). https://doi.org/10.1029/2006RG000199.1.INTRODUCTION

Buarque, D.C., Clarke, R.T., Mendes, C.A.B.: Spatial correlation in precipitation trends in the Brazilian Amazon. J. Geophys. Res. Atmos. **115**, 1–14 (2010). https://doi.org/10.1029/2009JD013329

Caminade, C., Terray, L.: Twentieth century Sahel rainfall variability as simulated by the ARPEGE AGCM, and future changes. Climate Dynamics **35**, 75–94 (2010)

Carvalho, L.M.V., Jones, C., Liebmann, B.: The South Atlantic convergence zone: intensity, form, persistence, and relationships with intraseasonal to interannual activity and extreme rainfall. J. Clim. **17**, 88–108 (2004)

Chao, Y., Ghil, M., McWilliams, J.C.: Pacific interdecadal variability in this century's sea surface temperatures. Geophys. Res. Lett. **27**, 2261–2264 (2000). https://doi.org/10.1029/1999GL011324

Charney, J.G.: Dynamics of deserts and drought in the Sahel. Q. J. R. Meteorol. Soc. **101**, 193–202 (1975). https://doi.org/10.1002/qj.49710142802

Chavez, F.P., Ryan, J., Lluch-Cota, S.E., ?iquen, M.: From anchovies to sardines and back: multidecadal change in the Pacific Ocean. Science **299**, 217–221 (2003)

Chiang, J.C., Kushnir, Y., Giannini, A.: Deconstructing atlantic intertropical convergence zone variability: influence of the local cross-equatorial sea surface temperature gradient and remote forcing from the eastern equatorial Pacific. J. Geophys. Res. Atmos. **107** (2002)

Chou, C., Tu, J.-Y., Tan, P.-H.: Asymmetry of tropical precipitation change under global warming. Geophys. Res. Lett. **34** (2007)

Clement, A., Bellomo, K., Murphy, L.N., Cane, M.A., Mauritsen, T., Rädel, G., Stevens, B.: The Atlantic Multidecadal Oscillation without a role for ocean circulation. Science **350**, 320–324 (2015)

Compo, G.P., Sardeshmukh, P.D.: Oceanic influences on recent continental warming. Clim. Dyn. **32**, 333–342 (2009). https://doi.org/10.1007/s00382-008-0448-9

Cook, K.H.: Climate science: the mysteries of Sahel droughts. Nat. Geosci. **1**, 647–648 (2008). https://doi.org/10.1038/ngeo320

Cook, K.H., Vizy, E.K.: Coupled model simulations of the West African monsoon system: twentieth- and twenty-first-century simulations. J. Clim. **19**, 3681–3703 (2006)

Coriolis, G.G.: Mémoire sur les équations du mouvement relatif des systèmes de corps. Journal de l'Ecole polytechnique, Paris (1835)

da Silva, V.D.P.R.: On climate variability in Northeast of Brazil. J. Arid. Environ. **58**, 575–596 (2004). https://doi.org/10.1016/j.jaridenv.2003.12.002

Dai, A.: Drought under global warming: a review. Wiley Interdiscip. Rev. Clim. Chang. **2**, 45–65 (2011). https://doi.org/10.1002/wcc.81

Dai, A.: The influence of the inter-decadal Pacific oscillation on US precipitation during 1923–2010. Clim. Dyn. **41**, 633–646 (2013). https://doi.org/10.1007/s00382-012-1446-5

de Albuquerque Cavalcanti, I.F.: The influence of extratropical Atlantic Ocean region on wet and dry years in North-Northeastern Brazil. Front. Environ. Sci. **3**, 1–10 (2015). https://doi.org/10.3389/fenvs.2015.00034

Dee, D.P., Uppala, S., Simmons, A., Berrisford, P., Poli, P., Kobayashi, S., Andrae, U., Balmaseda, M., Balsamo, G., Bauer, D.P., et al.: The ERA-interim reanalysis: configuration and performance of the data assimilation system. Q. J. R. Meteorol. Soc. **137**, 553–597 (2011)

Delworth, T.L., Greatbatch, R.J.: Multidecadal thermohaline circulation variability driven by atmospheric surface flux forcing. J. Clim. **13**, 1481–1495 (2000)

Delworth, T.L., Mann, M.E.: Observed and simulated multidecadal variability in the Northern Hemisphere. Clim. Dyn. **16**, 661–676 (2000). https://doi.org/10.1007/s003820000075

Deser, C., Alexander, M.A., Xie, S.-P., Phillips, A.S.: Sea surface temperature variability: patterns and mechanisms **2** (2010). https://doi.org/10.1146/annurev-marine-120408-151453

Deser, C., Phillips, A.S., Hurrell, J.W.: Pacific interdecadal climate variability: linkages between the tropics and the North Pacific during boreal winter since 1900. J. Clim. **17**, 3109–3124 (2004)

Dettinger, M., Battisti, D., McCabe, G., Bitz, C., Garreaud, R.: Interhemispheric effects of interannual and decadal ENSO-like climate variations on the Americas. In: Interhemispheric Climate Linkages: Present and Past Climates in the Americas and their Societal Effects, pp. 1–16 (2001)

Doblas-Reyes, F.J., Andreu-Burillo, I., Chikamoto, Y., García-Serrano, J., Guemas, V., Kimoto, M., Mochizuki, T., Rodrigues, L.R.L., van Oldenborgh, G.J.: Initialized near-term regional climate change prediction. Nat. Commun. **4**, 1715 (2013). https://doi.org/10.1038/ncomms2704

Dong, B., Sutton, R.: Dominant role of greenhouse-gas forcing in the recovery of Sahel rainfall. Nat. Clim. Chang. **5**, 757–760 (2015)

Edwards, P.N.: History of climate modeling. Wiley Interdiscip. Rev. Clim. Chang. **2**, 128–139 (2011)

Enfield, D.B., Mestas-Nuñez, A.M., Trimble, P.J.: The Atlantic multidecadal oscillation and its relation to rainfall and river flows in the continental U.S. Geophys. Res. Lett. **28**, 2077–2080 (2001). https://doi.org/10.1029/2000GL012745

Espinoza Villar, J.C., Ronchail, J., Guyot, J.L., Cochonneau, G., Naziano, F., Lavado, W., De Oliveira, E., Pombosa, R., Vauchel, P.: Spatio-temporal rainfall variability in the Amazon basin countries (Brazil, Peru, Bolivia, Colombia, and Ecuador). Int. J. Climatol. **29**, 1574–1594 (2009)

Farneti, R., Molteni, F., Kucharski, F.: Pacific interdecadal variability driven by tropical-extratropical interactions. Clim. Dyn. **42**, 3337–3355 (2014). https://doi.org/10.1007/s00382-013-1906-6

Fernandes, K., Giannini, A., Verchot, L., Baethgen, W., Pinedo-Vasquez, M.: Decadal covariability of Atlantic SSTs and western amazon dry-season hydroclimate in observations and CMIP5 simulations. Geophys. Res. Lett. **42**, 6793–6801 (2015)

Flato, G., Marotzke, J., Abiodun, B., Braconnot, P., Chou, S.C., Collins, W.J., Cox, P., Driouech, F., Emori, S., Eyring, V., et al.: Evaluation of climate models. In: Climate Change 2013: The Physical Science Basis. Contribution of Working Group I to the Fifth Assessment Report of the Intergovernmental Panel on Climate Change, vol. 5, pp. 741–866 (2013)

Folland, C.K., Colman, A.W., Rowell, D.P., Davey, M.K.: Predictability of northeast Brazil rainfall and real-time forecast skill, 1987–98. J. Clim. **14**, 1937–1958 (2001)

Folland, C., Palmer, T., Parker, D.: Sahel rainfall and worldwide sea temperatures, 1901–85. Nature **320**, 602–607 (1986)

Friedman, A.R., Hwang, Y.-T., Chiang, J.C., Frierson, D.M.: Interhemispheric temperature asymmetry over the twentieth century and in future projections. J. Clim. **26**, 5419–5433 (2013)

Gaetani, M., Mohino, E.: Decadal prediction of the Sahelian precipitation in CMIP5 simulations. J. Clim. **26**, 7708–7719 (2013). https://doi.org/10.1175/JCLI-D-12-00635.1

Gaetani, M., Flamant, C., Bastin, S., Janicot, S., Lavaysse, C., Hourdin, F., Braconnot, P., Bony, S.: West African monsoon dynamics and precipitation: the competition between global SST warming and CO_2 increase in CMIP5 idealized simulations. Clim. Dyn. **48**, 1353–1373 (2017)

Gallego, D., Ordóñez, P., Ribera, P., Peña-Ortiz, C., García-Herrera, R.: An instrumental index of the West African Monsoon back to the nineteenth century. Q. J. R. Meteorol. Soc. **141**, 3166–3176 (2015)

García-Serrano, J., Rodríguez-Fonseca, B., Bladé, I., Zurita-Gotor, P., de la Cámara, A.: Rotational atmospheric circulation during North Atlantic-European winter: the influence of ENSO. Clim. Dyn. **37**, 1727–1743 (2011). https://doi.org/10.1007/s00382-010-0968-y

García-Serrano, J., Guemas, V., Doblas-Reyes, F.: Added-value from initialization in predictions of Atlantic multi-decadal variability. Clim. Dyn. **44**, 2539–2555 (2015)

Giannini, A., Saravanan, R., Chang, P.: Oceanic forcing of Sahel rainfall on interannual to interdecadal time scales. Science **302**, 1027–1030 (2003). https://doi.org/10.1126/science.1089357

Giannini, A., Salack, S., Lodoun, T., Ali, A., Gaye, A.T., Ndiaye, O.: A unifying view of climate change in the Sahel linking intra-seasonal, interannual and longer time scales. Environ. Res. Lett. **8**, 24010 (2013). https://doi.org/10.1088/1748-9326/8/2/024010

Gill, A.E.: Some simple solutions for heat-induced tropical circulation. Q. J. R. Meteorol. Soc. **106**, 447–462 (1980). https://doi.org/10.1002/qj.49710644905

Goldenberg, S.B.: The recent increase in Atlantic hurricane activity: causes and implications. Science **293**, 474–479 (2001). https://doi.org/10.1126/science.1060040

Good, P., Lowe, J.A., Collins, M., Moufouma-Okia, W.: An objective tropical Atlantic sea surface temperature gradient index for studies of south Amazon dry-season climate variability and change. Philos. Trans. R. Soc. Lond. Ser. B Biol. Sci. **363**, 1761–1766 (2008). https://doi.org/10.1098/rstb.2007.0024

Gray, S.T.: A tree-ring based reconstruction of the Atlantic Multidecadal Oscillation since 1567 A.D. Geophys. Res. Lett. **31**(L12), 205 (2004). https://doi.org/10.1029/2004GL019932

Grimm, A.M., Saboia, J.P.: Interdecadal variability of the South American precipitation in the monsoon season. J. Clim. **28**, 755–775 (2015)

Grodsky, S.A., Carton, J.A.: The intertropical convergence zone in the South Atlantic and the equatorial cold tongue. J. Clim. **16**, 723–733 (2003)

Grodsky, S.A., Carton, J.A., Nigam, S.: Near surface westerly wind jet in the Atlantic ITCZ. Geophys. Res. Lett. **30**, 3–6 (2003)

Gu, G., Adler, R.F.: Interdecadal variability/long-term changes in global precipitation patterns during the past three decades: global warming and/or pacific decadal variability? Clim. Dyn. **40**, 3009–3022 (2013)

Hadley, G.: Concerning the cause of the general trade-winds: By Geo. Hadley, Esq; FRS. Philos. Trans. **39**, 58–62 (1735)
Harris, P.P., Huntingford, C., Cox, P.M.: Amazon Basin climate under global warming: the role of the sea surface temperature. Philos. Trans. R. Soc. Lond. B Biol. Sci. **363**, 1753–1759 (2008)
Hartmann, D.L.: Global Physical Climatology, 1st edn. Academic Press, INC, San Diego, California (1994). https://doi.org/10.1016/B978-0-12-328531-7.00011-6
Hastenrath, S., Heller, L.: Dynamics of climatic hazards in northeast Brazil (1977). https://doi.org/10.1002/qj.49710343505
Held, I.M., Delworth, T.L., Lu, J., Findell, K.L., Knutson, T.R.: Simulation of Sahel drought in the 20th and 21st centuries. In: Proceedings of the National Academy of Sciences of the United States of America, vol. 102, 17 891–17 896 (2005). https://doi.org/10.1073/pnas.0509057102
Hoerling, M., Hurrell, J., Eischeid, J., Phillips, A.: Detection and attribution of twentieth-century northern and southern African rainfall change. J. Clim. **19**, 3989–4008 (2006). https://doi.org/10.1175/jcli3842.1
Holton, J.: An Introduction to Dynamical Meteorology, 4th edn. Elsevier Academic Press (2004)
Houghton, J.T., Ding, Y., Griggs, D.J., Noguer, M., van der Linden, P.J., Dai, X., Maskell, K., Johnson, C.: Climate Change 2001: The Scientific Basis. The Press Syndicate of the University of Cambridge (2001)
Hourdin, F., Musat, I., Grandpeix, J.-Y., Polcher, J., Guichard, F., Favot, F., Marquet, P., Boone, A., Lafore, J.-P., Redelsperger, J.-L., et al.: AMMA-model intercomparison project. Bull. Am. Meteorol. Soc. **91**, 95–104 (2010)
Hurrell, J.W., Kushnir, Y., Ottersen, G.: An overview of the North Atlantic oscillation. In: The North Atlantic Oscillation, Climatic Significance and Environmental Impact, pp. 1–35 (2003). https://doi.org/10.1029/134GM01
Janicot, S., Moron, V., Fontaine, B.: Sahel droughts and ENSO dynamics. Geophys. Res. Lett. **23**, 515–518 (1996)
Janicot, S., Harzallah, A., Fontaine, B., Moron, V.: West African monsoon dynamics and eastern equatorial Atlantic and Pacific SST anomalies (1970–88). J. Clim. **11**, 1874–1882 (1998). https://doi.org/10.1175/1520-0442(1998)011<1874:WAMDAE>2.0.CO;2
Janicot, S., Trzaska, S., Poccard, I.: Summer Sahel-ENSO teleconnection and decadal time scale SST variations. Clim. Dyn. **18**, 303–320 (2001). https://doi.org/10.1007/s003820100172
Jin, F.-F., Kim, S.T., Bejarano, L.: A coupled-stability index for ENSO. Geophys. Res. Lett. **33** (2006)
Joly, M.: Le rôle des océans dans la variabilité climatique de la mousson africaine. Université Paris-Est, Theses (2008)
Karlsson, J., Svensson, G., Rodhe, H.: Cloud radiative forcing of subtropical low level clouds in global models. Clim. Dyn. **30**, 779–788 (2008)
Kayano, M.T., Andreoli, R.V.: Relationships between rainfall anomalies over northeastern Brazil and the El Niño-Southern Oscillation. J. Geophys. Res. Atmos. **111**, 1–11 (2006). https://doi.org/10.1029/2005JD006142
Keenlyside, N.S., Ba, J., Mecking, J., Omrani, N.-E., Latif, M., Zhang, R., Msadek, R.: North Atlantic multi-decadal variability-mechanisms and predictability. In: Climate Change: Multi-decadal and Beyond, p. 141 (2015)
Keenlyside, N.S., Latif, M., Jungclaus, J., Kornblueh, L., Roeckner, E.: Advancing decadal-scale climate prediction in the North Atlantic sector. Nature **453**, 84–88 (2008). https://doi.org/10.1038/nature06921
Kerr, R.A.: A North Atlantic climate pacemaker for the centuries. Science **288**, 1984–1985 (2000). https://doi.org/10.1126/science.288.5473.1984
Kim, H.M., Webster, P.J., Curry, J.A.: Evaluation of short-term climate change prediction in multi-model CMIP5 decadal hindcasts. Geophys. Res. Lett. **39**, 1–7 (2012). https://doi.org/10.1029/2012GL051644
Kirtman, B., Power, S.B., Adedoyin, J.A., Boer, G.J., Bojariu, R., Camilloni, I., Doblas-Reyes, F.J., Fiore, A.M., Kimoto, M., Meehl, G.A., Prather, M., Sarr, A., Schär, C., Sutton, R., van Olden-

borgh, G.J., Vecchi, G., Wang, H.-J.: Near-term climate change: projections and predictability. In: Climate Change 2013: The Physical Science Basis. Contribution of Working Group I to the Fifth Assessment Report of the Intergovernmental Panel on Climate Change, pp. 953–1028 (2013). https://doi.org/10.1017/CBO9781107415324.023
Klein, S.A., Soden, B.J., Lau, N.-C.: Remote sea surface temperature variations during ENSO: evidence for a tropical atmospheric bridge. J. Clim. **12**, 917–932 (1999)
Knight, J.R.: A signature of persistent natural thermohaline circulation cycles in observed climate. Geophys. Res. Lett. **32**(L20), 708 (2005). https://doi.org/10.1029/2005GL024233
Knight, J.R., Folland, C.K., Scaife, A.A.: Climate impacts of the Atlantic multidecadal oscillation. Geophys. Res. Lett. **33**(L17), 706 (2006). https://doi.org/10.1029/2006GL026242
Knudsen, M.F., Seidenkrantz, M.-S., Jacobsen, B.H., Kuijpers, A.: Tracking the Atlantic Multidecadal Oscillation through the last 8,000 years. Nat. Commun. **2**, 178 (2011). https://doi.org/10.1038/ncomms1186
Kousky, V.E.: Frontal influences on northeast Brazil. Mon. Weather. Rev. **107**, 1140–1153 (1979)
Krishnan, R., Sugi, M.: Pacific decadal oscillation and variability of the Indian summer monsoon rainfall. Clim. Dyn. **21**, 233–242 (2003). https://doi.org/10.1007/s00382-003-0330-8
Kucharski, F., Molteni, F., King, M.P., Farneti, R., Kang, I.-S., Feudale, L.: On the need of intermediate complexity general circulation models: a SPEEDY example. Bull. Am. Meteorol. Soc. **94**, 25–30 (2013)
Kwon, Y.O., Frankignoul, C.: Stochastically-driven multidecadal variability of the Atlantic meridional overturning circulation in CCSM3. Clim. Dyn. **38**, 859–876 (2012). https://doi.org/10.1007/s00382-011-1040-2
Latif, M., Keenlyside, N.S.: A perspective on decadal climate variability and predictability. Deep. Sea Res. Part II Top. Stud. Ocean. **58**, 1880–1894 (2011). https://doi.org/10.1016/j.dsr2.2010.10.066
Latif, M., Collins, M., Stouffer, R., Pohlmann, H., Keenlyside, N.: The physical basis for prediction of Atlantic sector climate on decadal timescales. CLIVAR Exch. **31**, 6–8 (2004)
Lau, K., Yang, S.: Walker circulation. Encycl. Atmos. Sci., 2505–2510 (2003)
Lebel, T., Ali, A.: Recent trends in the Central and Western Sahel rainfall regime (1990–2007). J. Hydrol. **375**, 52–64 (2009)
Lenters, J.D., Cook, K.H.: On the origin of the Bolivian high and related circulation features of the South American climate. J. Atmos. Sci. **54**, 656–678 (1997)
Li, G., Xie, S.-P.: Tropical biases in CMIP5 multimodel ensemble: the excessive equatorial Pacific cold tongue and double ITCZ problems. J. Clim. **27**, 1765–1780 (2014)
Liebmann, B., Mechoso, C.R.: South American Monsoon System. The Global Monsoon System, pp. 137–157 (2011)
Lin, J.-L.: The double-ITCZ problem in IPCC AR4 coupled GCMs: Ocean-atmosphere feedback analysis. J. Clim. **20**, 4497–4525 (2007)
Liou, K.-N.: An Introduction to Atmospheric Radiation, vol. 84 (2002)
López-Parages, J., Villamayor, J., Gómara, I., Losada, T., Martín-Rey, M., Mohíno, E., Polo, I., Rodríguez-Fonseca, B., Suárez, R.: Nonstationary interannual teleconnections modulated by multidecadal variability/No-estacionariedad de teleconexiones interanuales modulada por variabilidad multi-decadal. Física de la Tierra **25**, 11 (2013)
Lu, J., Delworth, T.L.: Oceanic forcing of the late 20th century Sahel drought. Geophys. Res. Lett. **32** (2005)
Lu, J.: The dynamics of the Indian Ocean sea surface temperature forcing of Sahel drought. Clim. Dyn. **33**, 445–460 (2009)
Luo, J.-J., Masson, S., Roeckner, E., Madec, G., Yamagata, T.: Reducing climatology bias in an ocean-atmosphere CGCM with improved coupling physics. J. Clim. **18**, 2344–2360 (2005)
Lyman, J.M., Good, S.A., Gouretski, V.V., Ishii, M., Johnson, G.C., Palmer, M.D., Smith, D.M., Willis, J.K.: Robust warming of the global upper ocean. Nature **465**, 334–337 (2010). https://doi.org/10.1038/nature09043

MacDonald, G.M., Case, R.A.: Variations in the Pacific Decadal Oscillation over the past millennium. Geophys. Res. Lett. **32**(L08), 703 (2005). https://doi.org/10.1029/2005GL022478

Mann, M.E., Zhang, Z., Rutherford, S., Bradley, R.S., Hughes, M.K., Shindell, D., Ammann, C., Faluvegi, G., Ni, F.: Global signatures and dynamical origins of the Little Ice Age and Medieval Climate Anomaly. Science **326**, 1256–1260 (2009)

Mantua, N.J., Hare, S.R.: The Pacific decadal oscillation. J. Ocean. **58**, 35–44 (2002). https://doi.org/10.1023/a:1015820616384

Mantua, N.J., Hare, S.R., Zhang, Y., Wallace, J.M., Francis, R.C.: A Pacific interdecadal climate oscillation with impacts on salmon production. Bull. Am. Meteorol. Soc. **78**, 1069–1079 (1997)

Marengo, J.A., Jones, R., Alves, L., Valverde, M.: Future change of temperature and precipitation extremes in South America as derived from the PRECIS regional climate modeling system. Int. J. Climatol. **29**, 2241–2255 (2009)

Marengo, J.A.: Interdecadal variability and trends of rainfall across the Amazon basin. Theor. Appl. Climatol. **78**, 79–96 (2004). https://doi.org/10.1007/s00704-004-0045-8

Marengo, J.A.: Vulnerabilidade, impactos e adaptação à mudança do clima no semi-árido do Brasil. Parcerias estratégicas **13**, 149–176 (2008)

Marengo, J.A.: Long-term trends and cycles in the hydrometeorology of the Amazon basin since the late 1920s. Hydrol. Process. **23**, 3236–3244 (2009). https://doi.org/10.1002/hyp.7396

Marengo, J.A., Tomasella, J., Uvo, C.R.: Trends in streamflow and rainfall in tropical South America: Amazonia, eastern Brazil, and northwestern Peru. J. Geophys. Res. **103**, 1775–1783 (1998)

Marengo, J.A., Soares, W.R., Saulo, C., Nicolini, M.: Climatology of the low-level jet east of the Andes as derived from the NCEP-NCAR reanalyses: characteristics and temporal variability. J. Clim. **17**, 2261–2280 (2004)

Marengo, J.A., Liebmann, B., Grimm, A.M., Misra, V., Dias, P.L.S., Cavalcanti, I.F.A., Carvalho, L.M.V., Berbery, E.H., Ambrizzi, T., Vera, C.S., Saulo, A.C., Nogues-paegle, J., Zipser, E., Seth, A., Alves, L.M.: Recent developments on the South American monsoon system. Int. J. Climatol. **32**, 1–21 (2012). https://doi.org/10.1002/joc.2254

Marengo, J.A., Torres, R.R., Alves, L.M.: Drought in Northeast Brazil-past, present, and future. Theor. Appl. Climatol. **129**, 1–12 (2016). https://doi.org/10.1007/s00704-016-1840-8

Martin, E.R., Thorncroft, C.D.: The impact of the AMO on the West African monsoon annual cycle. Q. J. R. Meteorol. Soc. **140**, 31–46 (2014)

Martin, E.R., Thorncroft, C., Booth, B.B.B.: The multidecadal Atlantic SST-Sahel rainfall teleconnection in CMIP5 simulations. J. Clim. **27**, 784–806 (2014). https://doi.org/10.1175/JCLI-D-13-00242.1

McCabe, G.J., Palecki, M.A., Betancourt, J.L.: Pacific and Atlantic Ocean influences on multidecadal drought frequency in the United States. In: Proceedings of the National Academy of Sciences of the United States of America, vol. 101, pp. 4136–4141 (2004). https://doi.org/10.1073/pnas.0306738101

Mechoso, C.R., Robertson, A.W., Barth, N., Davey, M., Delecluse, P., Gent, P., Ineson, S., Kirtman, B., Latif, M., Treut, H.L., et al.: The seasonal cycle over the tropical Pacific in coupled ocean-atmosphere general circulation models. Mon. Weather. Rev. **123**, 2825–2838 (1995)

Medhaug, I., Furevik, T.: North Atlantic 20th century multidecadal variability in coupled climate models: sea surface temperature and ocean overturning circulation (2011)

Meehl, G.A., Hu, A., Arblaster, J.M., Fasullo, J., Trenberth, K.E.: Externally forced and internally generated decadal climate variability associated with the interdecadal pacific oscillation. J. Clim. **26**, 7298–7310 (2013). https://doi.org/10.1175/JCLI-D-12-00548.1

Meehl, G.A., Hu, A.: Megadroughts in the Indian monsoon region and southwest North America and a mechanism for associated multidecadal Pacific Sea surface temperature anomalies. J. Clim. **19**, 1605–1623 (2006). https://doi.org/10.1175/JCLI3675.1

Meehl, G.A., Washington, W.M., Collins, W.D., Arblaster, J.M., Hu, A., Buja, L.E., Strand, W.G., Teng, H.: How much more global warming and sea level rise? Science **307**, 1769–1772 (2005)

Meehl, G.A., Goddard, L., Murphy, J., Stouffer, R.J., Boer, G., Danabasoglu, G., Dixon, K., Giorgetta, M.A., Greene, A.M., Hawkins, E., Hegerl, G., Karoly, D., Keenlyside, N., Kimoto, M.,

Kirtman, B., Navarra, A., Pulwarty, R., Smith, D., Stammer, D., Stockdale, T.: Decadal prediction. Bull. Am. Meteorol. Soc. **90**, 1467–1485 (2009). https://doi.org/10.1175/2009BAMS2778.1

Menary, M.B., Hodson, D.L., Robson, J.I., Sutton, R.T., Wood, R.A., Hunt, J.A.: Exploring the impact of CMIP5 model biases on the simulation of North Atlantic decadal variability. Geophys. Res. Lett. **42**, 5926–5934 (2015)

Miller, A.J., Schneider, N.: Interdecadal climate regime dynamics in the North Pacific Ocean: theories, observations and ecosystem impacts. Prog. Ocean. **47**, 355–379 (2000). https://doi.org/10.1016/S0079-6611(00)00044-6

Minobe, S.: Resonance in bidecadal and pentadecadal climate oscillations over the North Pacific: Role in climatic regime shifts. Geophys. Res. Lett. **26**, 855–858 (1999). https://doi.org/10.1029/1999GL900119

Mohino, E., Janicot, S., Bader, J.: Sahel rainfall and decadal to multi-decadal sea surface temperature variability. Clim. Dyn. **37**, 419–440 (2011a). https://doi.org/10.1007/s00382-010-0867-2

Mohino, E., Rodríguez-Fonseca, B., Mechoso, C.R., Gervois, S., Ruti, P., Chauvin, F.: Impacts of the tropical Pacific/Indian oceans on the seasonal cycle of the west african monsoon. J. Clim. **24**, 3878–3891 (2011b). https://doi.org/10.1175/2011JCLI3988.1

Mohino, E., Keenlyside, N., Pohlmann, H.: Decadal prediction of Sahel rainfall: where does the skill (or lack thereof) come from? Clim. Dyn. **47**, 3593–3612 (2016)

Molion, L.C.B., Lucio, P.S.: A note on Pacific Decadal Oscillation, El Nino Southern Oscillation, Atlantic Multidecadal Oscillation and the Intertropical Front in Sahel, Africa. Atmos. Clim. Sci. **03**, 269–274 (2013). https://doi.org/10.4236/acs.2013.33028

Monerie, P.A., Fontaine, B., Roucou, P.: Expected future changes in the African monsoon between 2030 and 2070 using some CMIP3 and CMIP5 models under a medium-low RCP scenario. J. Geophys. Res. Atmos. **117**, 1–12 (2012). https://doi.org/10.1029/2012JD017510

Moura, A.D., Shukla, J.: On the dynamics of droughts in Northeast Brazil: observations, theory, and numerical experiments with a general circulation model (1981)

Newman, M., Compo, G.P., Alexander, M.A.: ENSO-forced variability of the Pacific decadal oscillation. J. Clim. **16**, 3853–3857 (2003)

Newman, M., Alexander, M.A., Ault, T.R., Cobb, K.M., Deser, C., Di Lorenzo, E., Mantua, N.J., Miller, A.J., Minobe, S., Nakamura, H., Schneider, N., Vimont, D.J., Phillips, A.S., Scott, J.D., Smith, C.A.: The Pacific decadal oscillation, revisited. J. Clim. **29**, 4399–4427 (2016). https://doi.org/10.1175/JCLI-D-15-0508.1

Nicholson, S.: On the question of the "recovery" of the rains in the West African Sahel. J. Arid. Environ. **63**, 615–641 (2005)

Nicholson, S.E.: A revised picture of the structure of the "monsoon" and land ITCZ over West Africa. Clim. Dyn. **32**, 1155–1171 (2009)

Nicholson, S.E.: The West African sahel: a review of recent studies on the rainfall regime and its interannual variability. ISRN Meteorol. **2013**, 32 (2013). https://doi.org/10.1155/2013/453521

Nicholson, S.E., Klotter, D., Dezfuli, A.K.: Spatial reconstruction of semi-quantitative precipitation fields over Africa during the nineteenth century from documentary evidence and gauge data. Quat. Res. **78**, 13–23 (2012)

Nobre, P., Shukla, J.: Variations of sea surface temperature, wind stress, and rainfall over the tropical atlantic and South America. J. Clim. **9**, 2464–2479 (1996)

Nobre, P., Marengo, J., Cavalcanti, I., Obregon, G., Barros, V., Camilloni, I., Campos, N., Ferreira, A.: Seasonal-to-decadal predictability and prediction of South American climate. J. Clim. **19**, 5988–6004 (2006)

Oueslati, B., Bellon, G.: The double ITCZ bias in CMIP5 models: interaction between SST, large-scale circulation and precipitation. Clim. Dyn. **44**, 585–607 (2015)

Panthou, G., Vischel, T., Lebel, T.: Recent trends in the regime of extreme rainfall in the central sahel. Int. J. Climatol. **4006**, 3998–4006 (2014). https://doi.org/10.1002/joc.3984

Park, J.-Y., Bader, J., Matei, D.: Northern-hemispheric differential warming is the key to understanding the discrepancies in the projected Sahel rainfall. Nat. Commun. **6**, 5985 (2015). https://doi.org/10.1038/ncomms6985

Philander, S.G.: El Niño, La Niña, and the Southern Oscillation, International Geophys Series, vol. 46. Academic Press, INC, San Diego, California (1990)

Pohlmann, H., Botzet, M., Latif, M., Roesch, A., Wild, M., Tschuck, P.: Estimating the decadal predictability of a coupled AOGCM. J. Clim. **17**, 4463–4472 (2004). https://doi.org/10.1175/3209.1

Power, S., Casey, T., Folland, C., Colman, A., Mehta, V.: Inter-decadal modulation of the impact of ENSO on Australia, Clim. Dyn. **15**, 319–324 (1999). https://doi.org/10.1007/s003820050284

Rahmstorf, S.: Thermohaline Ocean circulation. Encycl. Quat. Sci., 1–10 (2006). https://doi.org/10.1016/B0-44-452747-8/00014-4

Rao, V.B., Hada, K.: Characteristics of rainfall over Brazil: annual variations and connections with the Southern Oscillation. Theor. Appl. Climatol. **42**, 81–91 (1990). https://doi.org/10.1007/BF00868215

Richter, I.: Climate model biases in the eastern tropical oceans: causes, impacts and ways forward. Wiley Interdiscip. Rev. Clim. Chang. **6**, 345–358 (2015)

Richter, I., Xie, S.-P.: On the origin of equatorial Atlantic biases in coupled general circulation models. Clim. Dyn. **31**, 587–598 (2008)

Richter, I., Xie, S.-P., Behera, S.K., Doi, T., Masumoto, Y.: Equatorial Atlantic variability and its relation to mean state biases in CMIP5. Clim. Dyn. **42**, 171–188 (2014)

Robertson, A.W., Mechoso, C.R.: Interannual and decadal cycles in river flows of southeastern South America. J. Clim. **11**, 2570–2581 (1998)

Rodrigues, R.R., Haarsma, R.J., Campos, E.J.D., Ambrizzi, T.: The impacts of inter-El Niño variability on the tropical Atlantic and northeast Brazil climate. J. Clim. **24**, 3402–3422 (2011). https://doi.org/10.1175/2011JCLI3983.1

Rodríguez-Fonseca, B., Mohino, E., Mechoso, C.R., Caminade, C., Biasutti, M., Gaetani, M., Garcia-Serrano, J., Vizy, E.K., Cook, K., Xue, Y., et al.: Variability and predictability of West African droughts: a review on the role of sea surface temperature anomalies. J. Clim. **28**, 4034–4060 (2015)

Rodwell, M.J., Rowell, D.P., Folland, C.K.: Oceanic forcing of the wintertime North Atlantic Oscillation and European climate. Nature **398**, 320–323 (1999). https://doi.org/10.1038/18648

Rotstayn, L.D., Lohmann, U.: Tropical rainfall trends and the indirect aerosol effect. J. Clim. **15**, 2103–2116 (2002)

Rowell, D.P.: Assessing potential seasonal predictability with an ensemble of multidecadal GCM simulations. J. Clim. **11**, 109–120 (1998)

Rowell, D.P., Folland, C.K., Maskell, K., Ward, M.N.: Variability of summer rainfall over tropical North Africa (1906–92): Observations and modelling. Q. J. R. Meteorol. Soc. **121**, 669–704 (1995)

Ruiz-Barradas, A., Carton, J.A., Nigam, S.: Structure of interannual-to-decadal climate variability in the tropical Atlantic sector. J. Clim. **13**, 3285–3297 (2000)

Salinger, M., Renwick, J., Mullan, A.: Interdecadal Pacific oscillation and south Pacific climate. Int. J. Climatol. **21**, 1705–1721 (2001)

Sanchez-Gomez, E., Cassou, C., Ruprich-Robert, Y., Fernandez, E., Terray, L.: Drift dynamics in a coupled model initialized for decadal forecasts. Clim. Dyn. **46**, 1819–1840 (2016)

Schlesinger, M.E., Ramankutty, N.: An oscillation in the global climate system of period 65–70 years. Nature **367**, 723–726 (1994). https://doi.org/10.1038/367723a0

Schneider, N., Cornuelle, B.D.: The forcing of the Pacific Decadal Oscillation. J. Clim. **18**, 4355–4373 (2005). https://doi.org/10.1175/JCLI3527.1

Shakun, J.D., Shaman, J.: Tropical origins of North and South Pacific decadal variability. Geophys. Res. Lett. **36**(L19), 711 (2009). https://doi.org/10.1029/2009GL040313

Smith, D.M., Eade, R., Dunstone, N.J., Fereday, D., Murphy, J.M., Pohlmann, H., Scaife, A.A.: Skilful multi-year predictions of Atlantic hurricane frequency. Nat. Geosci. **3**, 846–849 (2010). https://doi.org/10.1038/ngeo1004

Soden, B.J., Held, I.M.: An assessment of climate feedbacks in coupled ocean-atmosphere models. J. Clim. **19**, 3354–3360 (2006)

Souza, E.B., Ambrizzi, T.: ENSO impacts on the South American rainfall during 1980s: Hadley and Walker circulation. Atmósfera **15**, 105–120 (2002)

Srokosz, M.A., Bryden, H.L.: Observing the Atlantic Meridional Overturning Circulation yields a decade of inevitable surprises. Science **348**, 1255575 (2015). https://doi.org/10.1126/science.1255575

Stocker, T.: Climate Change 2013: The Physical Science Basis: Working Group I Contribution to the Fifth Assessment Report of the Intergovernmental Panel on Climate Change. Cambridge University Press (2014)

Stroeve, J.C., Kattsov, V., Barrett, A., Serreze, M., Pavlova, T., Holland, M., Meier, W.N.: Trends in Arctic sea ice extent from CMIP5, CMIP3 and observations. Geophys. Res. Lett. **39**, 1–7 (2012). https://doi.org/10.1029/2012GL052676

Suárez-Moreno, R., Rodríguez-Fonseca, B.: S4CAST v2.0: sea surface temperature based statistical seasonal forecast model. Geosci. Model. Dev. **8**, 3639–3658 (2015). https://doi.org/10.5194/gmd-8-3639-2015

Sutton, R.T., Hodson, D.L.R.: North Atlantic forcing of North American and European summer climate. Science **309**, 115–118 (2005). https://doi.org/10.1126/science.1109496

Sutton, R.T., Dong, B., Gregory, J.M.: Land/sea warming ratio in response to climate change: IPCC AR4 model results and comparison with observations. Geophys. Res. Lett. **34**, 2–6 (2007). https://doi.org/10.1029/2006GL028164

Svendsen, L., Hetzinger, S., Keenlyside, N., Gao, Y.: Marine-based multiproxy reconstruction of Atlantic multidecadal variability. Geophys. Res. Lett. **41**, 1295–1300 (2014). https://doi.org/10.1002/2013GL059076

Taylor, K.E., Penner, J.E.: Response of the climate system to atmospheric aerosols and greenhouse gases. Nature **369**, 734–737 (1994). https://doi.org/10.1038/369734a0

Terray, L.: Evidence for multiple drivers of North Atlantic multi-decadal climate variability. Geophys. Res. Lett. **39**, 6–11 (2012). https://doi.org/10.1029/2012GL053046

Thompson, D.W.J., Wallace, J.M., Kennedy, J.J., Jones, P.D.: An abrupt drop in Northern Hemisphere sea surface temperature around 1970. Nature **467**, 444–447 (2010). https://doi.org/10.1038/nature09394

Ting, M., Kushnir, Y., Seager, R., Li, C.: Forced and internal twentieth-century SST trends in the North Atlantic*. J. Clim. **22**, 1469–1481 (2009). https://doi.org/10.1175/2008JCLI2561.1

Ting, M., Kushnir, Y., Seager, R., Li, C.: Robust features of Atlantic multi-decadal variability and its climate impacts. Geophys. Res. Lett. **38**, 1–6 (2011). https://doi.org/10.1029/2011GL048712

Tourre, Y.M., Rajagopalan, B., Kushnir, Y., Barlow, M., White, W.B.: Patterns of coherent decadal and interdecadal climate signals in the Pacific Basin during the 20th century (2001). https://doi.org/10.1029/2000GL012780

Trenberth, K.E., Branstator, G.W., Karoly, D., Kumar, A., Lau, N.C., Ropelewski, C.: Progress during TOGA in understanding and modeling global teleconnections associated with tropical sea surface temperatures. J. Geophys. Res. Ocean. **103**, 14 291–14 324 (1998). https://doi.org/10.1029/97jc01444

Trenberth, K.E., Jones, P., Ambenje, P., Bojariu, R., Easterling, D., Klein Tank, A., Parker, D., Rahimzadeh, F., Renwick, J., Rusticucci, M., et al.: Observations: surface and atmospheric climate change. In: Climate Change 2007: the Physical Science Basis. Contribution of Working Group I to the Fourth Assessment Report of the Intergovernmental Panel on Climate Change (2007)

Trenberth, K.E.: Atmospheric moisture recycling: role of advection and local evaporation. J. Clim. **12**, 1368–1381 (1999)

Trenberth, K.E., Hurrell, J.W.: Decadal atmosphere-ocean variations in the Pacific. Clim. Dyn. **9**, 303–319 (1994). https://doi.org/10.1007/BF00204745

Trenberth, K.E., Shea, D.J.: Atlantic hurricanes and natural variability in 2005. Geophys. Res. Lett. **33**, 1–4 (2006). https://doi.org/10.1029/2006GL026894

van Oldenborgh, J.G., Doblas-Reyes, F.J., Wouters, B., Hazeleger, W.: Decadal prediction skill in a multi-model ensemble. Clim. Dyn. **38**, 1263–1280 (2012). https://doi.org/10.1007/s00382-012-1313-4

Vellinga, M., Wood, R.A.: Global climatic impacts of a collapse of the Atlantic thermohaline circulation. Clim. Change **54**, 251–267 (2002)

Vera, C., Baez, J., Douglas, M., Emmanuel, C.B., Marengo, J., Meitin, J., Nicolini, M., Nogues-Paegle, J., Paegle, J., Penalba, O., Salio, P., Saulo, C., Silva Dias, M.A., Silva Dias, P., Zipser, E.: The South American low-level jet experiment. Bull. Am. Meteorol. Soc. **87**, 63–77 (2006a). https://doi.org/10.1175/BAMS-87-1-63

Vera, C., Higgins, W., Amador, J., Ambrizzi, T., Garreaud, R., Gochis, D., Gutzler, D., Lettenmaier, D., Marengo, J., Mechoso, C.R., Nogues-Paegle, J., Silva Dias, P.L., Zhang, C.: Toward a unified view of the American monsoon systems. J. Clim. **19**, 4977–5000 (2006b). https://doi.org/10.1175/JCLI3896.1

Wahl, S., Latif, M., Park, W., Keenlyside, N.: On the tropical Atlantic SST warm bias in the Kiel climate model. Clim. Dyn. **36**, 891–906 (2011)

Wainer, I., Soares, J.: North northeast Brazil rainfall and its decadal-scale relationship to wind stress and sea surface temperature. Geophys. Res. Lett. **24**, 277–280 (1997)

Wang, B., Ding, Q., Liu, J.: Concept Of global monsoon. In: The Global Monsoon System: Research and Forecast, pp. 3–14 (2011)

Wang, B., Ding, Q.: Global monsoon: Dominant mode of annual variation in the tropics. Dyn. Atmos. Ocean. **44**, 165–183 (2008). https://doi.org/10.1016/j.dynatmoce.2007.05.002

Wang, B., Liu, J., Kim, H.-J., Webster, P.J., Yim, S.-Y., Xiang, B.: Northern Hemisphere summer monsoon intensified by mega-El Nino/southern oscillation and Atlantic multidecadal oscillation. Proc. Natl. Acad. Sci. **110**, 5347–5352 (2013). https://doi.org/10.1073/pnas.1219405110

Wang, C., Zhang, L., Lee, S.-K., Wu, L., Mechoso, C.R.: A global perspective on CMIP5 climate model biases. Nat. Clim. Chang. **4**, 201–205 (2014). https://doi.org/10.1038/nclimate2118

Webster, P.J., Fasullo, J.T.: Dynamical theory. In: Holton, J., Curry, J.A. (eds.) Encyclopedia of Atmospheric Sciences, pp. 1370–1385 (2003). Chapter Monsoon: D

Xie, S.P., Deser, C., Vecchi, G.A., Ma, J., Teng, H.Y., Wittenberg, A.T.: Formation, global warming pattern: sea surface temperature and rainfall. J. Clim. **23**, 966–986 (2010). https://doi.org/10.1175/2009jcli3329.1

Yoon, J.H., Zeng, N.: An Atlantic influence on Amazon rainfall. Clim. Dyn. **34**, 249–264 (2010). https://doi.org/10.1007/s00382-009-0551-6

Zeng, N., Neelin, J.D.: A land-atmosphere interaction theory for the tropical deforestatin problem. J. Clim. **12**, 857–872 (1999)

Zhang, R., Delworth, T.L.: Impact of Atlantic multidecadal oscillations on India/Sahel rainfall and Atlantic hurricanes. Geophys. Res. Lett. **33**(L17), 712 (2006). https://doi.org/10.1029/2006GL026267

Zhang, R., Delworth, T.L., Sutton, R., Hodson, D.L.R., Dixon, K.W., Held, I.M., Kushnir, Y., Marshall, J., Ming, Y., Msadek, R., Robson, J., Rosati, A.J., Ting, M., Vecchi, G.A.: Have aerosols caused the observed Atlantic multidecadal variability? J. Atmos. Sci. **70**, 1135–1144 (2013b). https://doi.org/10.1175/JAS-D-12-0331.1

Zhang, R., Sutton, R., Danabasoglu, G., Delworth, T.L., Kim, W.M., Robson, J., Yeager, S.G.: Comment on "The Atlantic Multidecadal Oscillation without a role for ocean circulation". Science **352**, 1527–1527 (2016). https://doi.org/10.1126/science.aaf1660

Zhou, J., Lau, K.M.: Principal modes of interannual and decadal variability of summer rainfall over South America. Int. J. Climatol. **21**, 1623–1644 (2001). https://doi.org/10.1002/joc.700

Chapter 2
Data and Methodology

Abstract The relationship between the main modes of SST variability and rainfall at decadal time scales is addressed in this Ph.D. Thesis through the analysis of simulations of several GCMs. In addition, one AGCM is also used for the realization of experiments designed to study the occurrence of a long rainy period in the late-19th century. In parallel, observational data from gridded data bases and reanalysis are also used to make a comparison with the results obtained from the simulations. To analyze all the climatic variables and distinguish the links among them, discriminant analysis and other statistical techniques are used. This chapter describes the main characteristics of the GCMs and simulations analyzed, the experimental design, the observational data and the methodology used throughout this Thesis.

2.1 The CMIP5

The fifth phase of the Coupled Model Intercomparison Project (CMIP5) emerged in 2008 as a result of an agreement among several of the most important modeling groups in the world. The purpose of this agreement is to coordinate the experiments carried out by each institution so as to make an intercomparison between the most updated models to date and to try to solve certain issues that arose in the preparation of the last Intergovernmental Panel on Climate Change (IPCC) report published in 2007 on the understanding of past and future climate change (Taylor et al. 2012). Hence, the CMIP5 provides a multi-model context aiming at:

- Assess the GCMs skill to reproduce the recent past.
- Evaluate the uncertain mechanisms involved in the carbon cycle feedbacks that produce different effects among GCMs.
- Examine the predictability of climate and the predictive skill of forecasting systems at decadal time scales.

CMIP5 includes two types of climate change modeling using GCMs. On the one hand, there are short-term simulations, which are initialized with observed ocean

J. Villamayor, *Influence of the Sea Surface Temperature Decadal Variability on Tropical Precipitation: West African and South American Monsoon*, Springer Theses, https://doi.org/10.1007/978-3-030-20327-6_2

and sea ice conditions and only simulate 10–30 years (Meehl et al. 2009). These simulations are also called decadal prediction experiments as their aim is to predict short-time climate change. On the other, the long-term experiments simulate time scales of one or more centuries. CMIP5 includes several types of these long-term simulations, like the historical, covering much of the industrial period (since the mid-19th century), and the pre-industrial control (piControl), which aim at simulating climate prior to the industrial revolution. The long-term experiments are initialized from piControl simulations. While there is internal climate variability in these simulations, it does not have to be in phase with the observed one. For this reason, in the long-term simulations, the simulated temporal sequence and duration of the periodic climate events do not necessarily coincide with observations. This must be taken into account when analyzing the model simulations by comparing them with the observations and not attributing these discrepancies to errors in the model (Taylor et al. 2012).

The external forcing used in historical simulations includes natural factors of variability, such as solar radiation or volcanic eruptions, and anthropogenic factors, such as the burning of fossil fuels, methane emissions or the changes in land uses. In addition to this forcing response, the climate shows independent variations that are caused by the internal interactions of the complex nonlinear climate system (Taylor et al. 2012). The piControl simulation is therefore useful to analyze these internal climate variations separately, which the forced CMIP5 experiments reproduce together with the externally forced variability.

In this work, the results of the historical and piControl simulations of a set of CMIP5 models are used because of their temporal extension. So, it is possible to study the climate variability at decadal-to-multidecadal time scales and to analyze the influence of the external forcing on the results by comparing the forced and unforced simulations.

With the objective of speculating on the future relationship between SST and precipitation at decadal time scales, also the CMIP5 future projections have been analyzed. These are the Representative Concentration Pathway 8.5 (hereinafter RCP8.5), which are typically run from 2006 to 2100 with a radiative forcing rising up to $8.5\,\mathrm{W/m^2}$ in 2100 induced by the emission of global warming gases (Riahi et al. 2007).

Detection and attribution of climate change simulations provided by some CMIP5 models have also been analyzed to shed some light on the role of the different components of the external forcing inducing uneven SST variability between the forced and unforced experiments. These simulations impose the piControl conditions but with the greenhouse forcing evolution of the historical experiment (hereinafter historicalGHG), excluding other factors of radiative forcing.

2.1.1 CMIP5 Data

Four different simulations from a set of 17 CMIP5 models have been analyzed. The names of these models, the number of years, the simulated period and the number of realizations run in each experiment are listed in Table 2.1.

The analyzed outputs are monthly data that are interpolated to a common grid of 2.8° in latitude and longitude and 17 vertical levels. Surface temperature data, precipitation and atmospheric variables, such as horizontal wind components, specific humidity and surface pressure, are used. Surface temperature is, on the continents, the soil temperature in radiative equilibrium (called skin temperature) and, on the oceans and seas, it is the SST. Land area fraction data have also been used to separate the SST from the continental surface temperature. Precipitation data include both the liquid and solid phases of all types of clouds (large-scale and convective systems).

2.2 The LMDZ Model

A set of simulations have been performed in this Thesis using a particular AGCM: the fifth version of the *Laboratoire de Météorologie Dynamique* model (LMDZ, where "Z" stands for "zoom"). This AGCM is developed in the *Institut Pierre Simon Laplace* (IPSL) and this version is the atmospheric component of the IPSL-CM5A coupled model (Hourdin et al. 2013) used in the CMIP5. The LMDZ is coupled to the latest version of the land surface model ORCHIDEE (Krinner et al. 2005). The configuration of the model used has a spatial resolution with 2.5° and 1.25° in longitude and latitude, respectively, and 39 vertical levels.

2.2.1 Experimental Design

Two kind of experiments have been performed with the LMDZ model to study the Sahel rainfall variability during the second half of the 19th century. Firstly, a set of 19 members starting from a long-term simulation have been run for the period 1854–2000 with the same boundary conditions but different initial state of the atmosphere. The boundary conditions imposed to the model to perform this simulation are the sea ice cover, the SST and the atmospheric gases that affect the external climate forcing. The sea ice cover data used are from HadISST1 (Rayner et al. 2003), extending back to year 1854 the monthly climatological data provided for the period 1870–1900. The configuration of GHGs concentration is the same as used in the historical simulations of CMIP5 (Dufresne et al. 2013). This has been combined with stratospheric aerosols effects estimated from registers of volcanic eruptions (Sato et al. 1993). The ERSST.v4 data base (Huang et al. 2015; Liu et al. 2015) has been used for the SST forcing, which provides data from 1854.

Table 2.1 List of CMIP5 models used, the modeling groups responsible for their development and number of years (#years) or period analyzed and number of realizations (#rea) of each simulation. All data available at http://pcmdi9.llnl.gov

	Model name	Modeling group	piControl	historical		RCP8.5		historicalGHG	
			#years	period	#rea	period	#rea	period	#rea
1.	bcc-csm1-1	Beijing Climate Center, China Meteorological Administration	500	1850–2012	3	2006–2100	1	1850–2012	1
2.	CanESM2	Canadian Center for Climate Modeling and Analysis	996	1850–2005	5	2006–2100	5	1850–2005	5
3.	CCSM4	National Center for Atmospheric Research	501	1850–2005	6	2006–2100	6	1850–2005	3
4.	CNRM-CM5	Centre National de Recherches Météorologiques/Centre Européen de Recherche et Formation Avancée en Calcul Scientifique	850	1850–2005	10	2006–2100	5	1850–2005	6
5.	CSIRO-Mk3–6-0	Commonwealth Scientific and Industrial Research Organization in collaboration with Queensland Climate Change Centre of Excellence	500	1850–2005	10	2006–2100	10	1850–2005	5
6.	FGOALS-g2	LASG, Institute of Atmospheric Physics, Chinese Academy of Sciences and CESS, Tsinghua University	700	1850–2005	4	2006–2100	1	1850–2005	1
7.	GISS-E2-H	NASA Goddard Institute for Space Studies	540	1850–2005	5	2006–2100	5	1850–2005	5
8.	GISS-E2-R	NASA Goddard Institute for Space Studies	550	1850–2005	6	2006–2100	5	1850–2005	5
9.	HadGEM2-CC	Met Office Hadley Centre	240	1860–2004	1	2006–2099	3	Not available	
10.	HadGEM2-ES	Met Office Hadley Centre	575	1860–2004	5	2006–2100	4	1860–2004	4
11.	inmcm4	Institute for Numerical Mathematics	500	1850–2005	1	2006–2100	1	Not available	
12.	IPSL-CM5A-LR	Institut Pierre-Simon Laplace	1000	1850–2005	6	2006–2100	4	1850–2005	5
13.	MIROC5	Japan Agency for Marine-Earth Science and Technology, Atmosphere and Ocean Research Institute and National Institute for Environmental Studies	670	1850–2012	5	2006–2100	3	Not available	
14.	MIROC-ESM-CHEM	Japan Agency for Marine-Earth Science and Technology, Atmosphere and Ocean Research Institute and National Institute for Environmental Studies	255	1850–2005	1	2006–2100	1	1850–2005	1
15.	MPI-ESM-LR	Max Planck Institute for Meteorology	1000	1850–2005	3	2006–2100	3	Not available	
16.	MRI-CGCM3	Meteorological Research Institute	500	1850–2005	5	2006–2100	1	1850–2005	1
17.	NorESM1-M	Norwegian Climate Centre	501	1850–2005	3	2006–2100	1	1850–2005	1

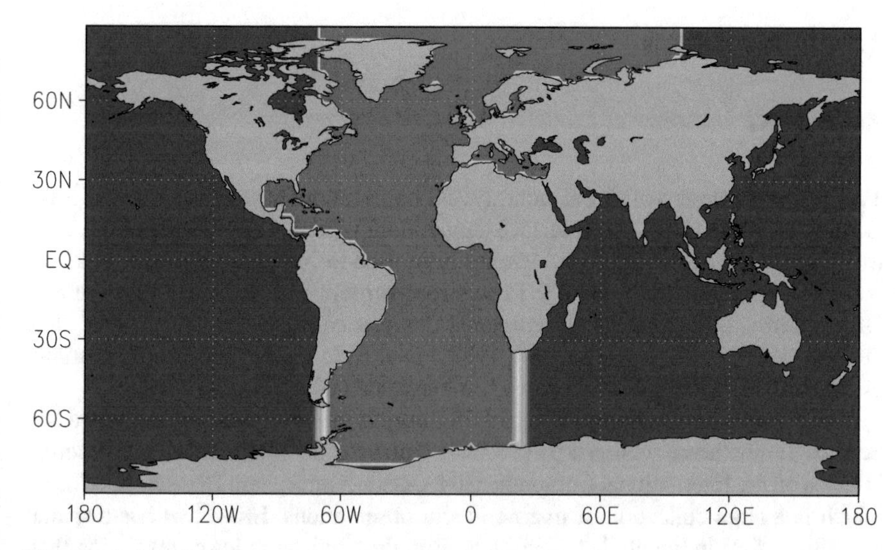

Fig. 2.1 Red and purple areas are, respectively, the Atlantic and Indo-Pacific sectors defined for the sensitivity experiments. Rainbow color bands indicate the buffered boundaries between the two oceanic areas

Secondly, a set of sensitivity experiments are performed with an ensemble-size of 5 members.[1] The two different sensitivity experiments have been performed by varying the superimposed state of the global SST for the period 1854–1910. In one of the sensitivity experiments (ATLVAR), the boundary condition imposed in the Atlantic sector is the observed evolution of the SSTs, as in the previous simulation of reference (REF). Outside this area, the climatological annual cycle of SSTs relative to 1854–1910 is used. Outputs of this experiment are noted with "atl" subscript. In the other experiment (INPVAR), the observed SSTs evolution is imposed in the Indo-Pacific ocean surface while the Atlantic SSTs are fixed to the climatological annual cycle. Outputs of this experiment are noted with "inp" subscript. The Atlantic sector spans its entire basin, including the Mediterranean Sea and part of the Southern Ocean to the south. In order to smooth the transition between the two ocean regions, we use two meridional buffer zones in 65°–75°W and 7°–17°E, south of the South American and African continents, separating the Atlantic sector from the Pacific and the Indian Ocean, respectively (Fig. 2.1). In these oceanic boundaries, the SSTA added to the climatology in one region or another are buffered through a linear relaxation along 5 grid points. In the Northern hemisphere, the ice cover separate the oceans and no buffer zones are used.

[1]The initial ensemble has been used to show that 5 realizations are enough to extract a significant signal from the internal weather noise on the ensemble mean (see Chap. 7).

2.3 Observations

2.3.1 SST

Two different data sets of global monthly SST observations have been used. One is the Hadley Center sea ice and sea surface temperature version 1 (HadISST1, from 1870 to 2009) (Rayner et al. 2003), which has been used to compare the results obtained from CMIP5 simulations. HadISST1 is a reconstruction of SST data from the Met Office Marine Data Bank, the International Comprehensive Ocean-Atmosphere Data Set and satellite measurements from 1982 onwards, interpolated to a fully spatially distributed grid with a resolution of 1° in longitude and latitude.

The other observational data base of SST used is the Extended Reconstructed Sea Surface Temperature version 4 (ERSST.v4) (Huang et al. 2015). This reconstruction is based on the International Comprehensive Ocean-Atmosphere Dataset (ICOADS), which is a large collection of marine in situ observations. ERSST.v4 has a spatial resolution of 2° in longitude and latitude. But, despite having lower resolution than HadISST1, ERSST.v4 provides data for a longer period back in time (from 1854 to 2015). Therefore, this data base has been used in this Thesis to perform climate simulations of the second half of the 19th century. In addition, this latest version of ERSST has improved metadata bias correction and completeness with respect to previous versions and has been assessed that it exhibits more realistic SST variability (Huang et al. 2015, 2016; Liu et al. 2015; Diamond and Bennartz 2015). The previous version of these data (ERSST.v3b) (Smith et al. 2008) was also used for the comparison with the multi-model analysis, providing similar results to those of HadISST1. Therefore, for the sake of simplicity, only the results from HadISST1 are shown in the report of the Thesis.

2.3.2 Precipitation

The available records of observed precipitation may be sparse and inaccurate in some cases, especially during the early 20th century and in uninhabited regions, such as the Amazonia (Marengo 2004) (Fig. 2.2). The precipitation gridded data use selected observations and provide long time series by spatially interpolating the available station records. But in turn, this artificial reconstruction generates high uncertainty of the resulting data in regions with scarce observations. Therefore, to gain confidence on the observational results, three gridded data sets dealing with different interpolation methods are analyzed. One is the Version-7 of the Global Precipitation Climatology Centre (GPCC v7, from 1901 to 2013) (Schneider et al. 2016), another one is the Climatic Research Unit time series version 3.24.01 (CRU TS3.24.01, from 1901 to 2015) (Harris et al. 2014) and the third one is the University of Delaware Air Temperature and Precipitation version 4.01 (UDEL v4.01, form 1900 to 2014) (Willmott et al. 2001). The three are monthly databases of continental

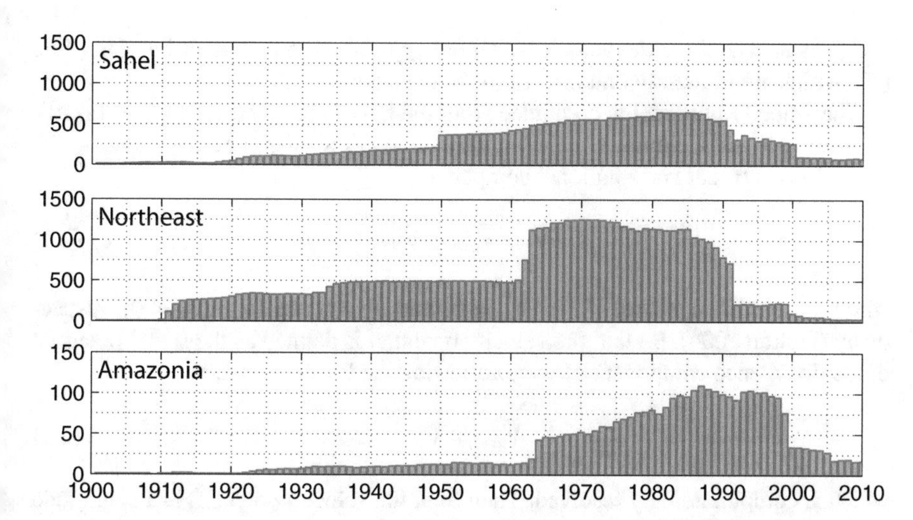

Fig. 2.2 Number of rain-gauge stations in time used in the GPCC v7 data base within the Sahel (between 17.5°W–10°E and 10°–17.5°N), the Northeast of Brazil (between 46°–35°W and 9°–2°S) and the Amazon region (between 10°S–5°N and 76°–55°W). Note the different scale used in the bottom plot with respect to the other two

coverage with a longitude and latitude resolution of 0.5° based on precipitation data from weather stations distributed around the world.

2.3.3 *Reanalysis*

In order to study the atmospheric dynamics, data from reanalyses are also used. These are based on the assimilation of observational data and, therefore, also have inherent uncertainties that are mainly attributed to the model and the observations used. To deal with it, we use two different reanalysis to compare the results: the European Center for Medium-Range Weather Forecasts reanalysis of the 20th Century (ERA-20C, from 1900 to 2010) (Poli et al. 2013) and the NOAA-CIRES 20th Century Reanalysis version 2c (20CRV2c, from 1851 to 2014) (Compo et al. 2011). Both reanalyses are performed by assimilating surface pressure from the International Surface Pressure Databank and the ERA-20C also includes wind observations from the International Comprehensive Ocean-Atmosphere Data Set. The outputs used from both reanalyses are monthly data with a horizontal resolution of 1° in longitude and latitude and 37 and 24 vertical levels, respectively for the ERA-20C and the 20CRV2c.

The variables used that are a direct product of the reanalyses are the vertical wind components (u and v), the specific humidity (q) and the surface pressure (sp). Other variables analyzed are the velocity potential (χ) and the moisture or humidity flux ($\vec{F}$), which are indirectly obtained from the reanalyses.

The velocity potential is computed from the horizontal wind field ($\vec{V} = (u, v)$). According to the Helmholtz theorem, the horizontal wind can be expressed as the sum of its divergent and rotational components:

$$\vec{V} = \vec{V}_{div} + \vec{V}_{rot}, \tag{2.1}$$

which have null rotational ($\nabla \times \vec{V}_{div} = 0$) and divergence ($\nabla \cdot \vec{V}_{rot} = 0$), respectively (Holton 2004). So that the velocity potential is defined as the scalar potential of the divergence component of wind following the Poison equation:

$$\vec{V}_{div} = \nabla \chi, \tag{2.2}$$

which is computationally resolved. Therefore, the velocity potential is a magnitude that measures the spreading of wind, such that the divergent wind flows out from centers of low velocity potential with a speed that is proportional to the velocity potential gradient.

The moisture flux is a useful variable to study the flow of the water vapor content in the atmosphere. To compute it, firstly the moist advection is calculated as the product of the specific humidity and horizontal wind. Then the moisture flux is obtained by vertically integrating the moist advection from the surface to high levels:

$$\vec{F} = -\frac{1}{g} \int_{sp}^{0} q\vec{V} dp, \tag{2.3}$$

where g is the gravity acceleration and p the atmospheric pressure. Particularly, in this Thesis the moisture flux is calculated by integrating from surface to 200 hPa level.

2.4 Methodology

2.4.1 Data Processing

Climate study involves the analysis of large amounts of data. The use of statistical tools is therefore required to efficiently process all these data, in order to extract and synthesize as much useful information as they can reveal. The methods used in this Thesis for the data processing are presented below. (For further details, the reader is referred to Wilks 2005).

2.4.1.1 Anomalies of a Field

Climate variability refers to departures of a certain climate characteristic from its long-term mean, also known as climatological value. Therefore, any climate characteristic can be expressed as a combination of its climatology plus time-dependent variations. Such variations are the so called anomalies. So, given a vector (if the climate characteristic is an index or a variable at a given location) or matrix (in case it is a field of a certain climate variable with spatial dimension) $\vec{X}$ describing the evolution of a climate characteristic over time t, the anomaly is obtained by subtracting its climatology ($\hat{\vec{\mu}}$):

$$\vec{X}'(t) = \vec{X}(t) - \hat{\vec{\mu}} \tag{2.4}$$

Particularly, in this Thesis we use data averaged over the whole year for SST (annual anomalies) and data averaged over particular seasons (seasonal anomalies) for the rest of variables.

2.4.1.2 Fourier Analysis

A time series is typically expressed in terms of the time domain, the same in which they are observed. But it is sometimes useful to use its expression in the frequency domain, in terms of contributions occurring at different characteristic frequencies of cosine waves (von Storch and Zwiers 2002). In the case of a discrete time series x_n with $n = 1, 2, ..., N$ data, this expression is as follows:

$$x_n = \sum_{k=0}^{N/2} C_k \cdot cos(\omega_k n - \phi_k), \tag{2.5}$$

in case N is pair (if N is odd, the summation extends from $k = 0$ to $(N-1)/2$). C_k is the amplitude and ϕ_k the phase of each kth cosine wave of characteristic angular frequency $\omega_k = 2\pi \mathrm{k/N}$. The values of these magnitudes are encoded in the expression of the time series in the frequency domain ($\hat{x}_k$), which is given by the Fourier transform:

$$\hat{x}_k = \sum_{n=0}^{N-1} x_n \cdot e^{-i\omega_k n}, \tag{2.6}$$

such that, the amplitudes C_k and phases ϕ_k can be expressed in terms of the real and the imaginary components of $\hat{x}_k$, which is a complex number, as follows:

$$C_k = \sqrt{Re(\hat{x}_k)^2 + Im(\hat{x}_k)^2} \tag{2.7}$$

$$\phi_k = tan^{-1}\frac{Im(\hat{x}_k)}{Re(\hat{x}_k)}. \tag{2.8}$$

The power spectrum of a signal quantifies how much of the variance of the data is contributed by a particular frequency. From the Fourier transform, the power spectrum is C_k^2, a function of ω_k which is typically normalized by $N/2\sigma^2$, where σ^2 is the variance of the time series x_n.

2.4.1.3 Frequency Filter

Climate shows variations at diverse time scales due to the numerous physical processes that modulate it. The signals of such variations are all mixed in the time series of the climate variables. The variability associated with certain physical processes can be separated from others that are not of our interest by filtering the time series. Such a filtering is performed through the use of time filters that allow decomposing the original signal in other time series with stationary variability. For example, a time series $Y(t)$ can be expressed as the combination of high (H) and low (L) variability components:

$$Y(t) = Y(t)^H + Y(t)^L \tag{2.9}$$

One way to filter a time series is using the Fourier analysis. Since it computes the amplitudes at all frequencies contributing to the total signal, we can modify them by means of a transfer function $H(\omega_k)$ that weights the amplitudes C_k such that in the frequency domain it is zero for frequencies above (below) a cutoff value ω_k, typically referred to as cutoff frequency, in the case of a low-pass (high-pass) filter.

$$x_{n,filtered} = \sum_{k=0}^{N/2} C_k H(\omega_k) \cdot cos(\omega_k n - \phi_k) \tag{2.10}$$

$$H(\omega_k) = \frac{C_{k,filtered}}{C_k}. \tag{2.11}$$

However, this method has serious shortcomings since the reconstructed time series may not resemble the original one and it requires complete records of data, hence the problems are worst near the ends of the finite time series.

In order to avoid these problems, there are filters designed so that the filtered series resembles the original one. These filters have a weighting function in the time domain with a certain number of weights (m) to achieve the desired transfer function. The number of weights determines the degree of the filter, so that the higher the degree of the filter is, the sharper the transition of the transfer function from 1 to 0 in the cutoff frequency is. Nevertheless, in the case of recursive filters, whose filtered value depends on weighted values of the original time series and of previous filtered values to be computationally more efficient, too large degree of the filter may result in an unstable filtered signal.

In this Thesis we use the Butterworth filter (Butterworth 1930), which is a recursive one that is commonly used in climate studies (e.g. Parker et al. 2007; Mohino et al. 2011a). The transfer function of the Butterworth low-pass filter is expressed as:

$$\mid H(\omega_k) \mid = \frac{1}{\sqrt{1 + \left(\frac{\omega_k}{\omega_c}\right)^{2m}}}, \tag{2.12}$$

where m the order of the filter and ω_c the cutoff angular frequency (i.e., 2π times the ordinary frequency) such that the amplitude of the transfer function is $1/\sqrt{2}$. The Butterworth filter can be modified to a high-pass filter, using $1 - H(\omega_k)$ as transfer function, or add others in series forming a band-pass filter.

In this Thesis we use an implementation of the Butterworth filter described by (Mann, 2004) which deals with the constraint of the time series near the boundaries by completing the series implementing different approximations. Particularly, we use both low- and high-pass Butterworth filters of order 10 with the *minimum slope* boundary constraint, which fills the time series in a way that the tendency of the filtered one is minimized near the boundary. For a given periodicity T_c below or above which we want to smooth the time series (depending on whether we low- or high-pass filter), the cut-off frequency is defined as:

$$f_c = \frac{1}{T_c}. \tag{2.13}$$

2.4.1.4 Empirical Orthogonal Function

The Empirical Orthogonal Function (EOF) analysis, also referred to as Principal Component Analysis, is a multivariate statistical technique that is used in data analysis and statistical forecasting. It is widely used in atmospheric and oceanic sciences as a discriminant analysis to identify the most prominent recurrent patterns of simultaneous variation and their time evolution in a field of a given climate variable. The term EOF was firstly introduced in meteorology by (Lorenz, 1956) and the method is fully described by books devoted to climate statistics (e.g., Wilks 2005; von Storch and Zwiers 2002).

Climate fields generally consist of large number of data with substantial correlation among each other. Through the EOF analysis, such fields can be decomposed into a set of spatial structures termed as EOFs and other complementary temporal ones called Principal Components (PCs). Each pair of EOFs and PCs describes the modes of variability of the field. These modes account for a fraction of the total variance of the original field, so that it can be reconstructed as a linear combination of the EOFs times the PCs.

Specifically, lets consider a field of anomalies $\vec{X}'$ with time (t) and space (m) dimensions (Eq. 2.4). The relation between the variance of the field in two different points is represented by the covariance matrix C calculated in the temporal dimension:

$$C = \frac{1}{N-1} \sum_{t} \vec{X}'_m(t) \cdot \vec{X}'_m(t)^T; \qquad t = 1, ..., N. \tag{2.14}$$

The objective of the EOF analysis is then to decompose the matrix C into eigenvectors and eigenvalues, such that:

$$C \cdot E = D \cdot E, \tag{2.15}$$

where the matrix E contains the eigenvectors $\vec{e}^k$ associated with the eigenvalues λ^k, which are arranged in the diagonal matrix D.

Eigenvectors $\vec{e}^k$ represent the EOFs describing the spatial distributions of the variability modes and satisfy orthogonality among themselves. The associated PCs are expressed as the projection of the anomalous field $\vec{X}'$ onto the transposed eigenvectors $\vec{e}^k$:

$$\alpha^k(t) = \vec{e}^{kT} \cdot \vec{X}'(t) = \sum_{m=1}^{M} e^k_m \cdot \vec{X}'_m(t) \tag{2.16}$$

The original field of anomalies can be then reconstructed from a number of K modes that describe a large proportion of the total variance

$$\vec{X}'(t) \simeq \sum_{k=1}^{K} \alpha^k(t) \cdot \vec{e}^k, \tag{2.17}$$

and minimize the mean squared error (MSE) (von Storch and Frankignoul 1998):

$$MSE = \sum_{t=1}^{N} \left[\vec{X}'(t) - \sum_{k=1}^{K} \alpha^k(t) \cdot \vec{e}^k \right] \tag{2.18}$$

The modes represented by the kth pair of EOFs $(\vec{e}^k)$ and PCs $(\alpha^k(t))$ explain a fraction of variance $(fvar)$ of the original field that is proportional to the corresponding eigenvalues (λ^k):

$$fvar = \frac{\lambda^k}{\sum_{k=1}^{K} \lambda^k} \cdot 100\% \tag{2.19}$$

The PCs are typically sorted such that the first PC (PC1) corresponds to the mode that explains more of the total variance of the field $\vec{X}'(t)$ and so on. That is, in descending order of the corresponding eigenvalue ($\lambda^1 > \lambda^2 > ... > \lambda^K$).

Therefore, the EOF analysis helps to identify modes of climate variability. Though the analysis does not provide information on the physical processes that generate these modes of variability. This requires careful interpretation of the resulting EOFs and PCs.

2.4.1.5 Linear Regression

In simple terms, the linear regression is a technique that allows to quantify the linear relation between two variables (x, y) that are represented in a plot of scattered points (scatterplot), seeking to summarize such a relationship by a linear straight:

$$\hat{y} = \alpha + \beta \cdot x, \tag{2.20}$$

where $\hat{y}$ is the estimation of the variable y and α and β are the intercept and the slope of the straight, respectively.

The slope β represents the regression coefficient between both variables, which quantifies how much of the dependent variable y linearly varies in relation to the independent one x. The regression coefficient is the ratio of the covariance between x and y and the variance[2] of the independent variable:

$$\beta = \frac{\sigma_{xy}^2}{\sigma_x^2}, \tag{2.21}$$

$$\sigma_{xy}^2 = \frac{1}{n-1} \sum_{i=1}^{n} (x_i - \overline{x})(y_i - \overline{y}), \tag{2.22}$$

$$\sigma_x^2 = \frac{1}{n-1} \sum_{i=1}^{n} (x_i - \overline{x})^2, \tag{2.23}$$

where $\overline{x}$ and $\overline{y}$ represent the average values of the variables. σ_x^2 is the variance of x, a measure of the spread of a set of numbers out from their average value. In turn, the standard deviation of a variable is the square root of its variance (σ_x).

In this Thesis we frequently use linear regression to show spatial patterns of the regression coefficients between a given field of a variable and a temporal index, to

[2] The covariance and the variance are indistinctly denoted in this Thesis as σ_{xy}^2 and σ_x^2 or as $Cov[x, y]$ and $Var[x]$, respectively.

which we will refer as regression maps or patterns. In matrix form, the field and the index can be expressed as $\mathbf{Y}(m, t)$ and $\mathbf{X}(t, 1)$, respectively, where m refers to the spatial dimension and t to the temporal one. If the index $\mathbf{X}$ is standardized, the expression of the regression maps can be simplified to:

$$\beta(m, 1) = \frac{\mathbf{X}'(m, t) \cdot \mathbf{Y}'(t, 1)}{N - 1}; \qquad t = 1, ..., N. \tag{2.24}$$

In some cases we may also be interested in removing the linear trend of a time series by detrending it. To obtain a detrended time series, the linear estimation calculated by linear regression is simply removed from the original one.

2.4.1.6 Linear Correlation

The correlation estimates the degree of linear dependence that exists between two variables, that is, how they covariate one with respect to the other. In this way one can study to what extent the linear dependence of one variable with another is significant.

In particular, the linear correlation evaluates to what degree a set of scattered points (x, y) approaches to a straight, which is tightly related to the linear regression study. Thus, the closer the points to the line, the greater the correlation absolute value is. If the variable y increases with x, the slope of the regression line and the correlation will have positive values. If, on the contrary, y decreases with x, the slope and the correlation will be negative.

The correlation coefficient is expressed as the ratio of the covariance and the standard deviation (square root of the variance) of the variables x and y:

$$R_{xy} = \frac{\sigma_{xy}}{\sigma_x \sigma_y} \tag{2.25}$$

From the value of the correlation between two variables one can also obtain the coefficient of determination (R^2), which indicates the proportion of variance of the dependent variable that is accounted for by the independent one.

2.4.1.7 Multi-linear Regression Analysis

While the simple linear regression allows to estimate a dependent variable (predictand) from another independent one (predictor), the multi-linear regression offers the possibility to use an unlimited number (k = 1, ..., K) of predictors (x_k). The estimated y variable using these predictors is given by:

$$\hat{y} = \beta_0 + \beta_1 \cdot x_1 + \beta_2 \cdot x_2 + ... + \beta_K \cdot x_K, \tag{2.26}$$

and therefore the predictand variable can be expressed as:

$$y = \beta_0 + \beta_1 \cdot x_1 + \beta_2 \cdot x_2 + ... + \beta_K \cdot x_K + \epsilon, \tag{2.27}$$

where β_0 is the intercept, β_k the regression coefficients corresponding to each predictor and ϵ is the residual term or error of the estimation ($\epsilon = y - \hat{y}$).

Through the multi-linear regression analysis, the estimation of the dependent variable can be improved with respect to the simple linear regression one, if it depends on more than one potential predictor. In addition, we can evaluate which component is explained by each one of these predictors. The variance of the variable y can be therefore expressed in terms of the regression coefficients (Mohino et al. 2016). For the sake of simplicity, if we consider only two predictor variables (x_1 and x_2), the variance of the predictand (y) is:

$$Var[y] = \beta_1^2 + \beta_2^2 + 2\beta_1\beta_2 Cov[x_1, x_2] + Var[\epsilon], \tag{2.28}$$

where $Cov[x_1, x_2]$ stands for the covariance between x_1 and x_2.

In one particular case, we are going to consider three indices (x_1, x_2 and x_3) as predictors in a multi-linear analysis, one of which is orthogonal to the other two (i.e., the covariance between this index with the other two is $Cov[x_1, x_2] = 0$ and $Cov[x_1, x_3] = 0$). Therefore, the variance of the predictand can be expressed in terms of the regression coefficients of the three predictors as follows:

$$Var[y] = \beta_1^2 + \beta_2^2 + \beta_3^2 + 2\beta_2\beta_3 Cov[x_2, x_3] + Var[\epsilon]. \tag{2.29}$$

In this Thesis, the multi-linear regression analysis is used to study how a single time series depends linearly on a number of indices and to determine the contribution of each one to the total variance of the predictand time series.

2.4.2 *Statistical Significance*

One of the major challenges in climate studies is whether the results obtained through the analysis of the data sets are statistically significant. These data often provide a limited number of samples and this can lead to results that are not statistically robust. The evaluation of significance is done by a contrast of statistical hypotheses. It is based on the decision of whether or not to reject an assumption about some characteristic of the population of the data analyzed that is called null hypothesis H_0. The opposite assumption is the so called alternative hypothesis H_1 (von Storch and Zwiers 2002).

There are different statistical tests than can be applied to determine whether H_0 is rejected or accepted. The distribution of the data that we study determines the election of the hypothesis test to be applied. Each test has an associated statistic. This test statistic is a value that follows a probability density function, called null distribution,

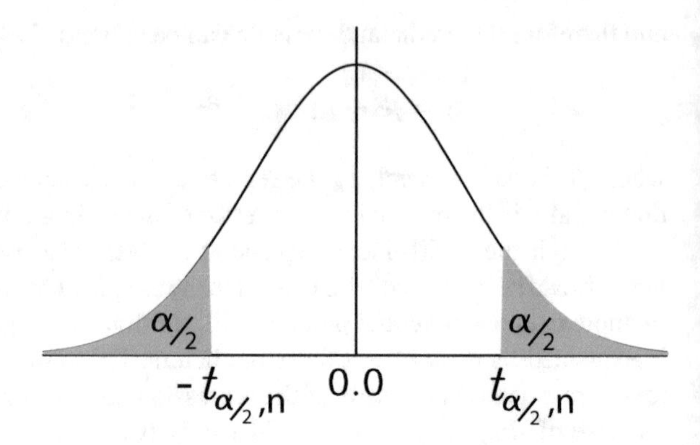

Fig. 2.3 Null distribution of a two-tailed Student t-test. The white area below the curve represents the *non-rejection* region and $t_{\alpha/2,n}$ stands for the test statistic value, corresponding to a α significant level and n degrees of freedom, from which H_0 can be rejected with a $1 - \alpha$ confidence level

which indicates the acceptance probability of H_0. So, given a significance level α, one may compute a statistic that determines whether H_0 can be rejected at a confidence level $1 - \alpha$ if its value is outside the *non-rejection* region of the null distribution (see one example in Fig. 2.3).

Regarding the null distribution associated with the different hypothesis tests, they can be classified into parametric and non-parametric tests. When the population to be studied is known and follows a normal distribution, parametric hypothesis tests are used assuming a theoretical probability density function. In contrast, when the distribution of the population is unknown and no assumptions about it can be made, a non-parametric test is used, for which a *ad-hoc* null distribution is built up (Wilks 2005).

2.4.2.1 Student T-Test

The Student t-test is a parametric hypothesis test in which the test statistic follows a Student's t-distribution under the null hypothesis (Fig. 2.3) (Gorgas et al. 2009). The t-test are typically used to determine whether two sets of data are significantly different from each other.

When we measure the degree of association between two variables through the correlation, we can estimate whether this relationship is statistically significant. To determine if the correlation is significant, a hypothesis test is made so that:

$$H_0 : r = 0, \text{ the two samples are linearly independent.}$$
$$H_1 : r \neq 0, \text{ the two samples are linearly dependent.}$$

In this case, the value of the t-Student statistic that determines whether H_0 must or must not be rejected is given by the expression:

$$t = \frac{r\sqrt{N-2}}{\sqrt{1-r^2}}, \tag{2.30}$$

which follows a null distribution of Student with N-2 degrees of freedom, where N stands for the number of data. Therefore, given a significance level α, if $t < t_{\alpha/2}$, then H_0 can be rejected with a $(1-\alpha)\cdot 100\%$ confidence level.

The Student t-test can be also applied to compare the mean value of a sample of data $\overline{x}$ with respect to a particular value μ_0. In this case, the hypothesis that we pose is:

$$H_0 : \overline{x} = \mu_0, \text{ the mean value is equal to} \mu_0$$
$$H_1 : \overline{x} \neq \mu_0, \text{ the mean value and } \mu_0 \text{ are different}$$

and the value of the statistic is given by this expression:

$$t = \frac{\overline{x} - \mu_0}{\sqrt{\frac{\sigma^2}{N}}}, \tag{2.31}$$

where σ^2 and N are the variance and the size of the sample x, respectively. Particularly, in this Thesis we want to know if the mean value of the anomalies of a variable over certain years is significantly different from zero (i.e., $\mu_0 = 0$).

Another application of the Student t-test is to assess the confidence intervals of the slope α of a linear regression line $\hat{y}$ estimated from the y dependent values of x (see Eq. 2.20). To determine if the slope is significantly close to zero, the following hypothesis is posed:

$$H_0 : \beta = 0, \text{ the slope is negligible.}$$
$$H_1 : \beta \neq 0, \text{ the slope is different from zero.}$$

In this case, the statistic is given by the expression:

$$t = \beta\sqrt{\frac{(N-1)Var[y]}{Var[y-\hat{y}]}}, \tag{2.32}$$

which follows a Student distribution with N-2 degrees of freedom.

2.4.2.2 "Random-Phase" Test of Ebisuzaki

In this Thesis, we frequently compare the relation between time series of data that are low-pass filtered by means of regression and correlation. These data may then present high autocorrelation and it cannot be assumed that the samples follow a normal distribution. Hence, a non-parametric hypothesis test is better suited.

A "random-phase" test, based on Ebisuzaki (1997), is used to statistically assess the significance of the regression and correlation values. This test is based on the comparison between the regression or correlation and a null distribution of regression or correlation coefficients (100 values in our case), respectively, obtained from random series designed to preserve the autocorrelation of the original ones. These synthetic time series are generated by randomly altering the phase of the original ones (ϕ_k in Eq. 2.5), using Fourier transforms, and maintaining their periodicities. Therefore, our null hypothesis supposes that the relation between the two time series is obtained by chance.

For the multi-model analysis held in this Thesis, model-mean regression patterns across the 17 models are composed to show common relations reproduced by the CMIP5 models. To evaluate the statistical significance of these model-mean patterns, the "random-phase" test is adapted. In this case, the averaged regression is compared with a null distribution constructed from mean regression coefficients out of 17 pairs of random time series generated from the original ones of each model.

2.4.2.3 ANOVA

The 1-way analysis of variance (ANOVA) compares the means of two or more groups of samples by assessing the equality of their variances. ANOVA uses the F-statistic of Ronald Fisher to statistically test this equality, hence it is a parametric hypothesis test. In climate studies, ANOVA can be used to assess how many realizations of a GCM simulation are needed to neglect the internal weather noise of the model with respect to the simulated climate response to the imposed forcing (e.g. Caminade and Terray 2010). If we consider that the only sources of variance of one model simulation are its internal weather noise and the variance induced by the imposed forcing, we can use ANOVA to evaluate the variance between different realizations (attributed to weather noise) with respect to the variance within them (obtained from the common externally forced variability) (Zwiers 1996). ANOVA provides the probability value that the variance between the different realizations is significantly lower than the common variance.

In this Thesis the ANOVA is used to evaluate the confidence of the set of realizations of the experiments performed to study the Sahel rainfall response to the imposed forcings (see Sect. 2.2.1).

2.4.3 Characterization of the Decadal-to-Multidecadal SST Modes

The characterization of the modes of variability is done by defining a pattern that describes the spatial distribution of the associated anomalies and an index that describes its temporal evolution. The method used to characterized the main modes of SST decadal-to-multidecadal variability is described below.

2.4.3.1 The GW Index

Considering that the GW is associated with global processes induced by the radiative forcing (Meehl et al. 2004), the GW index is defined as the global average of SSTA (Ting et al. 2009), calculated between 45°S–60°N to avoid problems with sea ice changes and regions with sparse coverage of observed SST data (Trenberth and Shea 2006; Baines and Folland 2007). Then, the resulting time series is low-pass filtered using a Butterworth filter with a 40-year cut-off period in order to isolate the higher-than-decadal signal of the SST from the variability of lower frequency. In this way we obtain the GW index. In many cases, the standardized GW index is used in this Thesis (i.e., divided by its standard deviation).

In the case of the CMIP5 simulations with imposed external forcing that provide more than one realization (see Table 2.1), the SSTA data is averaged among all the ensemble members prior to computing the GW index. Since each member is initialized from different states, the internal SST variability is not common among members of the ensemble, while the external forcing imposed is the same. Hence, the externally forced signal of the SSTA field can be isolated from the internal variability by computing the ensemble-mean. Note that the GW is not calculated from the piControl simulation of CMIP5 models since the radiative forcing imposed is constant in time and, therefore, no GW signal on SST is expected.

2.4.3.2 The "Residual" SST

Before calculating the AMV and IPO indices, the GW signal is removed from the original SSTA field (Trenberth and Shea 2006; Ting et al. 2009; Mohino et al. 2011a), instead of simply detrending the resulting indices as in other works (Enfield et al. 2001). Considering that the GW may not be spatially uniform throughout the global SST field, we follow the technique of (Mohino et al., 2011a) to compute a "residual" SSTA field to eliminate as much of the external forcing effects on the SST as possible.

Firstly, the GW spatial pattern of SSTA (GW_{patt}) is computed as the regression of the SSTA onto the GW index (GW_{ind}):

$$GW_{patt}(m, 1) = \frac{SSTA(m, t) \cdot GW_{ind}(t, 1)}{N - 1}; \qquad t = 1, ..., N \tag{2.33}$$

where m and t stand for the spatial and temporal dimensions, respectively.

Then, the "residual" SSTA ($SSTA_{res}$) is obtained by subtracting to the original SSTA a "GW-fitted" field, which is reconstructed as the product between the GW spatial pattern and its transposed index:

$$SSTA_{res}(m, t) = SSTA(m, t) - GW_{patt}(m, 1) \cdot GW^{t}_{ind}(1, t). \tag{2.34}$$

2.4.3.3 The AMV and IPO Indices

The AMV and IPO indices are computed as the first PC of an EOF analysis applied to the "residual" SSTA in the North Atlantic (between 0° and 60°N) and the basin-wide Pacific Ocean (between 45°S and 60°N), respectively. The "residual" SSTA field is previously area weighted and low pass-filtered through a Butterworth filter with a 13-year cut-off period. As an exception, in the case of the unforced piControl simulations the AMV and IPO indices are calculated from the original SSTA field instead of the "residual".

There are more straightforward ways to characterize the AMV index. For example, it is commonly obtained as the detrended index of SSTA averaged over the North Atlantic basin (e.g. Enfield et al. 2001; Knight 2005). However, the use of the EOF to define the AMV is also accepted by several authors (e.g. Mestas-Nuñez and Enfield 1999; Enfield et al. 2001; Baines and Folland 2007; Parker et al. 2007; Mohino et al. 2011a, 2016) and no relevant differences have been found between the results obtained using one definition or another (Martin et al. 2014). As for the IPO, the EOF analysis is the most commonly used approach to define its index (e.g. Zhang et al. 1997; Power et al. 1999; Arblaster et al. 2002; Meehl et al. 2013). Although there are also other definitions based on differences of SSTA averaged over the main centers of action of the IPO over the Pacific basin (Henley et al. 2015).

2.4.3.4 The AMV and IPO Patters

To characterize the SSTA spatial patterns associated with the AMV and IPO indices, as well as the GW, maps are obtained by regression of the anomalous field of any variable (without being previously filtered) onto the corresponding index. For the SSTA patterns, the annual values of the original data (not the "residual" SSTA) have been used for the regression. Whilst for the atmospheric variables, the regression maps have been computed using seasonal anomalies during JAS, for the Sahel, and DJFMAM, for the Amazon and Northeast regions. In the case of the Northeast region, its short rainy season is restricted to the months from March to May. However, the results concerning this region do not change substantially whether we use the anomalies averaged in DJFMAM or in its characteristic rainy season.

For those models with several realizations of the same experiment (see Table 2.1), their simulations are concatenated in time before the AMV and IPO estimation. This approach has been chosen to take advantage of the information provided by all the ensemble members together. Furthermore, from preliminary analysis of the methodology used in this Thesis, no differences were found between the resulting associated SSTA patterns and the time series obtained with this method and by analyzing the ensemble members separately, then averaging the patterns and putting the indices in series.

2.4.4 Indices from the LMDZ Simulations

From the resulting outputs of the simulations performed with the LMDZ model, two indices have been computed to analyze the results. One is the Sahel index of precipitation (also referred to as Sahel index, for simplicity) that is defined as the area-weighted average of the JAS seasonal precipitation between 17.5°W–10°E and 10°–17.5°N. The other index is the ASWI, which is computed as the proportion of days per month with predominant southwesterly wind between 29°–17°W and 7°–13°N and averaged in JAS, following (Gallego et al., 2015). The wind direction is obtained from the daily output of the model simulations of wind at 10 m above the surface. Both indices are filtered using a Butterworth filter with an 8-year cut-off period.

2.4.5 Limits of the Sahel, Amazon and Northeast Regions

In this Thesis, the Sahel refers to the region delimited between 10°–17.5°N and 17.5°W–10°E. The limits of the Amazonia are stated in 10°S–5°N and 76°–55°W and the Northeast of Brazil is assumed as the region spanned between 9°–2°S and 46°–35°W. These regions are delimited in some figures throughout the Thesis to guide the reader.

References

Arblaster, J., Meehl, G., Moore, A.: Interdecadal modulation of Australian rainfall. Clim. Dyn. **18**, 519–531 (2002)

Baines, P.G., Folland, C.K.: Evidence for a rapid global climate shift across the late 1960s. J. Clim. **20**, 2721–2744 (2007)

Butterworth, S.: On the theory of filter amplifiers. Wirel. Eng. **7**, 536–541 (1930)

Caminade, C., Terray, L.: Twentieth century Sahel rainfall variability as simulated by the ARPEGE AGCM, and future changes. Clim. Dyn. **35**, 75–94 (2010)

Compo, G.P., Whitaker, J.S., Sardeshmukh, P.D., Matsui, N., Allan, R.J., Yin, X., Gleason, B.E., Vose, R.S., Rutledge, G., Bessemoulin, P., et al.: The twentieth century reanalysis project. Quart. J. R. Meteorol. Soc. **137**, 1–28 (2011)

Diamond, M.S., Bennartz, R.: Occurrence and trends of eastern and central Pacific El Niño in different reconstructed SST data sets. Geophys. Res. Lett. **42** (2015)

Dufresne, J.-L., Foujols, M.-A., Denvil, S., Caubel, A., Marti, O., Aumont, O., Balkanski, Y., Bekki, S., Bellenger, H., Benshila, R., et al.: Climate change projections using the IPSL-CM5 earth system model: from CMIP3 to CMIP5. Clim. Dyn. **40**, 2123–2165 (2013)

Ebisuzaki, W.: A method to estimate the statistical significance of a correlation when the data are serially correlated. J. Clim. **10**, 2147–2153 (1997)

Enfield, D.B., Mestas-Nuñez, A.M., Trimble, P.J.: The Atlantic multidecadal oscillation and its relation to rainfall and river flows in the continental U.S. Geophys. Res. Lett. **28** 2077–2080 (2001). https://doi.org/10.1029/2000GL012745

Gallego, D., Ordóñez, P., Ribera, P., Peña-Ortiz, C., García-Herrera, R.: An instrumental index of the West African Monsoon back to the nineteenth century. Quart. J. R. Meteorol. Soc. **141**, 3166–3176 (2015)

Gorgas, J., Cardiel, N., Zamorano, J.: ESTADÍSTICA BÁSICA para estudiantes de Ciencias (2009)

Harris, I., Jones, P., Osborn, T., Lister, D.: Updated high-resolution grids of monthly climatic observations–the CRU TS3. 10 Dataset. Int. J. Climatol. **34**, 623–642 (2014)

Henley, B.J., Gergis, J., Karoly, D.J., Power, S., Kennedy, J., Folland, C.K.: A tripole index for the interdecadal Pacific oscillation. Clim. Dyn. **45**, 3077–3090 (2015)

Holton, J.: *An Introduction to Dynamical Meteorology*, 4th edn. Elsevier Academic Press (2004)

Hourdin, F., Foujols, M.-A., Codron, F., Guemas, V., Dufresne, J.-L., Bony, S., Denvil, S., Guez, L., Lott, F., Ghattas, J., et al.: Impact of the LMDZ atmospheric grid configuration on the climate and sensitivity of the IPSL-CM5A coupled model. Clim. Dyn. **40**, 2167–2192 (2013)

Huang, B., Banzon, V.F., Freeman, E., Lawrimore, J., Liu, W., Peterson, T.C., Smith, T.M., Thorne, P.W., Woodruff, S.D., Zhang, H.-M.: Extended reconstructed sea surface temperature version 4 (ERSST.v4). Part I: upgrades and intercomparisons. J. Clim. **28**, 911–930 (2015)

Huang, B., Thorne, P.W., Smith, T.M., Liu, W., Lawrimore, J., Banzon, V.F., Zhang, H.-M., Peterson, T.C., Menne, M.: Further exploring and quantifying uncertainties for extended reconstructed sea surface temperature (ERSST) version 4 (v4). J. Clim. **29**, 3119–3142 (2016)

Knight, J.R.: A signature of persistent natural thermohaline circulation cycles in observed climate. Geophys. Res. Lett. **32**, L20 708 (2005). https://doi.org/10.1029/2005GL024233

Krinner, G., Viovy, N., de Noblet-Ducoudré, N., Ogée, J., Polcher, J., Friedlingstein, P., Ciais, P., Sitch, S., Prentice, I.C.: A dynamic global vegetation model for studies of the coupled atmosphere-biosphere system. Global Biogeochem. Cycles **19** (2005)

Liu, W., Huang, B., Thorne, P.W., Banzon, V.F., Zhang, H.-M., Freeman, E., Lawrimore, J., Peterson, T.C., Smith, T.M., Woodruff, S.D.: Extended reconstructed sea surface temperature version 4 (ERSST.v4): Part II. Parametric and structural uncertainty estimations. J. Clim. **28**, 931–951 (2015)

Lorenz, E.N.: Empirical orthogonal functions and statistical weather prediction. Massachusetts Institute of Technology, Department of Meteorology (1956)

Mann, M.E.: On smoothing potentially non-stationary climate time series. Geophys. Res. Lett. **31** (2004)

Marengo, J.A.: Interdecadal variability and trends of rainfall across the Amazon basin. Theor. Appl. Climatol. **78**, 79–96 (2004). https://doi.org/10.1007/s00704-004-0045-8

Martin, E.R., Thorncroft, C., Booth, B.B.B.: The multidecadal Atlantic SST-Sahel rainfall teleconnection in CMIP5 simulations. J. Clim. **27**, 784–806 (2014). https://doi.org/10.1175/JCLI-D-13-00242.1

Meehl, G.A., Hu, A., Arblaster, J.M., Fasullo, J., Trenberth, K.E.: Externally forced and internally generated decadal climate variability associated with the interdecadal Pacific oscillation. J. Clim. **26**, 7298–7310 (2013). https://doi.org/10.1175/JCLI-D-12-00548.1

Meehl, G.A., Washington, W.M., Ammann, C.M., Arblaster, J.M., Wigley, T., Tebaldi, C.: Combinations of natural and anthropogenic forcings in twentieth-century climate. J. Clim. **17**, 3721–3727 (2004)

Meehl, G.A., Goddard, L., Murphy, J., Stouffer, R.J., Boer, G., Danabasoglu, G., Dixon, K., Giorgetta, M.A., Greene, A.M., Hawkins, E., Hegerl, G., Karoly, D., Keenlyside, N., Kimoto, M., Kirtman, B., Navarra, A., Pulwarty, R., Smith, D., Stammer, D., Stockdale, T.: Decadal prediction. Bull. Am. Meteorol. Soc. **90**, 1467–1485 (2009). https://doi.org/10.1175/2009BAMS2778.1

Mestas-Nuñez, A.M., Enfield, D.B.: Rotated global modes of non-ENSO sea surface temperature variability. J. Clim. **12**, 2734–2746 (1999)

Mohino, E., Janicot, S., Bader, J.: Sahel rainfall and decadal to multi-decadal sea surface temperature variability. Clim. Dyn. **37**, 419–440 (2011a). https://doi.org/10.1007/s00382-010-0867-2

Mohino, E., Keenlyside, N., Pohlmann, H.: Decadal prediction of Sahel rainfall: where does the skill (or lack thereof) come from? Clim. Dyn. **47**, 3593–3612 (2016)

Parker, D., Folland, C., Scaife, A., Knight, J., Colman, A., Baines, P., Dong, B.: Decadal to multi-decadal variability and the climate change background. J. Geophys. Res. Atmos. **112** (2007)
Poli, P., Hersbach, H., Tan, D., Dee, D., Thépaut, J.-n., Simmons, A., Peubey, C., Laloyaux, P., Komori, T., Berrisford, P., Dragani, R.: ERA report series, p. 59 (2013)
Power, S., Casey, T., Folland, C., Colman, A., Mehta, V.: Inter-decadal modulation of the impact of ENSO on Australia. Clim. Dyn. **15**, 319–324 (1999). https://doi.org/10.1007/s003820050284
Rayner, N., Parker, D.E., Horton, E., Folland, C., Alexander, L., Rowell, D., Kent, E., Kaplan, A.: Global analyses of sea surface temperature, sea ice, and night marine air temperature since the late nineteenth century. J. Geophys. Res. Atmos. **108** (2003)
Riahi, K., Grübler, A., Nakicenovic, N.: Scenarios of long-term socio-economic and environmental development under climate stabilization. Technol. Forecast. Soc. Change **74**, 887–935 (2007). https://doi.org/10.1016/j.techfore.2006.05.026
Sato, M., Hansen, J.E., McCormick, M.P., Pollack, J.B.: Stratospheric aerosol optical depths, 1850–1990. J. Geophys. Res. Atmos. **98** 22987–22994 (1993)
Schneider, U., Becker, A., Finger, P., Meyer-Christoffer, A., Rudolf, B., Ziese, M.: GPCC full data reanalysis version 7.0: monthly land-surface precipitation from rain gauges built on GTS based and historic data (2016). https://doi.org/10.5065/D6000072
Smith, T.M., Reynolds, R.W., Peterson, T.C., Lawrimore, J.: Improvements to NOAA's historical merged land-ocean surface temperature analysis (1880–2006). J. Clim. **21**, 2283–2296 (2008)
Taylor, K.E., Stouffer, R.J., Meehl, G.A.: An overview of CMIP5 and the experiment design. Bull. Am. Meteorol. Soc. **93**, 485–498 (2012)
Ting, M., Kushnir, Y., Seager, R., Li, C.: Forced and internal Twentieth-century SST trends in the North Atlantic*. J. Clim. **22**, 1469–1481 (2009). https://doi.org/10.1175/2008JCLI2561.1
Trenberth, K.E., Shea, D.J.: Atlantic hurricanes and natural variability in 2005. Geophys. Res. Lett. **33**, 1–4 (2006). https://doi.org/10.1029/2006GL026894
von Storch, H., Frankignoul, C.: Empirical modal decomposition in coastal oceanography. In: The Global Coastal Ocean: Processes and Methods, vol. 10, p. 419 (1998)
von Storch, H., Zwiers, F.W.: Statistical Analysis in Climate Research. Cambridge University Press, Cambridge (2002)
Wilks, D.S.: *Statistical Methods in the Atmospheric Sciences*, vol. 91 (2005)
Willmott, C.J., Matsuura, K., Legates, D.: Terrestrial air temperature and precipitation: monthly and annual time series (1950–1999). Center for climate research version 1 (2001)
Zhang, Y., Wallace, J.M., Battisti, D.S.: ENSO-like interdecadal variability: 1900–93. J. Clim. **10**, 1004–1020 (1997)
Zwiers, F.: Interannual variability and predictability in an ensemble of AMIP climate simulations conducted with the CCC GCM2. Clim. Dyn. **12**, 825–847 (1996)

Part II
Results I: Multi-Model Analysis

This part presents the results derived from the study of the influence of the main modes of decadal SST variability on the Sahel and northern Brazil rainfall using a set of simulations of several CMIP5 models. This analysis begins by characterizing the main mode of the global SST variation at longer-than-decadal time scales (the GW) and then the two leading modes of the decadal-to-multidecadal SST variability of the major ocean basins (the AMV and IPO). Afterwards, the influence of these modes on precipitation changes in the regions of interest are evaluated and the atmospheric mechanisms involved in the SST-precipitation teleconnection are identified. The role of the external radiative forcing in the AMV, the IPO and their impacts on rainfall, as well as the future evolution of the SST modes impacts, are assessed by comparing forced and unforced simulations and analyzing future projections of different GCMs of CMIP5. Finally, the contribution that each of the characterized modes of SST low-frequency variability have on the total decadal variance of rainfall in the regions of interest is calculated.

Chapter 3
Influence of the GW

Abstract The GW index and its corresponding SSTA pattern reproduced by the externally forced CMIP5 simulations and the observed one are characterized in this chapter. The impact of the GW patterns on precipitation in the Sahel and the northern Brazilian regions of the Amazon and the Northeast is also identified and analyzed. Historical simulations of the 20th century allow to compare the results with observations, while the analysis of future projections provide an insight of how the GW influence on rainfall in the regions of study might be along the current century.

3.1 The GW Index and Pattern

The GW index obtained from observational SSTA data broadly shows a warming trend throughout the 20th century (Fig. 3.1). But it also presents certain oscillatory variability. The minimum temperature of the observed period is around the 1910s. Then the global temperature steeply rises until the 1940s, where the warming decelerates and a slight relative cooling is observed throughout the 1950s until the 1960s. In the following decades, the global temperature increases again very rapidly and decelerates over the 2000s. The occurrence of these periods of GW deceleration agrees with other studies, which define them as hiatus decades and relate them with natural variability (e.g. Trenberth and Fasullo 2013; Meehl et al. 2013). Although it has to be noticed that the warming trend in the last part of the time series may be artificially decreased by the minimum slope condition at the edges used for the construction of the filtered GW index (Sect. 2.4).

Regarding CMIP5 models, the average of the GW indices among them show important similarities with the observed one (Fig. 3.1), being both highly correlated (Table 3.1). The model-mean GW presents a characteristic warming trend along the reproduced period with some oscillations. In contrast to observations, it shows a minimum peak in the 1890s. Then it rises up to the 1940s, decreases until the 1960s and rises again coinciding with the observed GW index. Both the observed and the model-mean GW indices indicate that the global SST experienced an increment of roughly 0.5 °C along the 20th century. Though the CMIP5 models tend to slightly

J. Villamayor, *Influence of the Sea Surface Temperature Decadal Variability on Tropical Precipitation: West African and South American Monsoon*, Springer Theses, https://doi.org/10.1007/978-3-030-20327-6_3

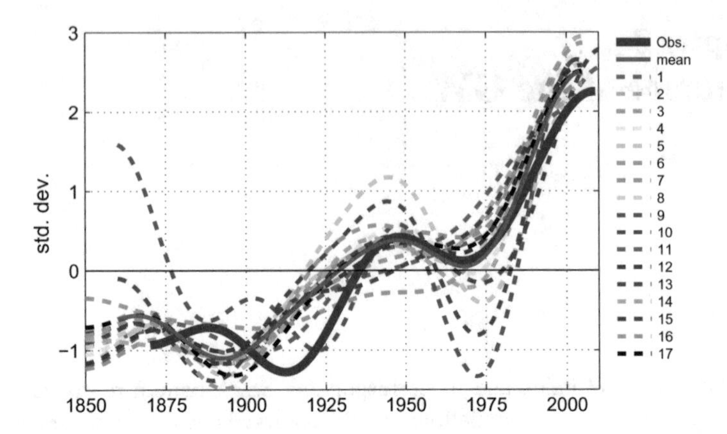

Fig. 3.1 Standardized GW indices from observations (HadISST1), the historical simulation of the CMIP5 models on average and individually. Each model is enumerated as in Table 2.1

overestimate the observed warming trend during the last two decades of the 20th century, coinciding with other works based on GCMs simulations with external natural and anthropogenic forcing imposed (Stott et al. 2000; Meehl et al. 2004; Hansen et al. 2005)

The individual model GW indices show certain discrepancies among themselves and with respect to observations. The correlation between the indices of each model and the one obtained from observations range between $R = 0.72$ and $R = 0.96$ (Table 3.1). The lowest score corresponds to the HadGEM2-CC model, which is followed by CSIRO-Mk3-6-0 and HadGEM2-ES. These models overestimate the amplitude of the observed global SSTA oscillations, particularly with a pronounced shift between the relative maximum and minimum of the 1940s and 1960s, respectively. Furthermore, the HadGEM2-CC model represents unrealistically warm global SST at the beginning of the simulation, prior to the 20th century.

In observations, the GW projects widespread intense warm SSTA roughly globally (Fig. 3.2a). This suggests that the vast global ocean surface has experienced a long-term warming trend over the last century, with greater intensity over western Pacific, the tropical and subtropical eastern Atlantic, the northern Indian basin and the Southern Ocean south of the Atlantic and Indian basins. A weak cooling is observed in the extratropical North Atlantic. This has been related to anomalous water freshening, as a consequence of reduced evaporation due to the air warming relative to SST, resulting in a weakening of the THC and the subarctic North Atlantic cooling (Kim and An 2013). The central and eastern tropical Pacific also presents a weak cooling associated with the GW. This is consistent with other works that attribute such a cooling to reinforced trade winds over the eastern Pacific promoted by the tropical Indian Ocean warming relative to the Pacific (Luo et al. 2012; Solomon and Newman 2012).

Table 3.1 Coefficients of the correlation (R) between the GW indices from observations and the CMIP5 models (including the model-mean) calculated for the common period among all models and observations of 1870–2004. All the correlation coefficients are statistically significant at a confidence level greater than 99%

Model-mean	0.95
1. bcc-csm1-1	0.95
2. CanESM2	0.91
3. CCSM4	0.95
4. CNRM-CM5	0.94
5. CSIRO-Mk3-6-0	0.80
6. FGOALS-g2	0.91
7. GISS-E2-H	0.94
8. GISS-E2-R	0.92
9. HadGEM2-CC	0.72
10. HadGEM2-ES	0.80
11. inmcm4	0.96
12. IPSL-CM5A-LR	0.95
13. MIROC5	0.91
14. MIROC-ESM-CHEM	0.94
15. MPI-ESM-LR	0.92
16. MRI-CGCM3	0.89
17. NorESM1-M	0.95

The CMIP5 models, on average, reproduce a SSTA pattern associated with the GW that presents a worldwide increase of the ocean temperatures with some regions warming more intensely than others (Fig. 3.2b). Broadly, the SST warming is more intense between the tropics than in the extratropical oceans. In particular, according to observations, the extratropical North Atlantic SST shows weak variability associated with the GW, while it is considerably more intense toward the equator and further south. Models also show more intense warming compared to the rest of the globe in the northern Indian basin and the Southern Ocean sector adjoining the Atlantic and Indian Ocean with high agreement among themselves. Similarly, CMIP5 models robustly reproduce weaker SSTA relative to the worldwide pattern over the extratropical North Pacific and to the south spanning the Southern Ocean, coinciding with observations. However, over the central and eastern equatorial Pacific, models robustly reproduce an intense warming that is opposite to observations. This unrealistic simulated warming is in accordance with other works that relate the inaccuracy of GCMs in reproducing the observed tropical Pacific cooling trend to biases concerning the way they simulate the warming contrast among the tropical ocean basins, local feedbacks and the surface net heat flux (Luo et al. 2017; Song and Zhang 2014). This discrepancy entails a significant difference, considering the important influence of the SST in this region on climate worldwide (Rodríguez-Fonseca et al. 2016), that

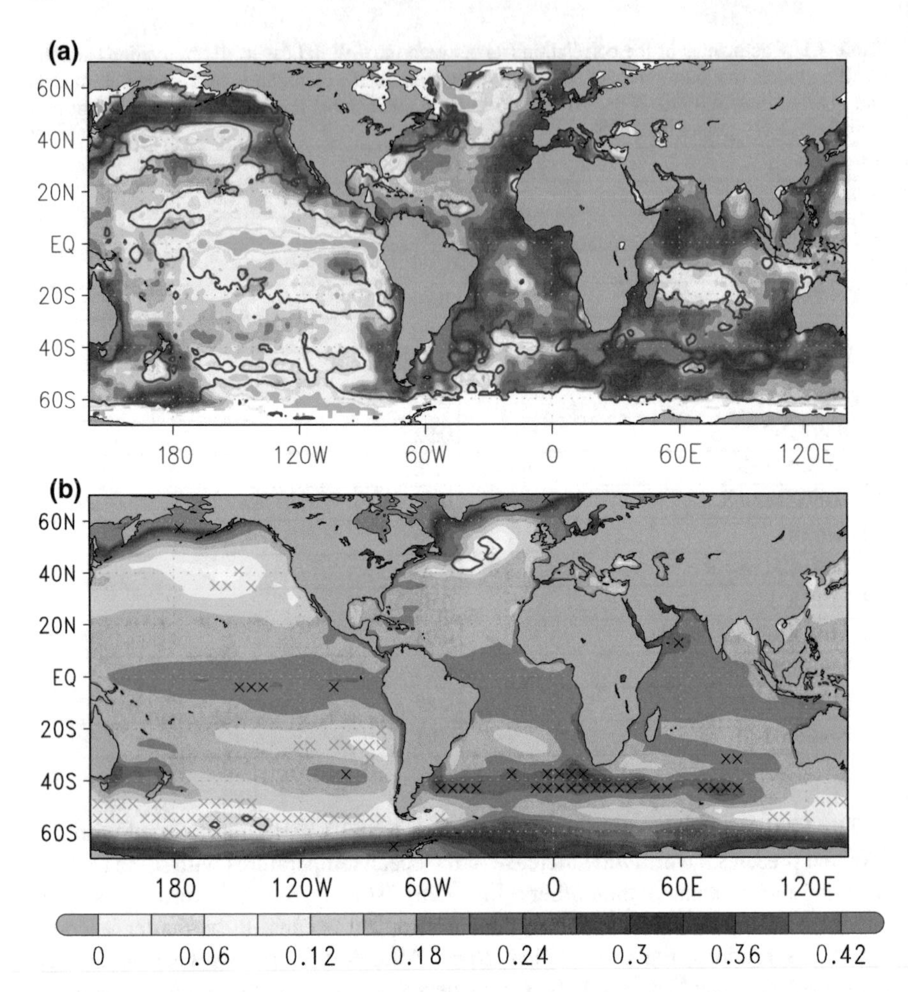

Fig. 3.2 Regression onto the standardized GW index of the unfiltered SSTA (K per standard deviation) from the HadISST1 observational data (**a**) and from the historical simulation averaged among the 17 CMIP5 models (**b**). Contours indicate the regions where the simple (**a**) or the averaged (**b**) regression is significant at the 5% level from a "random-phase" test. Black and blue crosses in (**b**) indicate points where the SSTA is, respectively, higher and lower than the global mean in at least 13 out of the 17 models analyzed

may affect the way in which models reproduce the GW impacts. In addition, there is broadly low agreement among models as to the regions with higher or lower warming relative to the global mean (sea crosses in Fig. 3.2b). This suggests that there may be certain discrepancies among CMIP5 models in simulating the distribution of SSTA gradients associated with the GW, which may induce model-dependency in terms of the GW impacts.

Regarding the RCP8.5 future projections from the CMIP5 models, the GW indices obtained from the reproduced SSTA show a practically linear warming trend over

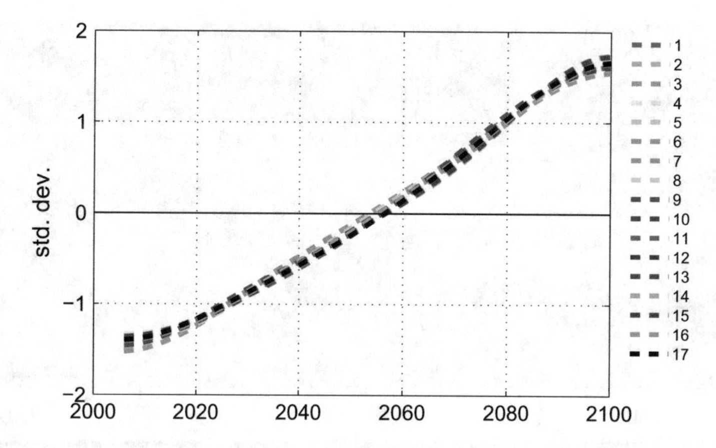

Fig. 3.3 Standardized GW indices from the RCP8.5 future projection of the CMIP5 models individually. Each model is enumerated as in Table 2.1

the 21st century with no clear oscillatory variability, in contrast to the historical simulation (Fig. 3.3). Such a quasi-linear increase of the global SST is consistent among all the models analyzed, without exception. However, there is certain discrepancy among them as to the total increase of the global SSTA, ranging from around 2–3.5 K over the simulated period (typically 2006–2100). This suggests that under the same radiative forcing conditions, the effects on the SST can be more or less intense depending on the model. The climate sensitivity of the models can vary among themselves due to differences in the climate feedbacks that they simulate, such as cloud radiative forcing (Andrews et al. 2012).

The model-mean SSTA pattern associated with the GW in the RCP8.5 future projection shows a widespread warming. This is consistent with the historical simulation, though the SSTA are notably more intense (note the different color scales used in Figs. 3.2 and 3.4). The SST rise is also more intense in some regions than in others, but they are differently distributed than in the historical simulations of the 20th century. The warming in the tropical and northern sector of the oceans is more intense than the global mean, with the only exception of the subpolar North Atlantic where it is weaker. While south of the tropics, the SSTA are mostly weaker than the global mean. Such an interhemispheric difference of the SST warming between the north, including the tropics, and the south is highly consistent among the CMIP5 models and may have important implications on rainfall changes in the regions of interest of this Thesis (e.g. Folland et al. 1986; Hastenrath and Greischar 1993; Folland et al. 2001; Harris et al. 2008; Giannini et al. 2013; Park et al. 2015).

In summary, CMIP5 models on average reproduce the observed GW time series and some of the basic aspects of its SSTA pattern in the historical simulation. However, they do not reproduce the same tropical Pacific SSTA and show certain disagreement among themselves as to the amplitude of the oscillatory variability of the GW index. Such discrepancies can be relevant to determine the way in which

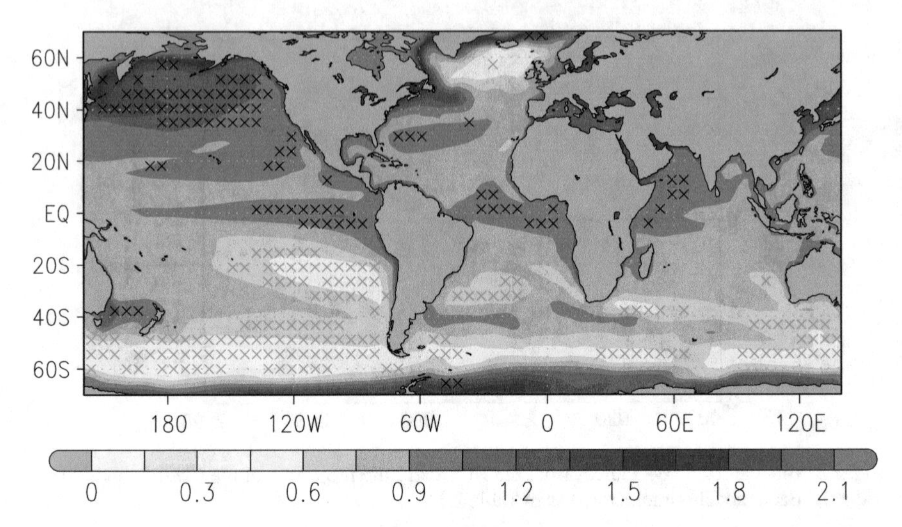

Fig. 3.4 Regression onto the standardized GW index of the unfiltered SSTA (K per standard deviation) from the RCP8.5 future projection averaged among the 17 CMIP5 models. The averaged regression is significant at the 5% level from a "random-phase" test throughout the global pattern. Black and blue crosses in indicate points where the SSTA is, respectively, higher and lower than the global mean in at least 13 out of the 17 models analyzed

CMIP5 models reproduce the GW impacts on average and individually. Robustness among models in reproducing the GW is higher in the RCP8.5 future projections, with steeper warming trend and more intense SSTA pattern than in the historical simulation. The different distribution of SSTA in the GW patterns of the RCP8.5 and the historical simulations suggests that a worldwide SST warming may have different impacts on climate in the future. Since the GW is an externally forced mode of variability, the role of the main components of radiative forcing have to be addressed in order to shed some light on the causes that lead the differences among models and between the historical simulation and the future projection.

3.1.1 The Role of the GHGs and Aerosols

The differences among the GW indices of the models in the historical simulation is outstanding, whereas in the RCP8.5 future projection the indices obtained are practically identical. This suggests that the response of the global SST to the radiative forcing imposed in the historical simulation is notably model-dependent (Rotstayn et al. 2015), in contrast to the future projection. Regarding the radiative forcing imposed in these simulations, in the RCP8.5 future projection large concentrations of GHGs, especially carbon dioxide, are imposed with respect to the historical simulation, while aerosol concentrations are much lower than in the 20th century (Riahi et al. 2007). Considering that the parametrization of the aerosols effects can be freely

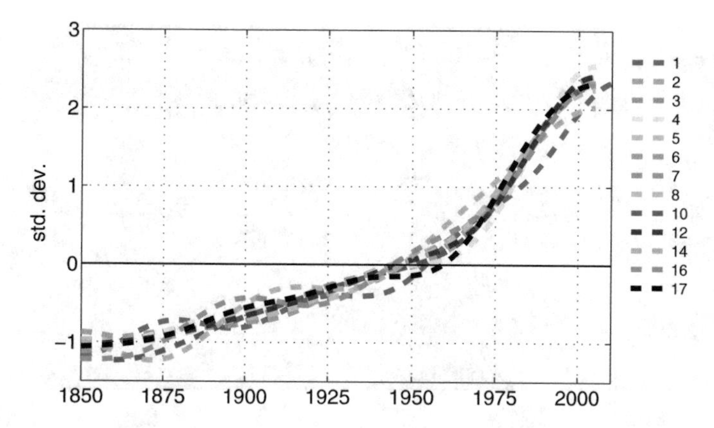

Fig. 3.5 Standardized GW indices from the historicalGHG detection and attribution experiment of the CMIP5 models individually. Each model is enumerated as in Table 2.1

set in each model (Taylor et al. 2012; Ekman 2014), it is reasonable to believe that they may induce the disparity among models in the historical simulation. This is in agreement with Rotstayn et al. (2015), who suggest that most of the inter-model variance in the simulated global-mean surface temperature is explained by changes in the radiative forcing induced by aerosols.

The GW indices obtained from the historicalGHG detection and attribution experiment reveal notably higher consistency among models than in the historical simulations (Fig. 3.5). In addition, these indices show almost no oscillatory variability but rather a linear warming trend of the global SST, in contrast to observations and consistently with RCP8.5 future projections. Considering that the only difference between the historical and historicalGHG simulations is the configuration of the radiative forcing imposed, this result suggest that the warming trend of the global SST described by the GW index is mainly induced by the effect of the GHGs. While the oscillatory variability of the GW index reproduced by the historical simulation is, by default, forced by aerosols effects. Therefore, these effects can be considered as the principal source of uncertainty among models regarding the accuracy with which they reproduce the observed GW evolution over the 20th century. It can also be deduced that models which overestimate the observed GW oscillations (such as CSIRO-Mk3-6-0, HadGEM2-CC and HadGEM2-ES), represent exaggerated aerosol effects on the global SST that lead to reproduce unrealistic variability at longer-than-decadal time scales.

Regarding the GW patterns of SSTA, the warming reproduced in the historical simulation by CMIP5 models, on average, is broadly more intense in the tropical band than in the extratropics (Fig. 3.2b). While the RCP8.5 future projection simulates larger anomalies between the tropics and the northern hemisphere than in the southern extratropics (Fig. 3.4). In turn, the historicalGHG detection and attribution simulation simulates a model-mean GW pattern with a worldwide warming, more intense between the tropics and further north than to the south (Fig. 3.6), which is

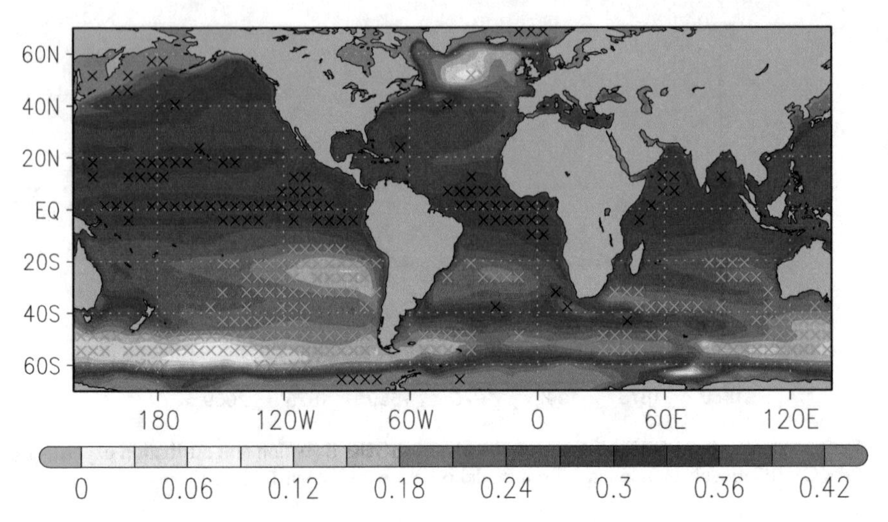

Fig. 3.6 Regression onto the standardized GW index of the unfiltered SSTA (K per standard deviation) from the historicalGHG detection and attribution experiment averaged among the 13 CMIP5 models for which these experiment is available (see Table 2.1). Contours indicate the regions where the averaged regression is significant at the 5% level from a "random-phase" test. Black and blue crosses in indicate points where the SSTA is, respectively, higher and lower than the global mean in at least 9 out of the 13 models analyzed

more akin to the RCP8.5 than the historical pattern. Although the warming is notably weaker than in the RCP8.5 future projections, according to the different GHGs concentration in both experiments (note the different scales used in the color shading in Figs. 3.4 and 3.6). The model-mean GW pattern of SSTA is also more robust in historicalGHG than in the historical simulation (see crosses in Figs. 3.2b and 3.6). While they may be some nonlinear effects of the GHGs and the aerosols on climate (Ming and Ramaswamy 2011), these results suggest that the simulated aerosol effects induce more uncertainties among models than the GHGs, as far as the GW pattern is concerned, and that the aerosols effect on SST is to dampen the warming induced by the GHGs increase in the northern hemisphere, in agreement with other works (Ekman 2014; Rotstayn and Lohmann 2002; Rotstayn et al. 2015; Ackerley et al. 2011). So, considering that the differential SST warming associated with the GW is crucial to explain some of its impacts on climate (e.g. Rotstayn and Lohmann 2002; Friedman et al. 2013; Park et al. 2015), the model-dependent parametrization of the aerosol effects is an important source of uncertainty among CMIP5 models that may hamper their ability to robustly simulate the GW impacts.

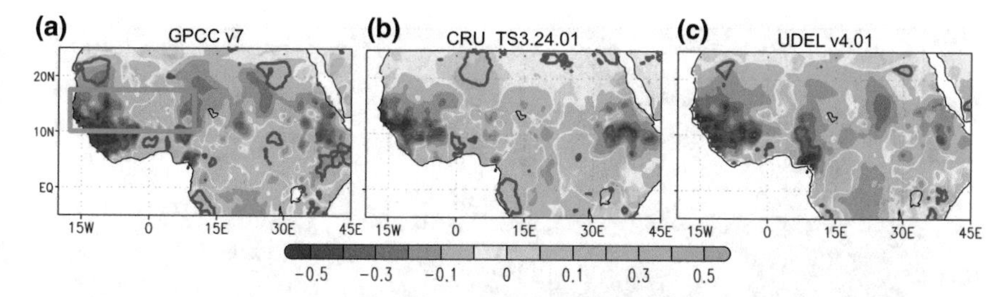

Fig. 3.7 Regression map of the unfiltered JAS precipitation anomaly from the (**a**) GPCC v7, (**b**) CRU TS3.24.01 and (**c**) UDEL v4.01 observational data bases onto the standardized GW index (units are mm day^{-1} per standard deviation). Contours indicate the regions where the regression is significant at the 5% level from a "random-phase" test. The orange box delimits the Sahel region

3.2 Sahel Rainfall Response

The observed precipitation response to the GW in West Africa during JAS is mostly negative (Fig. 3.7). The rainfall anomalies are barely statistically significant over most of West Africa and even locally with different opposite signs among the three observational data bases, specially in Central Africa. However, in the Sahel there are intense negative rainfall anomalies over the western half, extending north and south along the coast. In the rest of the Sahel region, there are mostly weak positive anomalies around the zero meridian and negatives close to the eastern boundary, more intense to the south. The anomalous drying to the west is robustly shown by all the sources of observations analyzed and is partially statistically significant toward the coast, where the drying is more prominent. Along the Gulf of Guinea coast there are two local areas with significant positive rainfall anomalies, one extending from Ghana to Nigeria and the other one in Gabon and its surrounding areas, which are robustly shown in all the observational data.

Consistently with the observed response of the JAS precipitation to the GW, the historical simulation of the CMIP5 models, on average, reproduce intense Sahel drying in the western half of the region (Fig. 3.8a). Such a drying is more intense along the coast, extending further south and is statistically significant and robust among models. In the eastern half of the Sahel, models also reproduce weaker or even positive rainfall anomalies and significant drying to the most southeastern part, similar to observations. Out of this region, models reproduce significant negative precipitation anomalies south of the Sahel, covering most of the Gulf of Guinea coast in contrast to observations.

In the RCP8.5 future projection of the CMIP5 models, on average, intense dying in western Sahel in response to the GW is also reproduced, in agreement with the historical simulation (Fig. 3.8b). Negative precipitation anomalies cover the coastal part of the Sahel, roughly west of 5°W and with high agreement among models, and there are opposite anomalies to the south. But in contrast to the historical simulation, there are intense positive precipitation anomalies in the rest of the Sahel region simulated by the future projection. These positive anomalies expand throughout

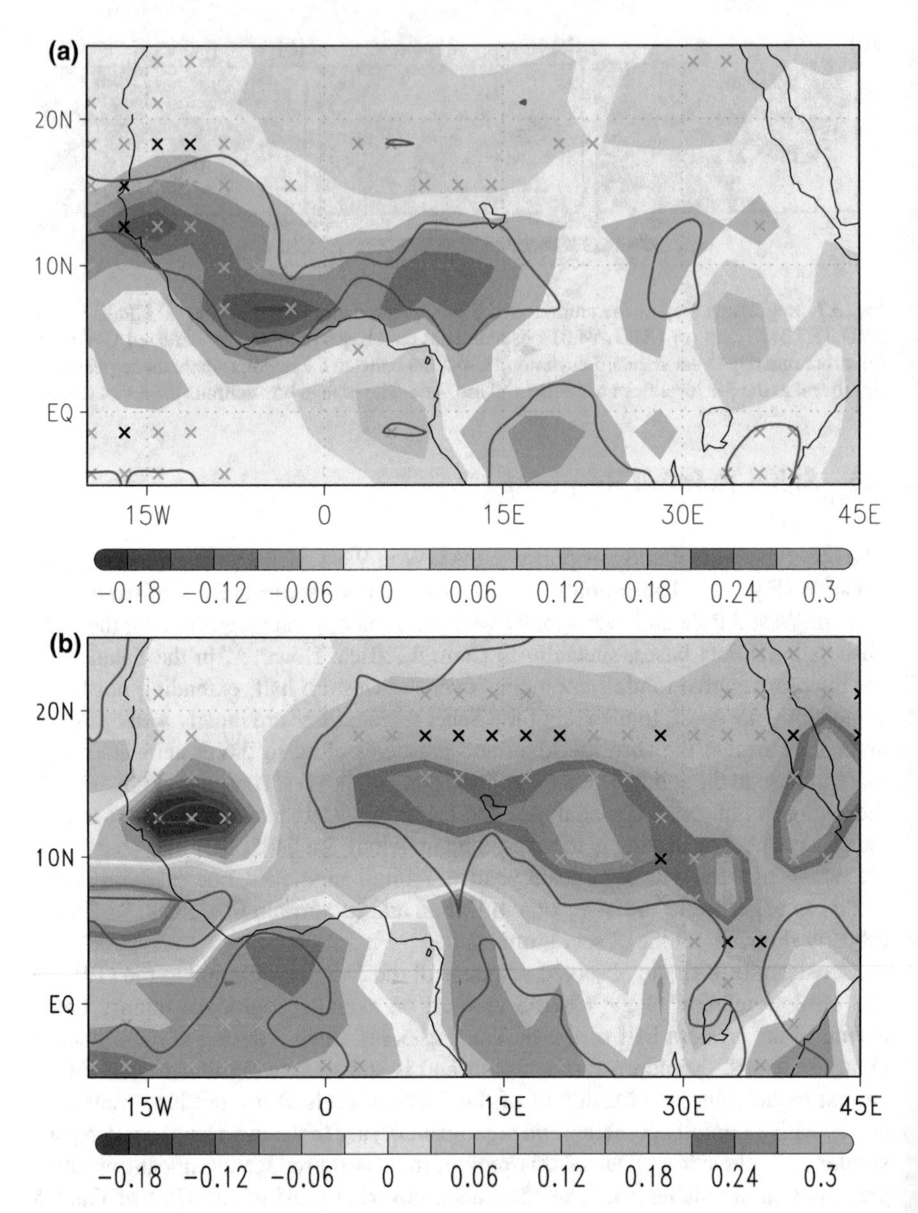

Fig. 3.8 Regression maps of the unfiltered JAS precipitation anomalies (units are mm day^{-1} per standard deviation) on the standardized GW index averaged among the 17 CMIP5 models in the historical (**a**) and the RCP8.5 (**b**) simulations. Black and gray crosses indicate points where the regression coefficient sign coincides in at least 15 and 13 out of the 17 models analyzed, respectively. Contours indicate the regions where the averaged regression is significant at the 5% level from a "random-phase" test

East Africa and show high statistical significance. Therefore, while in the historical simulation the response of the Sahel precipitation to a global increase of the SSTA is mostly a decrease, the RCP8.5 experiment, which presents more intense warming, reproduces a different precipitation response, with drought to the west and abundant rainfall to the east. The differential precipitation response to the GW between the west and the rest of the Sahel is consistent with other works addressing the Sahel rainfall trend projected for the future (Biasutti 2013; Seth et al. 2013; Monerie et al. 2017). They show that such a difference is due to a westward Sahel drying during the early monsoon season and abundant precipitation during the last moths of the rainy season, associated with a delay in the seasonal cycle of SST and the ITCZ attributed to the GHGs effect and local feedbacks involving the water recycling (Biasutti and Sobel 2009; Seth et al. 2013).

3.2.1 Inter-model Analysis

The historical simulation of the CMIP5 models reproduce, on average, GW time series similar to the observed one. They also simulate broadly the main features of the observed SSTA pattern and the same impact on Sahel rainfall. But individually, some models show important differences with respect to the others. The GW indices of some models have revealed certain deviation with respect to the model-mean, which is also reflected in their individual SSTA patterns (see Appendix A). Consistently, the precipitation response to the GW is not the same in all models.

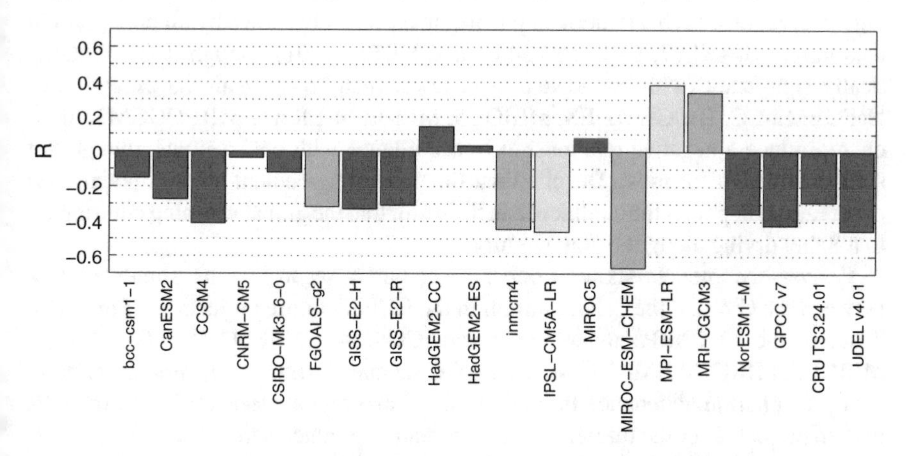

Fig. 3.9 Correlation (R) between the GW time series and the Sahel index of the JAS precipitation anomalies, 13-year low-pass filtered, from the historical simulation of the CMIP5 models individually and observations, using the three different precipitation data bases (GPCC v7, CRU TS3.24.01 and UDEL v4.01). Green, yellow and red colors indicate the correlation values that are significant at 95%, 90% or lower confidence level, respectively, following a "random-phase" test

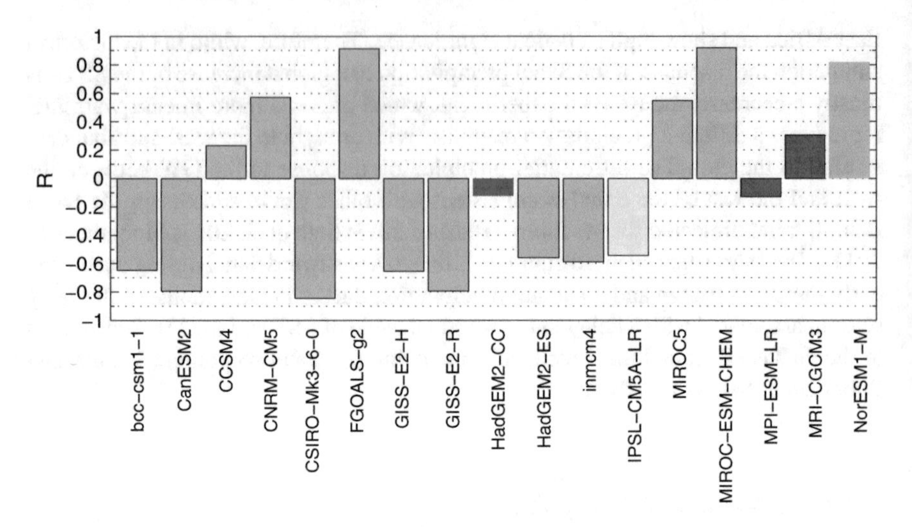

Fig. 3.10 Correlation (R) between the GW time series and the Sahel index of the JAS precipitation anomalies, 13-year low-pass filtered, from the RCP8.5 future projection of the CMIP5 models individually. Green, yellow and red colors indicate the correlation values that are significant at 95%, 90% or lower confidence level, respectively, following a "random-phase" test

In order to show the relationship between GW and the Sahel rainfall variability that the CMIP5 models individually reproduce, the correlation between the GW time series and the index of the Sahel rainfall anomalies in JAS, 13-year low-pass filtered, for each model is represented together with observations in Fig. 3.9. Regarding observations, such a relationship is negative, robustly shown by all precipitation data bases (GPCC v7, CRU TS3.24.01 and UDEL v4.01), though it is not statistically significant (with confidence level lower than 70%). With the exception of HadGEM2-CC, HadGEM2-ES, MIROC5, MPI-ESM-LR and MRI-CGCM3, models reproduce a negative relationship in accordance with observations, though not statistically sign for most. Despite this, the general agreement among models and observations supports the confidence in the result that the global warming contributed to a Sahel drying along the 20th century.

Conversely, models show stronger discrepancy regarding the simulated link between the GW and the Sahel rainfalll in the RCP8.5 future projections (Fig. 3.10). 7 out of the 17 CMIP5 models analyzed (CCSM4, CNRM-CM5, FGOALS-g2, MIROC5, MIROC-ESM-CHEM, MRI-CGCM3 and NorESM1-M) simulate anomalous precipitation abundance throughout the Sahel region associated with the GW pattern projected for the future. The rest, instead, reproduce Sahel drying. In general, the correlation between the GW and Sahel indices from the models in this case is higher than in the historical simulation and statistically significant. This suggests that the Sahel rainfall low-frequency variability may be more influenced by the GW in an hypothetical future with high GHGs emission than it was in the 20th century. However, due to the discrepancy among models, the GW effect on the Sahel rainfall projected for the future is largely uncertain. This result is in agreement with other works addressing the future evolution of the Sahel rainfall and showing that the future projections are highly uncertain since GCMs simulate different precipitation trends for the 21st century (Biasutti 2013; Park et al. 2015; Gaetani et al. 2017; Monerie et al. 2017; Biasutti and Giannini 2006; Giannini 2010; Cook and Vizy 2006).

3.2.2 *Causes of the GW Impact on Sahel Rainfall*

It has been shown that the GW induces a robust negative impact on Sahel rainfall in the historical simulations of CMIP5 as well as in observations. However, the projected Sahel precipitation response to the GW in the future is the opposite. This coincides

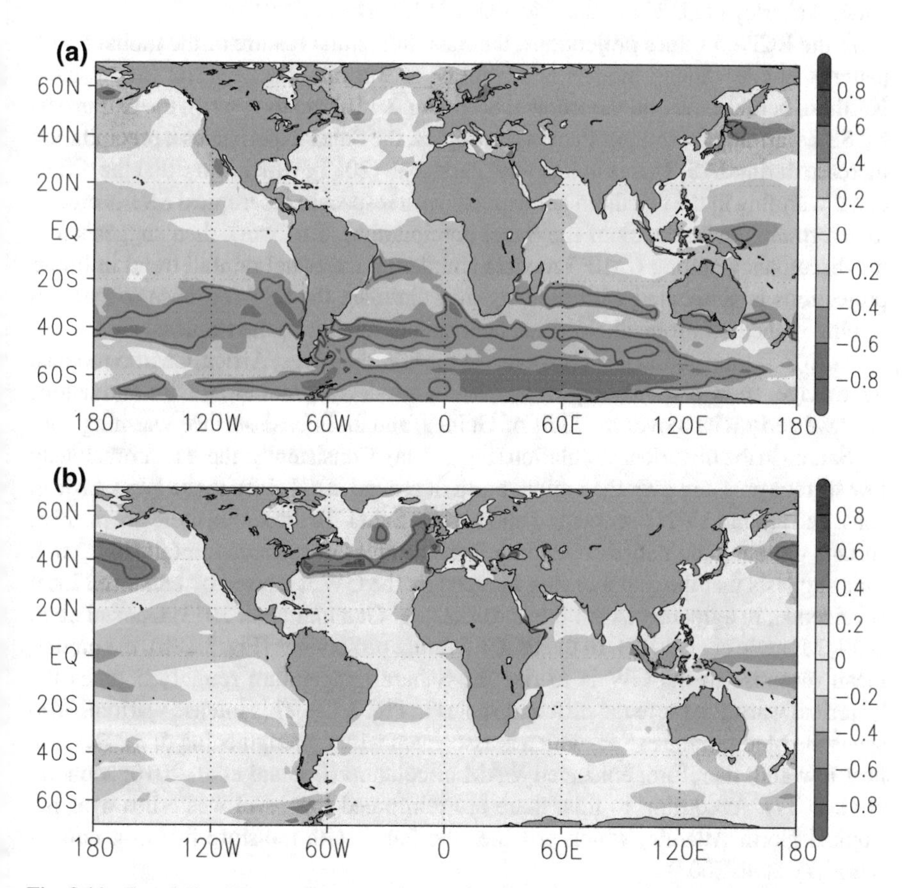

Fig. 3.11 Correlation between the regression coefficients of precipitation anomaly averaged over the Sahel (delimited area in Fig. 3.7a) and the regression patterns of GW of the 17 CMIP5 in series in the historical (**a**) and RCP8.5 (**b**) experiments (Figures A.1, A.2, A.3 and A.4 in Appendix A). Contours indicate areas where the correlation is significant at the 5% level according to a Student t-test

with the fact that the simulated GW patterns of SSTA are different in both the historical and the RCP8.5 experiments. Therefore, in order to understand the causes of the link between the GW and Sahel rainfall, it is interesting to identify the common areas of the global SST pattern that influence most on the precipitation response among all models. Figure 3.11 shows the correlation between the Sahel rainfall anomalies and the SSTA associated with the GW simulated by the CMIP5 models. It reveals that the main common feature of the GW pattern that causes the Sahel drying in the historical simulation among all CMIP5 models is an interhemispheric contrast of SST warming. The fact that the models simulate a relatively intense tropical and Southern hemisphere warming with respect to the extratropical Northern hemisphere favors that they reproduce the observed Sahel drying tendency over the historical period (Fig. 3.11a). This result is consistent with other works showing that latitudinal contrasts in the SST warming trend affect the Sahel precipitation changes (Folland et al. 1986; Ackerley et al. 2011; Hwang et al. 2013; Park et al. 2015).

In the RCP8.5 future projections, the most influential feature of the global SSTA patterns of GW among models is a contrast of warming between the extratropical Northern hemisphere and the tropical SST (Fig. 3.11b). So that, when the extratropical SST warming is stronger than in the tropics, the Sahel experiences a precipitation increase during JAS. This is in line with Park et al. (2015), which show that the differential warming in the Northern hemisphere with respect to the tropical SST is crucial to determine the tendency of the Sahel precipitation. This work then suggests that the discrepancy among CMIP5 models simulating the Sahel rainfall trend in future projections is associated with the way they simulate the SST response to external forcing in the extratropical Northern hemisphere relative to the tropics.

As to the atmospheric local response to the GW over West Africa, CMIP5 models, on average, show a weakening of the WAM low-level circulation, with anomalous northwesterly winds over the Gulf of Guinea, and an intense surface warming over the Sahara in the historical simulation (Fig. 3.12a). Consistently, there is anomalously low surface pressure over the region, north of around 20°N close to the West African coast and around 15°N over central and eastern Sahel. To the south, there are positive surface pressure anomalies coinciding with the areas of reduced rainfall (Fig. 3.8a). This suggests that the Sahel drying induced by the GW is caused by enhanced local subsidence, in agreement with other works (e.g. Giannini et al. 2013; Gaetani et al. 2017; Monerie et al. 2017). In the RCP8.5 future projections (Fig. 3.12b), the model-mean response to the GW is to rise the Sahara temperature roughly 5 times the historical warming (note the different scales in Fig. 3.12). This intense warming also spans the Mediterranean Sea, which is associated with a strengthening of the Saharan heat low and, therefore, enhanced WAM circulation (Gaetani et al. 2010; Monerie et al. 2017). According to this, there are reinforced low-level westerlies over the tropical North Atlantic, which provides the Sahel with moisture favoring rainfall (Grodsky et al. 2003).

Therefore, we find that during the historical period, CMIP5 models reproduce a robust Sahel drying associated with the GW that is consistent with observations. This is associated with a relatively weak warming in the extratropical Northern hemisphere compared to the rest of the global SST (Fig. 3.2b), which is likely a result of the aerosol

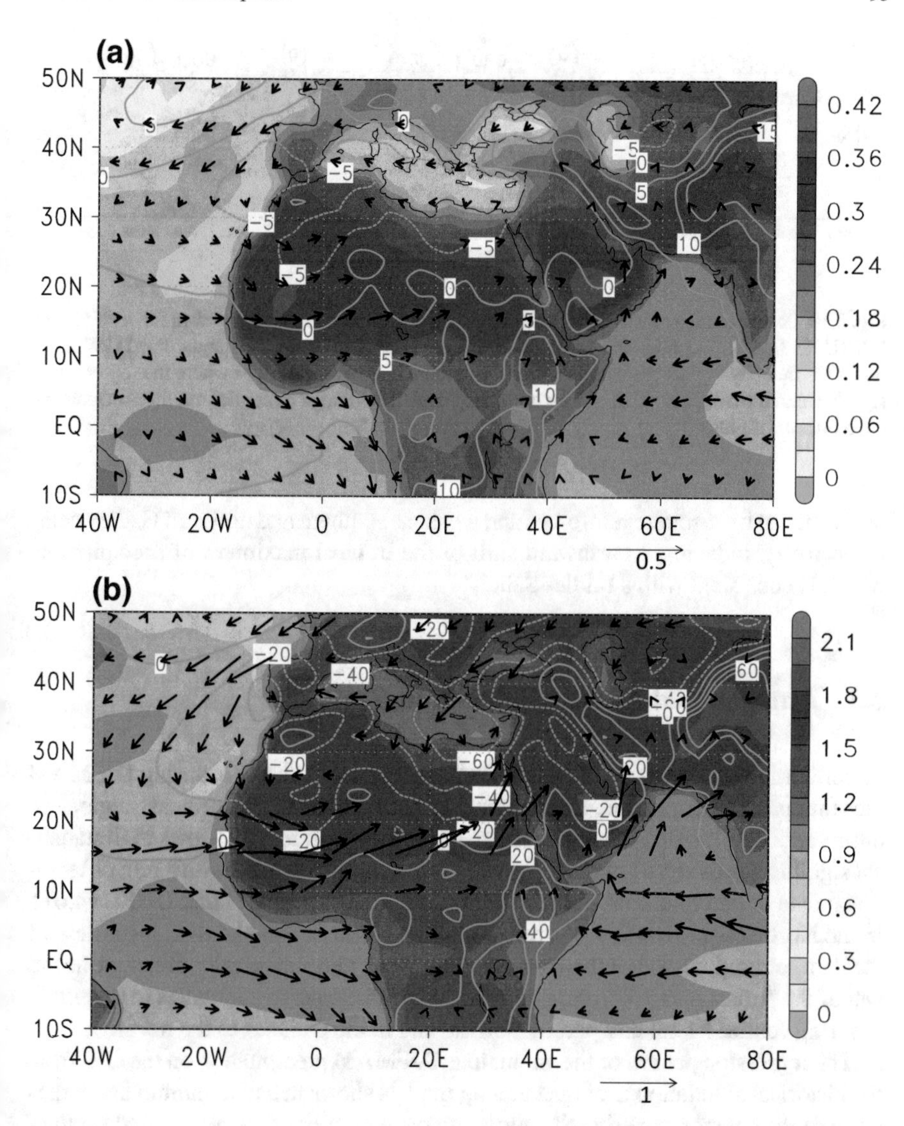

Fig. 3.12 Regression maps of the unfiltered JAS anomalies of surface temperature (shaded, units are K per standard deviation), surface pressure (contours, units are Pa per standard deviation) and horizontal wind at 925 hPa (vectors, units are m s^{-1} per standard deviation) on the standardized GW index averaged among the 17 CMIP5 models in the historical (**a**) and the RCP8.5 (**b**) simulations

effects. Conversely, in the future projections the radiative forcing is dominated by the GHGs effect, resulting in an interhemispheric warming contrast being more rapid in the north than to the south. Such a latitudinal contrast in the global SST has been associated with anomalous shifts of the global ITCZ and, hence, the tropical rainfall through changes in the Hadley circulation (Friedman et al. 2013): a faster warming of

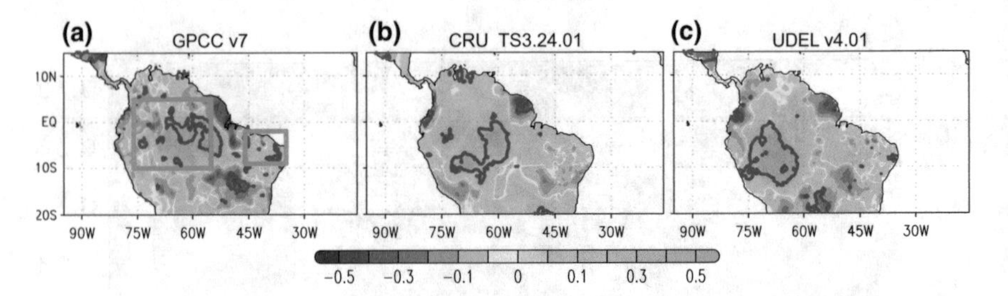

Fig. 3.13 Regression map of the unfiltered DJFMAM precipitation anomaly from the **a** GPCC v7, **b** CRU TS3.24.01 and **c** UDEL v4.01 observational data bases onto the standardized GW index (units are mm day^{-1} per standard deviation). Contours indicate the regions where the regression is significant at the 5% level from a "random-phase" test. The orange boxes delimit the Amazon and the Northeast regions

the Northern hemisphere reinforces and weakens southern and northern Hadley cells, respectively, inducing a northward shift of the tropical maximum of precipitation, which is consistent with a rainier Sahel.

3.3 Amazon and Northeast Rainfall Response

In northern South America, the precipitation response to the GW during DJFMAM presents positive anomalies throughout the Amazon and the Northeast regions in observations (Fig. 3.13). The increase in Amazon rainfall locally shows high statistical significance over part of the region and is consistent among the different observational data of precipitation analyzed (GPCC v7, CRU TS3.24.01 and UDEL v4.01). In the Northeast of Brazil, the same rainfall response is also obtained regardless of the data set used, although the anomalies are weak and statistically non-significant. Out of the Amazon and Northeast regions, there are negative anomalies to the south, covering central Brazil and part of Bolivia, and in the Guianas to the north.

The regression pattern of the anomalous DJFMAM precipitation on the GW from the historical simulation averaged among models shows negative rainfall anomalies north of the equator, covering the northernmost part of South America, and south of around 10°S, over central and eastern Brazil. Between roughly 0°–6°S there are positive anomalies in the Atlantic and Pacific sectors spanning the Northeast of Brazil and most of the Amazon region (Fig. 3.14a). This distribution of precipitation anomalies suggests a tropical rain-belt anomalously constrained to equatorial latitudes, coinciding with intense positive tropical SSTA. The simulated positive rainfall response to the GW in the Northeast is consistent with observations, showing high statistical significance and agreement among models to the north and west of the region. However, the reproduced Amazon rainfall anomalies associated with the GW are weak and statistically uncertain in contrast to observations.

The model-mean precipitation pattern of the RCP8.5 future projection shows more intense anomalies and higher statistical significance than the one of the historical sim-

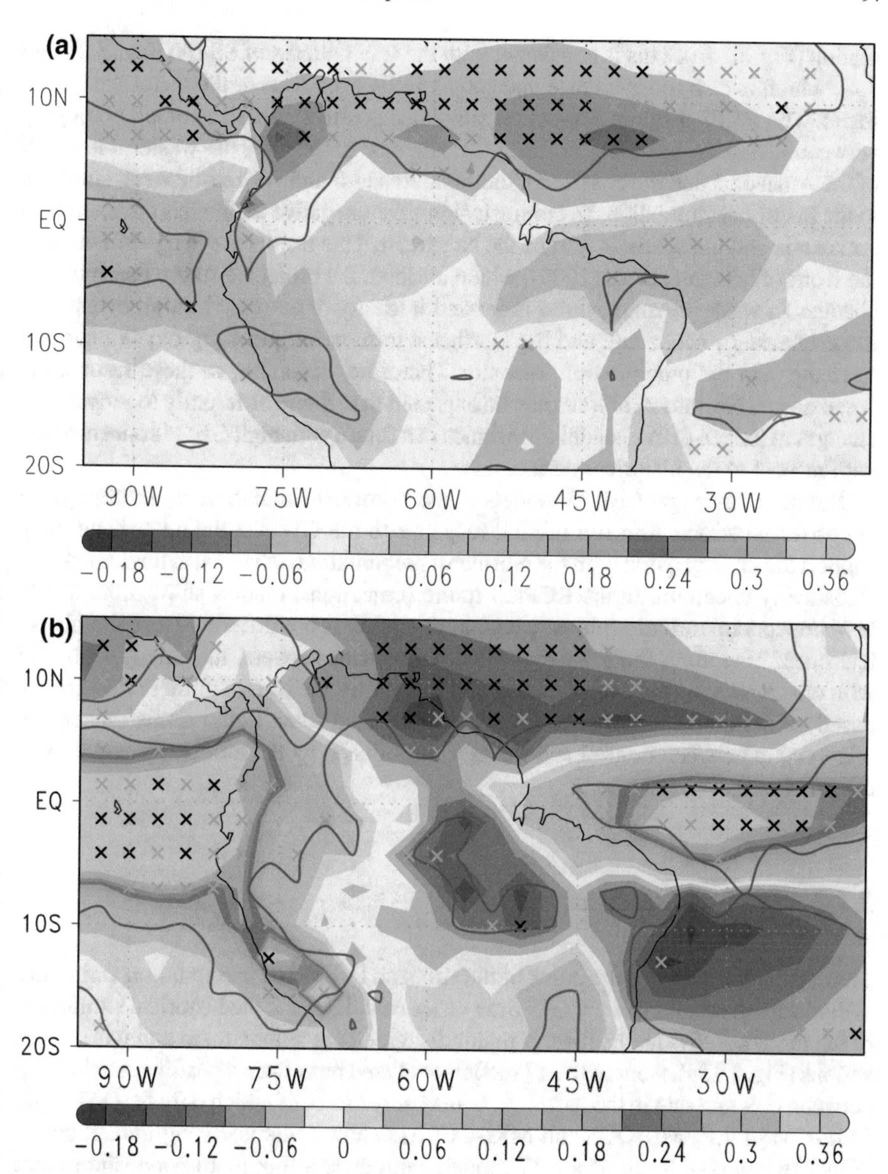

Fig. 3.14 Regression maps of the unfiltered DJFMAM precipitation anomalies (units are mm day^{-1} per standard deviation) on the standardized GW index averaged among the 17 CMIP5 models in the historical (**a**) and the RCP8.5 (**b**) simulations. Black and gray crosses indicate points where the regression coefficient sign coincides in at least 15 and 13 out of the 17 models analyzed, respectively. Contours indicate the regions where the averaged regression is significant at the 5% level from a "random-phase" test

ulation (Fig. 3.14b). This is consistent with the GW patterns of SSTA (Figs. 3.2b and 3.4), which is also more intense and robust in RCP8.5 than in the historical experiment. The RCP8.5 future projection simulates positive rainfall anomalies roughly between 8°S–5°N, spanning northern Brazil coast to the east, the westernmost part of the Amazonia and the coasts of Colombia, Ecuador and Peru to the west, similarly to the historical simulation. In contrast, significant negative anomalies cover eastern Amazonia and the southern half of the Northeast. This results are in agreement with the work of Marengo et al. (2009), which attribute the projected future precipitation changes in western Amazonia to increased intensity of extreme rainfall events and those in eastern Amazonia and the Northeast to more frequent dry days rather that to changes in the precipitation intensity. Therefore, according to these results, the Amazon and Northeast rainfall may be expected to respond differently to global SST changes at longer-than-decadal time scales in a future with high GHGs concentration with respect to the historical period.

Hence, on average, CMIP5 models in the historical simulation hardly reproduce the observed strong Amazon rainfall response to the GW. On the other hand, they support the observed link with the Northeast precipitation, which is, in turn, weak and statistically uncertain. In the RCP8.5 future projections, models show more robust GW impacts in northern South America, though they are different from the historical ones. Therefore, there are evident discrepancies between the experiments and with observations. However, it is difficult to identify the causes of these differences based on the model-mean analysis. In order to shed some light on this matter, the relationship between rainfall and the GW reproduced by the models individually is also analyzed.

3.3.1 Inter-model Analysis

The correlation between the index of the Amazon DJFMAM precipitation variability at decadal time scales and the GW time series calculated for the historical simulation of the models individually, reveals major disagreement among them and with observations (Fig. 3.15a). 9 out of the 17 models analyzed reproduce a positive relationship between GW and rain in the Amazon, as in observations, of which only FGOALS-g2, IPSL-CM5A-LR, MIROC5 and MRI-CGCM3 show statistically significant correlation. In contrast, 8 out of the 17 models reproduce a link that is opposite to that observed, which is significant in the case of the CanESM2, CNRM-CM5, GISS-E2-H and GISS-E2-R models. Therefore, approximately half of the models simulate a relationship between GW and the Amazon precipitation that is not significant and the rest show discrepant results. Hence, the result from the historical simulation is not consistent with observations.

Regarding the relationship between the Northeast rainfall changes at decadal time scales with the GW in the historical period, the observational results show a very weak link (correlation do not show statistical significance) (Fig. 3.15b). Most models (11 out of 17) individually reproduce a positive link in agreement with observations

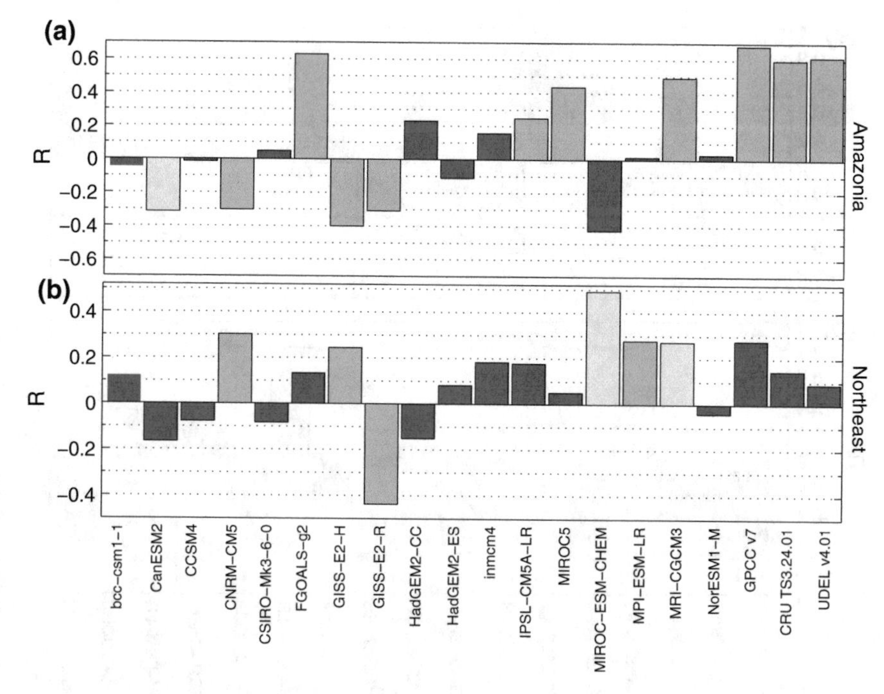

Fig. 3.15 Correlation (R) between the GW time series and the index of the DJFMAM precipitation anomalies, 13-year low-pass filtered, in the Amazonia (**a**) and the Northeast of Brazil (**b**) from the historical simulation of the CMIP5 models individually and observations, using the three different precipitation data bases (GPCC v7, CRU TS3.24.01 and UDEL v4.01). Green, yellow and red colors indicate the correlation values that are significant at 95%, 90% or lower confidence level, respectively, following a "random-phase" test

in the historical simulation, which is even statistically significant in the CNRM-CM5, GISS-E2-H, MIROC-ESM-CHEM, MPI-ESM-LR and MRI-CGCM3 models. As for the minority of models reproducing a relationship between the Northeast rainfall and the GW opposite to that observed, with the exception of the GISS-E2-R, the correlation that they show is small and statistically non-significant. Therefore, although the GW signal on the Northeast rainfall is weak and statistically uncertain, three different databases reveal a positive precipitation response that is supported by the historical simulation of most of the CMIP5 models.

In the RCP8.5 future projection, the models individually show a link between the GW and rainfall in the Amazon and Northeast regions that is notably more significant than in the historical simulations (Fig. 3.16). The index of the Amazon precipitation variability at decadal time scales is highly correlated with the GW time series (Fig. 3.16a). However, there is no clear consensus among models as to the sign of the precipitation anomalies, with 8 out of the 17 models showing a rainfall increase in response to the GW and 9 reproducing a decrease, coinciding with the historical simulation. The Northeast rainfall correlates negatively with the GW in most of the models (in 10 out of the 17 models analyzed) (Fig. 3.16b). This suggests

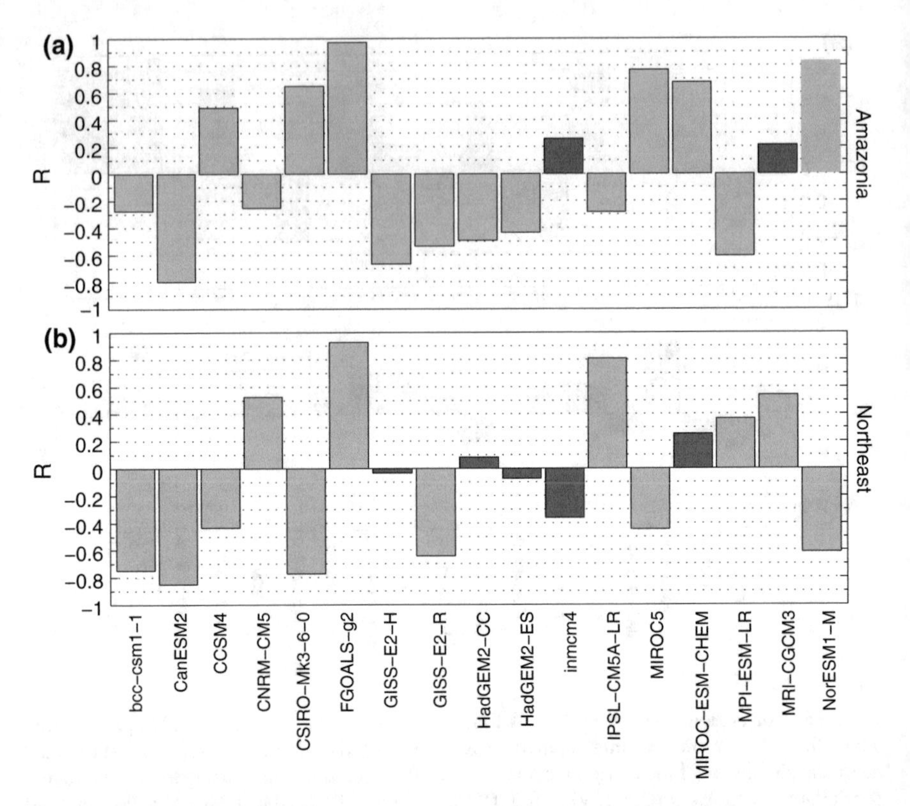

Fig. 3.16 Correlation (R) between the GW time series and the index of the DJFMAM precipitation anomalies, 13-year low-pass filtered, in the Amazonia (**a**) and the Northeast of Brazil (**b**) from the RCP8.5 future projection of the CMIP5 models individually. Green, yellow and red colors indicate the correlation values that are significant at 95%, 90% or lower confidence level, respectively, following a "random-phase" test

that the relationship between GW and precipitation in the Northeast could change in a hypothetical future with high concentrations of GHGs. But, due to the little agreement among models, it is difficult to confidently draw this conclusion from the RCP8.5 future projection.

3.3.2 *Causes of the GW Impact on Northern Brazil Rainfall*

The common characteristics among the CMIP5 models of the simulated GW pattern that influence most on the Amazon rainfall during DJFMAM are represented in Fig. 3.17. A differential warming between the Northern hemisphere, roughly north of 20°N, and the rest of the global SST may be a key feature of the GW pattern in determining the Amazon rainfall changes in historical simulations (Fig. 3.17a), as in the Sahel (Park et al. 2015). Nevertheless, this result is not statistically significant, suggesting that CMIP5 models do not show strong agreement among themselves

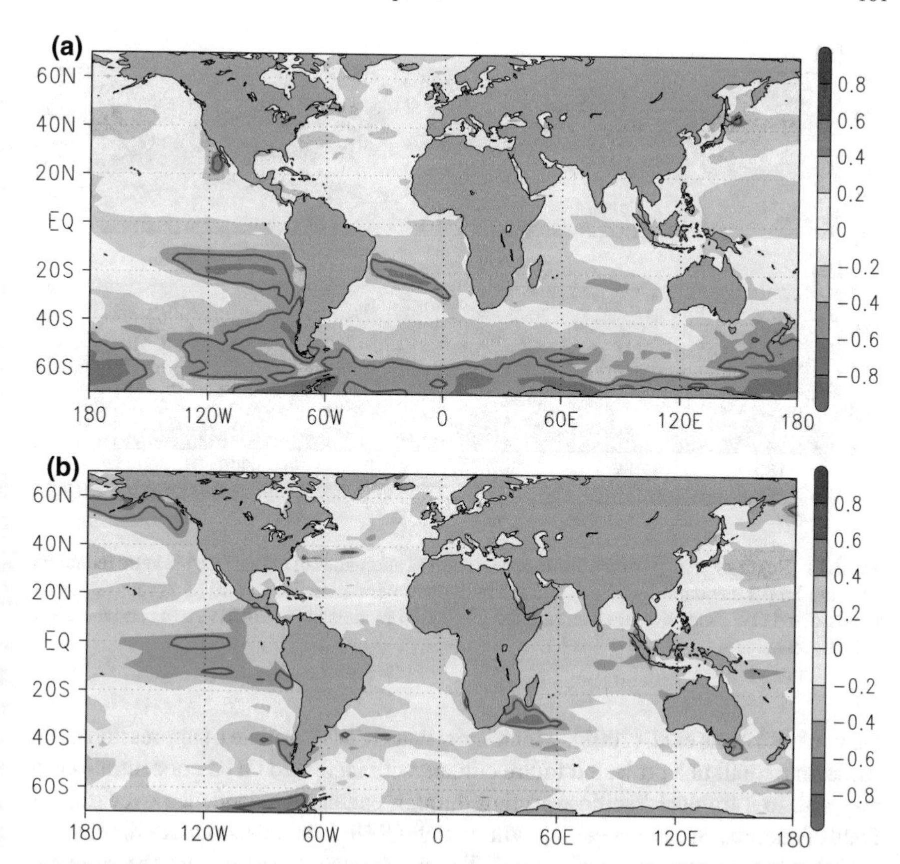

Fig. 3.17 Correlation between the regression coefficients of precipitation anomaly averaged over Amazonia (delimited area in Fig. 3.13a) and the regression patterns of GW of the 17 CMIP5 in series in the historical (**a**) and RCP8.5 (**b**) experiments (Figures A.1, A.2, A.5 and A.6 in Appendix A). Contours indicate areas where the correlation is significant at the 5% level according to a Student t-test

as to the distribution of the SSTA gradients associated with the GW that affect the Amazon rainfall. In addition, there seems to be also a relevant relationship between the Amazon precipitation response to the GW and the eastern tropical Pacific SSTA. Nevertheless, we have seen that CMIP5 models and observations disagree as to the tropical Pacific SST response to the GW (Fig. 3.2), which may be another cause of the discrepancy between observations and some models that simulate a negative Amazon rainfall response to the GW in the historical simulation. The fact that there is not an unique dominant feature of the GW pattern leading the Amazon rainfall, but more than one, may induce even more uncertainties among models.

In the RCP8.5 future projection, the link between rainfall and some specific wide pattern is mostly uncertain in the extratropics. The SSTA between the tropics, specially over eastern Pacific, seem to be more relevant for the future relationship between the Amazon precipitation and the GW (Fig. 3.17b). This is consistent with

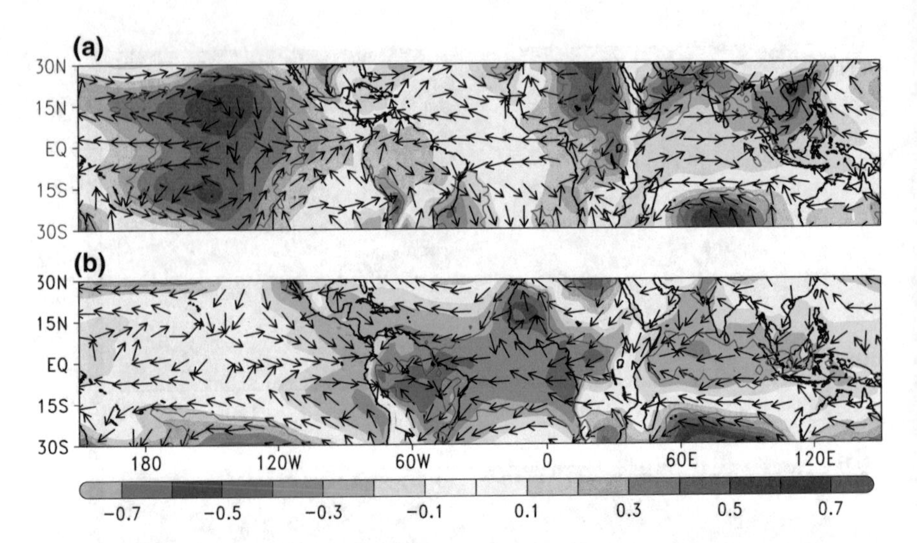

Fig. 3.18 Regression maps of the unfiltered DJFMAM anomalies of the surface pressure (contours, units are Pa per standard deviation) and the horizontal wind direction at 850 hPa (vectors) on the standardized GW index averaged among the 17 CMIP5 models in the 20CRV2c (**a**) and the ERA-20C (**b**) reanalyses

the work of Harris et al. (2008), which show that GCMs simulate a suppression of the Amazon rainfall in a projected future climate with increased GHG concentrations in response to a tropical Pacific warming through enhanced subsidence over tropical South America. So, the accuracy with which CMIP5 models simulate the tropical Pacific warming associated with the GW may explain, at least in part, the large discrepancy obtained among models in simulating the Amazon rainfall response in the RCP8.5 future projections (Fig. 3.16a). This discrepancy obtained among the CMIP5 models is also consistent with Joetzjer et al. (2013). This work relates the uncertainly projected Amazon rainfall to differences in the projected local moisture convergence and evapotranspiration among the different models, which are related with poorly constrained soil moisture feedbacks and may be another reason that explain the models disagreement. In addition to this, the Amazon precipitation response to the GW is projected to change in the future with respect to the historical period, with a strong gradient of precipitation anomalies within the region (Fig. 3.14b). The fact that the precipitation response is different within the same region makes the agreement among models more unlikely.

Although no clear link has been found between the Amazon rainfall anomalies and some specific characteristic of the GW SSTA pattern that CMIP5 models simulate, the tropical Pacific seems to be the most influential part. In addition, the model-mean DJFMAM rainfall response to the GW is robust among models over some parts of northern South America and the surroundings (Fig. 3.14b). Then, the model-mean anomalous atmospheric dynamics associated with the GW may provide further information about the link with precipitation and shed some light on the role of the tropical

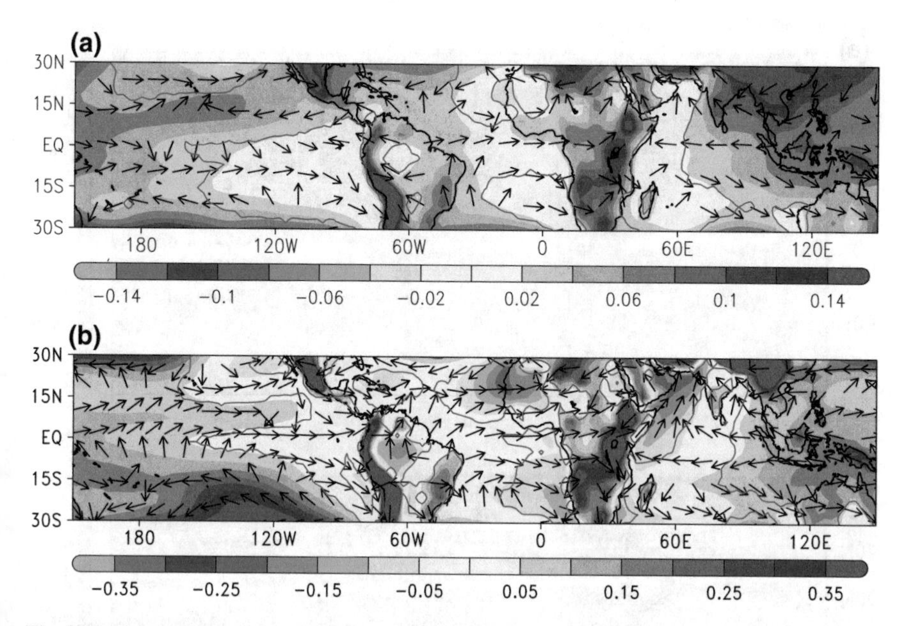

Fig. 3.19 Regression maps of the unfiltered DJFMAM anomalies of the surface pressure (contours, units are Pa per standard deviation) and the horizontal wind direction at 850 hPa (vectors) on the standardized GW index averaged among the 17 CMIP5 models in the historical (**a**) and the RCP8.5 (**b**) simulations

Pacific. In observations, the tropical surface pressure anomalies associated with the GW roughly show an enhancement over the Pacific and the Africa and a decrease in the Atlantic and the Indian Ocean, although with certain discrepancies between the two reanalyses used, specially over the continents (Fig. 3.18). Nevertheless, the low-level wind anomalies show more consistent response to the GW in both reanalyses, with anomalous wind divergence over the tropical Pacific and central Africa and convergence over northern South America and the Maritime Continent. This pattern suggest anomalous Walker circulation throughout the global tropics induced by the GW SST pattern. This is with enhanced subsidence over the tropical Pacific, coinciding with the observed cooling in this area (Fig. 3.2a), and over Africa and anomalous convection over the Amazonia, coinciding with the rainfall abundance (Fig. 3.13), and the Maritime Continent.

In contrast, CMIP5 models reproduce, on average, anomalous wind convergence over northwestern South America and western central Africa and divergence over eastern South America and the Maritime Continent in both historical and RCP8.5 simulations (Fig. 3.19). They also reproduce relatively low surface pressure over eastern tropical Pacific, eastern Atlantic and western Indian Ocean, coinciding with the regions of warmest SSTA (Figs. 3.2b and 3.4). This entails an anomalous Walker circulation in response to the GW pattern in the historical simulation that is different to the observed one. This difference is most likely related with the unrealistic tropical Pacific warming simulated by CMIP5 models. Hence, the fact that models fail in

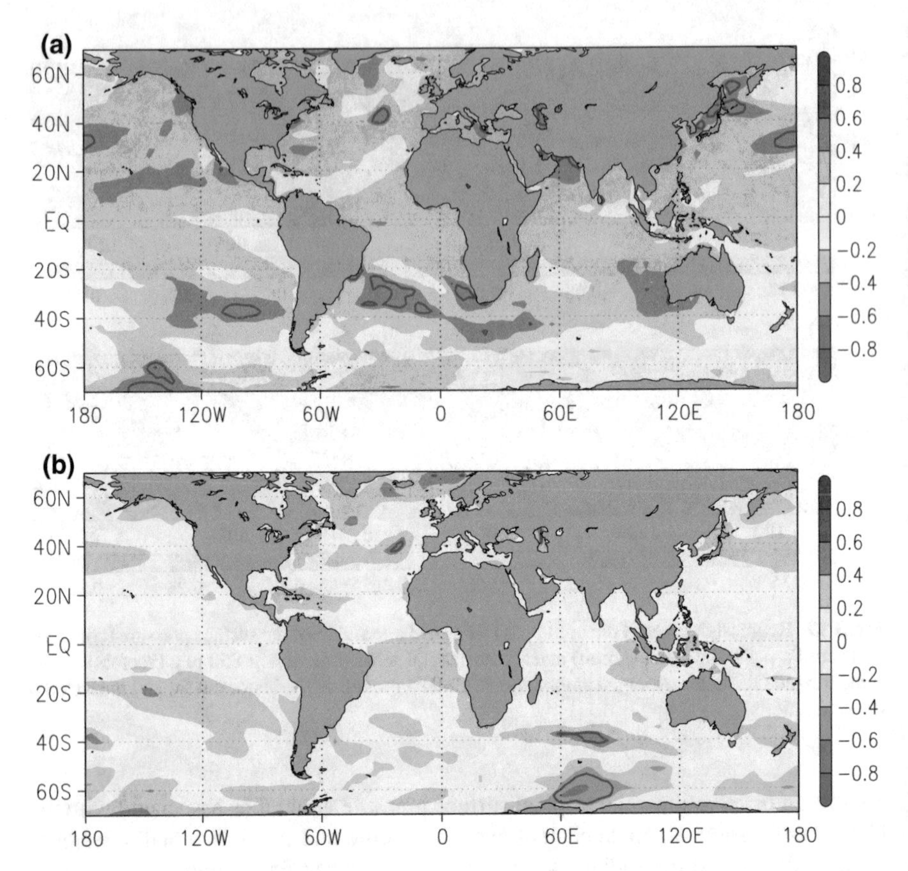

Fig. 3.20 Correlation between the regression coefficients of precipitation anomaly averaged over the Northeast (delimited area in Fig. 3.13a) and the regression patterns of GW of the 17 CMIP5 in series in the historical (**a**) and RCP8.5 (**b**) experiments (Figures A.1, A.2, A.5 and A.6 in Appendix A). Contours indicate areas where the correlation is significant at the 5% level according to a Student t-test

reproducing the observed tropical Pacific cooling associated with the GW affects the way in which they reproduce its impact on precipitation in northern South America. Due to the similarity of the results from the RCP8.5 with respect to the historical simulation, In addition, since RCP8.5 experiments show similar atmospheric response to the GW we cannot confidently interpret the projected rainfall changes for the future. Being, therefore, an important source of hampered skill in reproducing the observed GW impacts on northern South American rainfall.

Regarding the Northeast of Brazil, the simulated precipitation response to the GW is roughly associated with a global SST warming, but mostly not significant, in both the historical and the RCP8.5 experiments (Fig. 3.20). This suggests that there is no confident consensus among models as to the causes involved in the link between the GW pattern of SSTA and the Northeast rainfall response. However, Fig. 3.20a show that the simulated Northeast rainfall is related with a local gradient of SSTA in the GW pattern between the subtropical and the tropical South Atlantic, close to the coast of South America. It may be hypothesized that when the subtropical part is warmer than the tropical one, the thermal gradient may favor the ITCZ southward intrusion during the austral summer and bring more rainfall over the Northeast. This is consistent with the fact that models simulate positive rainfall anomalies in the historical simulations, where the maximum warming of the tropical Atlantic SST expands to around 15°S (Fig. 3.2b), and negative precipitation anomalies in the RCP8.5 future projection, where the gradient of SSTA over the tropical South Atlantic is closer to the equator (Fig. 3.4).

3.4 Summary of Main Findings

Addressing the objectives of the Thesis, in this Chapter we characterize the GW mode of variability, its impact on Sahel, Amazon and Northeast rainfall and the causes of such links. The results show that CMIP5 models, on average, reproduce the observed GW time evolution and most of the main features of its spatial pattern, except for the tropical Pacific SSTA. Individually, models show some uncertainties among themselves which have been attributed to the effects of the aerosol radiative forcing. Despite these uncertainties, the observed negative Sahel rainfall anomalies associated with the GW are robustly reproduced in the historical simulation. Such a rainfall decrease is associated with a weakened WAM circulation in response to the tropical SST warming. In the Northeast, models also reproduce the observed positive rainfall anomalies, while in the Amazonia the precipitation response to the GW is largely uncertain. The low skill of CMIP5 models in reproducing the tropical Pacific SSTA of the GW pattern has been associated with the simulated rainfall response in northern South America, which induces relevant differences between the associated atmospheric dynamics that models simulate and the observed one (this result will be published in Villamayor 2018).

The RCP8.5 future projections suggest that both the GW pattern and its impacts are expected to change in the future. Associated with a dominant effect of the GHGs with respect to the one of aerosols, the projected GW is more robust among models than in the historical simulation. Its SSTA pattern shows an intense Northern hemisphere warming with respect to the south. Consistently, the Sahel rainfall response to the GW is also projected to be different than in the historical period, with enhanced rainfall over most of the region except for the westernmost part. Same occurs with the Amazon and the Northeast, whose model-mean precipitation response show different anomalies within the regions and models have large discrepancies among themselves.

Summarizing, the main conclusions drawn from the results of this chapter are the followings:

- CMIP5 models reproduce the observed GW except for the associated tropical Pacific SSTA.

- The simulated aerosol effects are an important source of uncertainty among models that affect the way in which they simulate the GW.
- Models succeed in reproducing the observed Sahel rainfall response to the GW over the historical period.
- Models also reproduce the observed precipitation anomalies associated with the GW in the Amazon (although with high uncertainty among themselves) and in the Northeast. But this link is related to the tropical Pacific SSTA of the GW pattern, which is unrealistically reproduced. Hence, the reliability of these simulated impacts is little.
- RCP8.5 experiments project a different GW pattern dominated by the large GHGs concentration and, consistently, different impacts on precipitation.

References

Ackerley, D., Booth, B.B., Knight, S.H., Highwood, E.J., Frame, D.J., Allen, M.R., Rowell, D.P.: Sensitivity of twentieth-century Sahel rainfall to sulfate aerosol and CO_2 forcing. J. Clim. **24**, 4999–5014 (2011)

Andrews, T., Gregory, J.M., Webb, M.J., Taylor, K.E.: Forcing, feedbacks and climate sensitivity in CMIP5 coupled atmosphere-ocean climate models. Geophys. Res. Lett. **39** (2012)

Biasutti, M., Giannini, A.: Robust Sahel drying in response to late 20th century forcings, Geophys. Res. Lett. **33**, L11 706 (2006). https://doi.org/10.1029/2006GL026067

Biasutti, M., Sobel, A.H.: Delayed Sahel rainfall and global seasonal cycle in a warmer climate. Geophys. Res. Lett. **36** (2009)

Biasutti, M.: Forced Sahel rainfall trends in the CMIP5 archive. J. Geophys. Res. Atmos. **118**, 1613–1623 (2013)

Cook, K.H., Vizy, E.K.: Coupled model simulations of the West African monsoon system: twentieth- and twenty-first-century simulations. J. Clim. **19**, 3681–3703 (2006)

Ekman, A.M.: Do sophisticated parameterizations of aerosol-cloud interactions in CMIP5 models improve the representation of recent observed temperature trends? J. Geophys. Res. Atmos. **119**, 817–832 (2014)

Folland, C.K., Colman, A.W., Rowell, D.P., Davey, M.K.: Predictability of northeast Brazil rainfall and real-time forecast skill, 1987–98. J. Clim. **14**, 1937–1958 (2001)

Folland, C., Palmer, T., Parker, D.: Sahel rainfall and worldwide sea temperatures, 1901–85. Nature **320**, 602–607 (1986)

Friedman, A.R., Hwang, Y.-T., Chiang, J.C., Frierson, D.M.: Interhemispheric temperature asymmetry over the twentieth century and in future projections. J. Clim. **26**, 5419–5433 (2013)

Gaetani, M., Fontaine, B., Roucou, P., Baldi, M.: Influence of the Mediterranean Sea on the West African monsoon: intraseasonal variability in numerical simulations. J. Geophys. Res. Atmos. **115** (2010)

Gaetani, M., Flamant, C., Bastin, S., Janicot, S., Lavaysse, C., Hourdin, F., Braconnot, P., Bony, S.: West African monsoon dynamics and precipitation: the competition between global SST warming and CO_2 increase in CMIP5 idealized simulations. Clim. Dyn. **48**, 1353–1373 (2017)

Giannini, A., Salack, S., Lodoun, T., Ali, A., Gaye, A.T., Ndiaye, O.: A unifying view of climate change in the Sahel linking intra-seasonal, interannual and longer time scales. Environ. Res. Lett. **8**, 024010 (2013). https://doi.org/10.1088/1748-9326/8/2/024010

Giannini, A.: Mechanisms of climate change in the semiarid African Sahel: the local view. J. Clim. **23**, 743–756 (2010)

Grodsky, S.A., Carton, J.A., Nigam, S.: Near surface westerly wind jet in the Atlantic ITCZ. Geophys. Res. Lett. **30**, 3–6 (2003)
Hansen, J., Nazarenko, L., Ruedy, R., Sato, M., Willis, J., Del Genio, A., Koch, D., Lacis, A., Lo, K., Menon, S., et al.: Earth's energy imbalance: confirmation and implications. Science **308**, 1431–1435 (2005)
Harris, P.P., Huntingford, C., Cox, P.M.: Amazon Basin climate under global warming: the role of the sea surface temperature. Philos. Trans. R. Soc. Lond. B Biol. Sci. **363**, 1753–1759 (2008)
Hastenrath, S., Greischar, L.: Circulation mechanisms related to northeast Brazil rainfall anomalies. J. Geophys. Res. Atmos. **98**, 5093–5102 (1993)
Hwang, Y.-T., Frierson, D.M., Kang, S.M.: Anthropogenic sulfate aerosol and the southward shift of tropical precipitation in the late 20th century. Geophys. Res. Lett. **40**, 2845–2850 (2013)
Joetzjer, E., Douville, H., Delire, C., Ciais, P.: Present-day and future Amazonian precipitation in global climate models: CMIP5 versus CMIP3. Clim. Dyn. **41**, 2921–2936 (2013)
Kim, H., An, S.-I.: On the subarctic North Atlantic cooling due to global warming. Theor. Appl. Climatol. **114**, 9–19 (2013)
Luo, J.-J., Sasaki, W., Masumoto, Y.: Indian Ocean warming modulates Pacific climate change. In: Proceedings of the National Academy of Sciences, vol. 109, pp. 18701–18706 (2012)
Luo, J.-J., Wang, G., Dommenget, D.: May common model biases reduce CMIP5's ability to simulate the recent Pacific La Niña-like cooling? Clim. Dyn. 1–17 (2017)
Marengo, J.A., Jones, R., Alves, L., Valverde, M.: Future change of temperature and precipitation extremes in South America as derived from the PRECIS regional climate modeling system. Int. J. Climatol. **29**, 2241–2255 (2009)
Meehl, G.A., Hu, A., Arblaster, J.M., Fasullo, J., Trenberth, K.E.: Externally forced and internally generated decadal climate variability associated with the interdecadal pacific oscillation, J. Clim. **26**, 7298–7310 (2013). https://doi.org/10.1175/JCLI-D-12-00548.1
Meehl, G.A., Washington, W.M., Ammann, C.M., Arblaster, J.M., Wigley, T., Tebaldi, C.: Combinations of natural and anthropogenic forcings in twentieth-century climate. J. Clim. **17**, 3721–3727 (2004)
Ming, Y., Ramaswamy, V.: A model investigation of aerosol-induced changes in tropical circulation. J. Clim. **24**, 5125–5133 (2011)
Monerie, P.-A., Sanchez-Gomez, E., Boé, J.: On the range of future Sahel precipitation projections and the selection of a sub-sample of CMIP5 models for impact studies. Clim. Dyn. **48**, 2751–2770 (2017)
Park, J.-Y., Bader, J., Matei, D.: Northern-hemispheric differential warming is the key to understanding the discrepancies in the projected Sahel rainfall. Nat. Commun. **6**, 5985 (2015). https://doi.org/10.1038/ncomms6985
Riahi, K., Grübler, A., Nakicenovic, N.: Scenarios of long-term socio-economic and environmental development under climate stabilization. Technol. Forecast. Soc. Change **74**, 887–935 (2007). https://doi.org/10.1016/j.techfore.2006.05.026
Rodríguez-Fonseca, B., Suárez-Moreno, R., Ayarzagüena, B., López-Parages, J., Gómara, I., Villamayor, J., Mohino, E., Losada, T., Castaño-Tierno, A.: A review of ENSO influence on the North Atlantic. A non-stationary signal. Atmosphere **7**, 87 (2016)
Rotstayn, L.D., Lohmann, U.: Tropical rainfall trends and the indirect aerosol effect. J. Clim. **15**, 2103–2116 (2002)
Rotstayn, L.D., Collier, M.A., Shindell, D.T., Boucher, O.: Why does aerosol forcing control historical global-mean surface temperature change in CMIP5 models? J. Clim. **28**, 6608–6625 (2015)
Seth, A., Rauscher, S.A., Biasutti, M., Giannini, A., Camargo, S.J., Rojas, M.: CMIP5 projected changes in the annual cycle of precipitation in monsoon regions. J. Clim. **26**, 7328–7351 (2013)
Solomon, A., Newman, M.: Reconciling disparate twentieth-century Indo-Pacific ocean temperature trends in the instrumental record. Nat. Clim. Change **2**, 691–699 (2012)
Song, X., Zhang, G.J.: Role of climate feedback in El Niño-like SST response to global warming. J. Clim. **27**, 7301–7318 (2014)

Stott, P.A., Tett, S., Jones, G., Allen, M., Mitchell, J., Jenkins, G.: External control of 20th century temperature by natural and anthropogenic forcings. Science **290**, 2133–2137 (2000)
Taylor, K.E., Stouffer, R.J., Meehl, G.A.: An overview of CMIP5 and the experiment design. Bull. Am. Meteorol. Soc. **93**, 485–498 (2012)
Trenberth, K.E., Fasullo, J.T.: An apparent hiatus in global warming? Earth's Future **1**, 19–32 (2013)
Villamayor, J., Ambrizzi, T., Mohino, E.: Influence of decadal sea surface temperature variability on northern Brazil rainfall in CMIP5 simulations. Clim. Dyn. **51**, 563–579 (2018)

Chapter 4
Influence of the AMV

Abstract This Chapter presents the results from the characterization of the AMV in the CMIP5 simulations from several different models. Then, the precipitation response to the AMV in the regions of interest of this thesis is analyzed, as well as the atmospheric mechanisms responsible for such relationship. The same is done using observations in order to identify similarities or discrepancies. The results obtained from simulations that have the evolution of the observed external radiative forcing incorporated are compared with others that do not include such forcing. This comparison allows to find out whether the AMV has an externally forced component and how this effect can influence the impact on rainfall. Finally, the analysis of the RCP8.5 future projections provide an insight of the foreseeable evolution of both the AMV mode of SST and its impacts on rainfall in a hypothetical future with high amounts of GHGs emissions.

4.1 The AMV Index and Pattern

The evolution of the AMV index computed from observations (Fig. 4.1a) shows a negative phase during the first two decades of the 20th century followed by a positive one from the 1930s to the mid-1960s, when it turned to negative until the mid-1990s, in accordance with previous studies (e.g. Enfield et al. 2001; Sutton and Hodson 2005; Alexander et al. 2014). The associated AMV pattern (Fig. 4.1b) is characterized by a well-defined interhemisferic SSTA gradient in the Atlantic basin, with warm anomalies all across its northern half and cold ones in the southern part. The North Atlantic warming depicts a comma-shape pattern of SSTA. Anomalies are more intense in the northernmost part of the North Atlantic, south of Greenland,

---

The content of this Chapter referring to the AMV influence on northern South American rainfall is published in the following article:
Villamayor, J., Ambrizzi, T. & Mohino, E. (2018a): Influence of decadal sea surface temperature variability on northern Brazil rainfall in CMIP5 simulations. *Clim. Dyn.* https://doi.org/10.1007/s00382-017-3941-1.

J. Villamayor, *Influence of the Sea Surface Temperature Decadal Variability on Tropical Precipitation: West African and South American Monsoon*, Springer Theses, https://doi.org/10.1007/978-3-030-20327-6_4

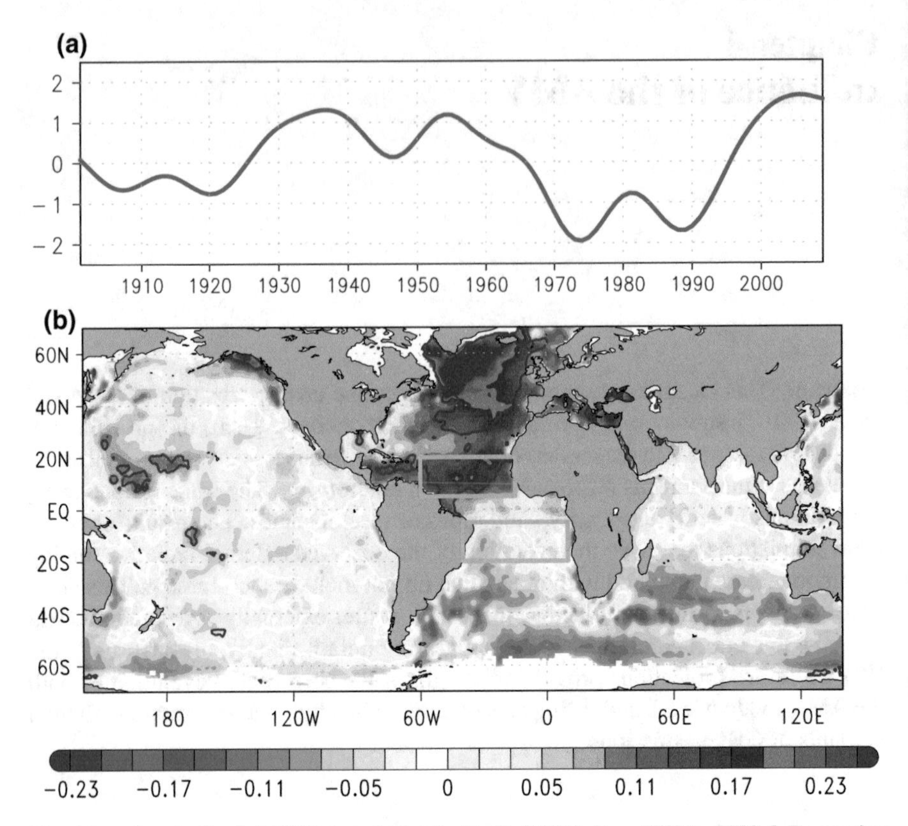

Fig. 4.1 **a** Standardized AMV index obtained with HadISST1 from 1901 to 2009. **b** Regression pattern of the unfiltered HadISST1 SSTA onto the standardized AMV index (units are K per standard deviation). Contours indicate the regions where the regression is significant at the 10% level from a "random-phase" test. Green boxes indicate the characteristic areas of the tropical Atlantic gradient. Adapted from Villamayor et al. (2018a)

and extend southward along the eastern part of the basin to the northern half of the tropical Atlantic.

A Fourier analysis of the AMV index is performed in order to find out whether it oscillates with one or more characteristic periodicities. The resulting power spectrum shows an outstanding and significant peak revealing that the AMV calculated from observations has a well-defined periodicity at around 65 years (Fig. 4.2). This result is consistent between two different observational data bases and coincides with other studies (e.g. Kerr 2000; Knight 2005).

Regarding state-of-the-art GCMs, the CMIP5 models reproduce an AMV pattern averaged across the 17 models that shows a coma-shape distribution of positive SSTA over the North Atlantic that is similar to the observed one in both the historical and piControl simulations (Fig. 4.3). Such a warming is highly consistent among the different models in the two simulations (crosses in Fig. 4.3 indicate the grid

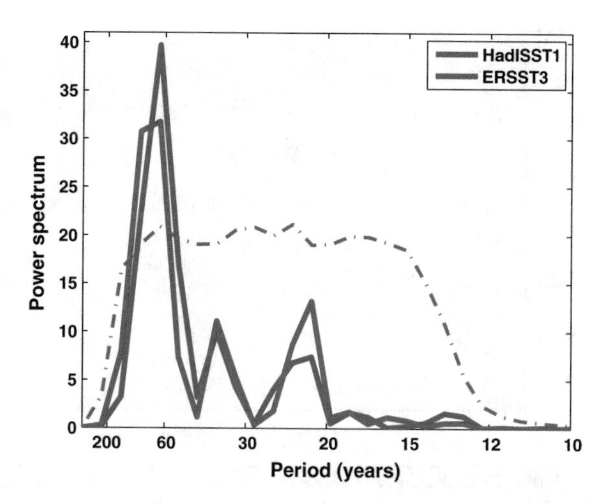

Fig. 4.2 Power spectrum of the Fourier transform of the detrended AMV indices from HadISST1 and ERSST3 data bases. Gray dashed line indicates the 95% confidence level following a non-parametric hypothesis test in which a probability density is built from the Fourier spectrum of 13-year low-pass filtered and detrended white noise time series

points where most of the models coincide in the sign of the regression coefficient). The simulated AMV time series show characteristic oscillation periodicities that are distributed in a range around 65 years that are consistent in both the historical and piControl simulations (Fig. 4.4) and is in agreement with observations (Fig. 4.2) (Kerr 2000; Knight 2005).

Compared to observations, CMIP5 models tend to underestimate the SSTA associated with the AMV, especially in the tropical and subtropical North Atlantic. There are also some differences between the forced and unforced experiments. Over the southern half of the Atlantic Ocean, the AMV pattern of the historical experiment presents weak and not significant anomalies. In contrast, the pattern of the piControl simulation shows significant cold SSTA south of the equator, with high consistence among the models and in agreement with the observed pattern. This result suggests that most of the models reproduce a more accurately defined SSTA interhemispheric gradient in the Atlantic in the piControl experiment than in the historical one. In the historical experiment, the AMV pattern shows mostly positive SSTA in the Indian Ocean, which are more consistent among the models in the northeastern part of the basin. It also shows significant warm anomalies in the northernmost part of the Pacific Ocean and weak ones in roughly the rest of the basin. On the other hand, the piControl AMV pattern shows weaker warm SSTA over the northern Indian Ocean and significant cooling to the south. In the Pacific, it also shows a significant extratropical warming to the north in agreement with the historical AMV pattern, but in contrast a cooling to the south (Fig. 4.3b).

4.2 The Sahel Rainfall Response

The observed precipitation response to the AMV is obtained through the regression of the JAS seasonal rainfall anomalies on the AMV index (Fig. 4.5). The resulting regression patterns reveal a significant increase of precipitation in West Africa

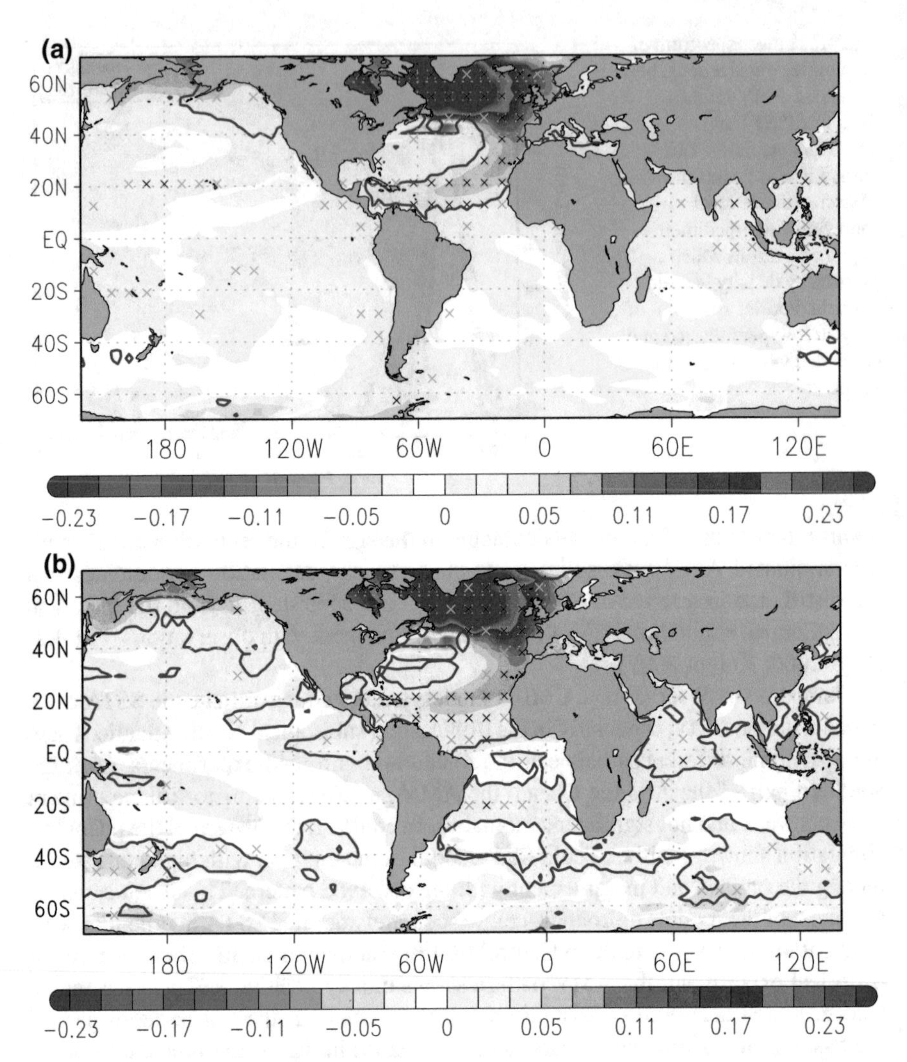

Fig. 4.3 Regression onto the AMV index of the unfiltered SSTA (K per standard deviation) averaged among the 17 CMIP5 models in the historical (**a**) and the piControl (**b**) simulations. Black and gray crosses indicate points where the sign of the regression coefficient coincides in at least 15 and 13 out of the 17 models analyzed, respectively. Contours indicate the regions where the averaged regression is significant at the 5% level from a "random-phase" test. Adapted from Villamayor et al. (2018a)

throughout a zonal band roughly between 10° and 17.5°N associated with positive AMV phases, with high agreement among the different observational data sets analyzed. This band covers the entire Sahel, taking maximum values over the center of the region. South of 10°N, the rainfall anomalies are mostly negatives along the Gulf of Guinea coast and the equatorial regions of Africa, although these anomalies have

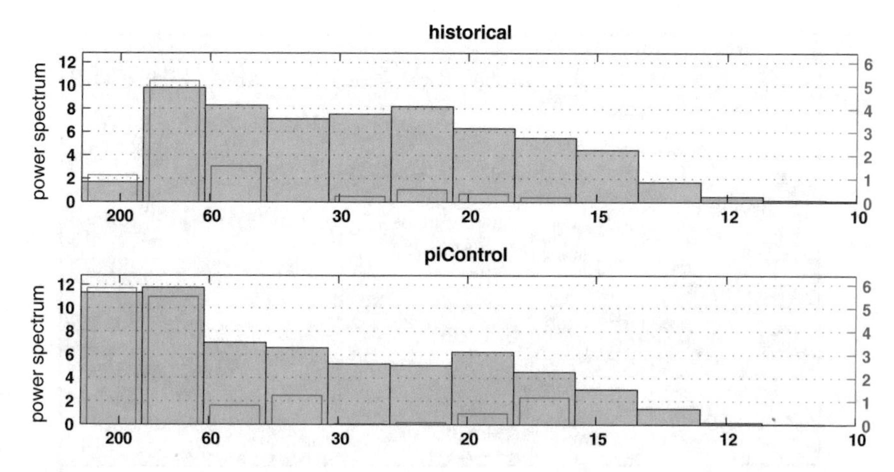

Fig. 4.4 Model-mean power spectra of the Fourier transform of the detrended AMV indices (blue bars) of the historical (top) and piControl (bottom) simulations. Red bars indicate the power spectra averaged over the 17 models but taking into account only 95% significant peaks for each of them. The statistical significance for each model's AMV is obtained following a non-parametric hypothesis test in which a probability density function is built from the Fourier spectrum of 13-year low-pass filtered and detrended white noise time series with the same time length as the simulation and averaged across the number of ensemble members in cases where more than one realization is available

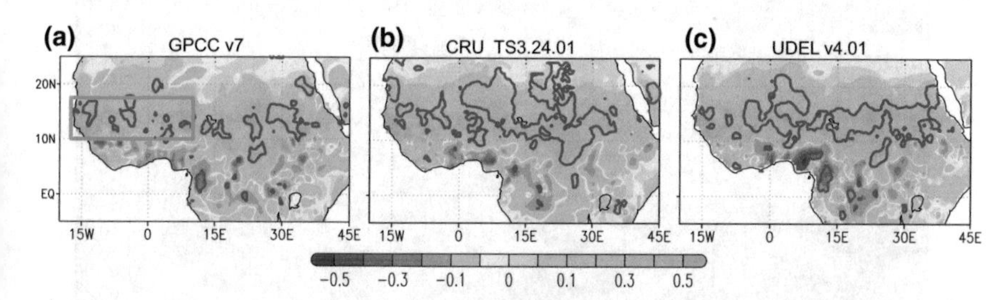

Fig. 4.5 Regression maps of the unfiltered JAS precipitation anomaly from the **a** GPCC v7, **b** CRU TS3.24.01 and **c** UDEL v4.01 observational data bases onto the standardized AMV index (units are mm day^{-1} per standard deviation). Contours indicate the regions where the regression is significant at the 5% level from a "random-phase" test. Orange box delimits the Sahel region

lower statistical significance. During negative phases of the AMV, the precipitation anomalies represented in the patterns are the opposite. Such a distribution of the rainfall anomalies represents a latitudinal shift of the tropical rain-belt associated with the ITCZ and is consistent with other studies (e.g. Zhang and Delworth 2006; Mohino et al. 2011a; Martin and Thorncroft 2014).

The AMV influence on West African rainfall that the CMIP5 models reproduce on average is broadly consistent with observations in both the historical and piControl simulations (Fig. 4.6). Both show enhanced precipitation over the Sahel associated with the AMV, being more intense towards the western coast. To the south, there are

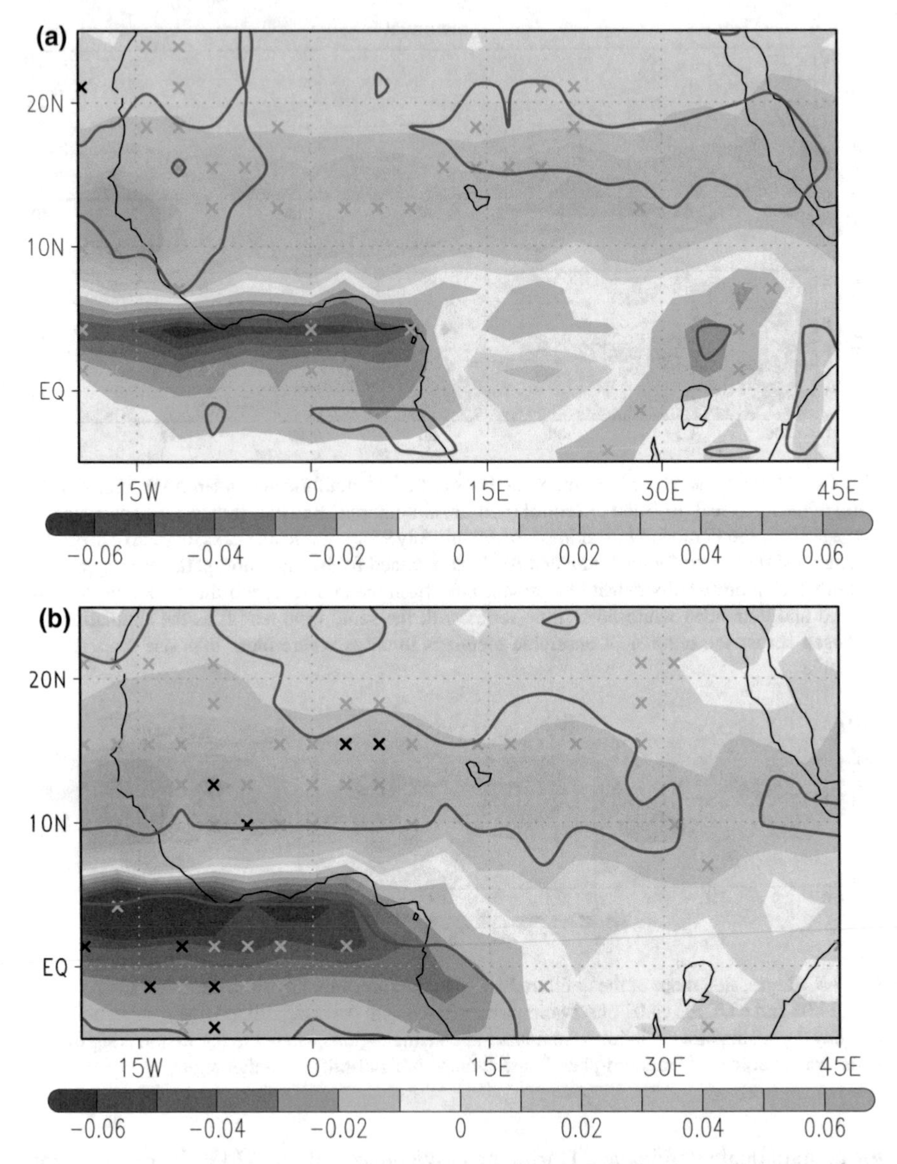

Fig. 4.6 Regression maps of the unfiltered JAS precipitation anomalies (units are mm day^{-1} per standard deviation) on the AMV index averaged among the 17 CMIP5 models in the historical (**a**) and the piControl (**b**) simulations. Black and gray crosses indicate points where the regression coefficient sign coincides in at least 15 and 13 out of the 17 models analyzed, respectively. Contours indicate the regions where the averaged regression is significant at the 5% level from a "random-phase" test

negative anomalies showing an anomalous shift of the tropical rain-belt over West Africa. Despite these similarities, the reproduced rainfall response to the AMV is notably weaker compared to observations (note that the scale of the color shading in the regression maps in Figs. 4.5 and 4.6 is different, with one ranging from −0.5 to 0.5 and another form −0.06 to 0.06 mm day^{-1} per standard deviation, respectively). They also underestimate the northward intrusion of rainfall over the Sahel with respect to observations: While the observed anomalies present a dipole around 10°N, the models show negative anomalies restricted to the Gulf of Guinea coast and positive ones roughly north of 6°N with maximum values confined to the southern part of the Sahel. The averaging among the 17 different models can explain part of the underestimation of the anomalies shown in the model-mean regression patterns. The regions where the models disagree in terms of the sign and amplitude of the regression coefficients will show low model-mean anomalies.

Regarding the differences between the forced and unforced simulations, the latter shows more consistency among the different models. While the distribution of rainfall anomalies are similar in both experiments, the intensity of the historical regression pattern is weaker than the piControl one. The former presents high statistical significance only in the western part of the Sahel, where the precipitation anomalies are more intense, and the model-mean anomalies of the latter are significant over the whole Sahel. This result suggests that the connection between Sahel rainfall and the AMV reproduced by historical simulations is lower than in the unforced one. This difference between both simulations is consistent with the associated SSTA patterns of the AMV, which are also more robust and show more intense contrast of SSTA over the Atlantic in piControl than in the historical experiment. A relationship between the way in which models simulate the Sahel precipitation and the Atlantic SSTA associated with the AMV may be therefore suggested.

4.2.1 Inter-model Analysis

In order to understand the reason why in the historical simulation the models reproduce a weaker link between the Sahel rainfall and the AMV than in piControl, a comparative analysis is made between both simulations of the models individually. Martin et al. (2014) conclude that CMIP5 models reproducing an AMV pattern with a well-defined interhemispheric SST gradient in the Tropical Atlantic can reproduce a significant link with the Sahel rainfall. Following this idea, Fig. 4.7 shows a scatter plot in which the amplitude of the Sahel rainfall anomalies associated with the AMV is plotted versus the intensity of the tropical Atlantic SSTA gradient, computed as the average over 5°–20°N and 60°–15°W minus 20°–5°S and 40°–10°W (see areas in Fig. 4.1b), for each model individually. The scatter plot evidences the underestimation of the Sahel rainfall response to the AMV that the CMIP5 models reproduce. The large majority of the models represent rainfall anomalies below 30% of the observed one, which is around 0.32 mm day^{-1} per standard deviation of the AMV index. MIROC-ESM-CHEM (number 14 in Fig. 4.7) is the model that approximates most to that value in the historical simulation, while in piControl it reproduces around half the observed precipitation response. The CSIRO-Mk3-6-0 model (number 5), instead, reproduces half the observed rainfall anomalies in piControl but very

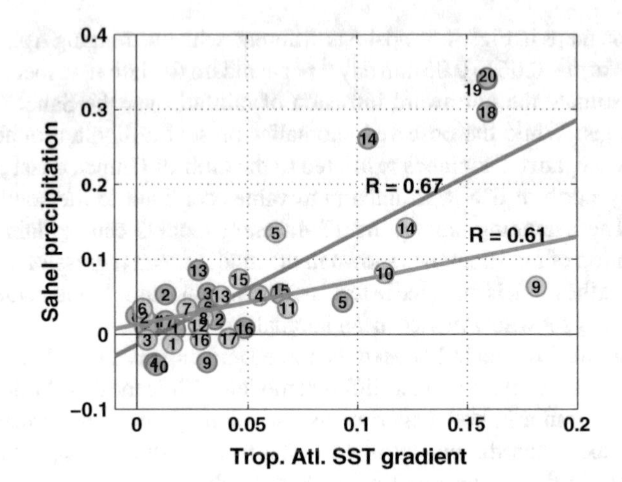

Fig. 4.7 **a** Scatter plot of the regression coefficient of precipitation anomaly over the Sahel (delimited region in Fig. 4.5a) and the SSTA tropical Atlantic gradient (difference between the delimited regions in Fig. 4.1b) relative to the AMV of each model in the historical (green) and the piControl (orange) simulations (Figs. B.1, B.2, B.3, and B.4 in Appendix B). The lines indicate the linear regression fitting of the corresponding colored points (R is the correlation coefficient). The numbers from 1 to 17 identify each model individually with the given number in Table 2.1. Numbers 18, 19 and 20 correspond to CRU TS3.24.01, GPCC v7 and UDEL v4.01 observed data (blue), respectively. Units for the horizontal and vertical axes are mm day^{-1} per standard deviation and K per standard deviation), respectively

low ones in historical simulation. The historical experiment in models CNRM-CM5, HadGEM2-CC and HadGEM2-ES (numbers 4, 9 and 10) and the piControl simulation in bcc-csm1-1, CCSM4, MRI-CGCM3 and NorESM1-M (numbers 1, 3, 16 and 17) even show negative anomalies associated with AMV, although very weak.

Consistent with the underestimation of rainfall intensity in response to the AMV, the interhemispheric SSTA gradient of the tropical Atlantic that the models reproduce is also lower than in observations, which is around 0.16 K per standard deviation. This gradient is weaker than 0.05 K per standard deviation in most models. Only 6 out of the 17 models show a higher value in piControl simulation and 3 in the historical one. The models approximating most to the observed value are HadGEM2-CC and HadGEM2-ES in piControl and MIROC-ESM-CHEM in both simulations (numbers 9, 10 and 14 in Fig. 4.7, respectively). In agreement with Martin et al. (2014), it is also also found that the first low-frequency variability mode of the North Atlantic SSTA reproduced by some models is not associated with an AMV-like SSTA pattern, i.e. with a well defined interhemispheric gradient of SSTA over the Atlantic (Figs. B.1 and B.2). Furthermore, the differences found between the historical and piControl experiments in the model-mean AMV patterns of SSTA are not appreciable in all the models individually, suggesting an important model dependence.

Regarding the relationship between the intensity of the interhemispheric gradient of the tropical Atlantic and the Sahel rainfall anomalies related to the AMV, in both, the historical and piControl simulations, there is strong linear correlation between both parameters (the correlation coefficients are $R = 0.67$ and $R = 0.61$, respectively, which are significant with a 95% confidence level according to a Student t-test). However, in the case of the historical simulation, this link is highly dependent on model MIROC-ESM-CHEM. If this model is not considered for the linear fitting, the significance of such a relationship is notably lower ($R = 0.36$, which is barely significant with a 90% confidence level according to a Student t-test). In contrast, in piControl simulations this linear correlation is more robust and does not depend so much on a particular model.

Following this results, it can be concluded that the models skill to simulate a realistic Sahel precipitation response to the AMV in the piControl simulation relies on their ability to reproduce an AMV pattern with a well-defined interhemispheric gradient of SSTA in the tropical Atlantic. However, such a relationship is not so clear in the historical experiment. Some models show remarkable differences between the forced and unforced simulations while others don not. For instance, HadGEM2-CC simulates a much more intense tropical Atlantic SST gradient in piControl than in the historical experiment, while MIROC-ESM-CHEM shows similarly intense gradients in both experiments. Such a different behavior among models suggest that there are some interactions between the external forcing and the AMV pattern they simulate that are model-dependent. This issue is more deeply addressed in Sect. 4.5.

4.2.2 Atmospheric Teleconnection Between AMV and Sahel Rainfall

Associated with the AMV, the surface pressure significantly weakens throughout the tropical North Atlantic sector, North Africa, North America and northwestern Europe during the boreal summer in observations (Fig. 4.8a). Focusing on the tropics, the weakened surface pressure over the tropical North Atlantic is consistent with anomalous low-level southerly winds close to the equator and westerlies around 10°N. Such an anomalous wind behavior implies the northward shift of the ITCZ (Knight et al. 2006) that favors rainy conditions. Inland, there is also significant weakening of the surface pressure over the Sahara creating a pressure gradient with respect to southern West Africa. The negative pressure anomalies in the Sahara represents the strengthening of the Saharan heat low, which is related with the Sahel precipitation (Biasutti et al. 2009; Lavaysse et al. 2009). These results are consistent with other studies addressing the AMV impact on the WAM based on observations (Martin and Thorncroft 2014).

The surface pressure and low-level wind response to the AMV reproduced by CMIP5 simulations show high consistency with observations. Associated with the AMV, the model-mean regression pattern of the JAS surface pressure presents negative and statistical significant anomalies throughout the North Atlantic, northern Africa and Europe in both the historical and piControl simulations (Fig. 4.9). In the

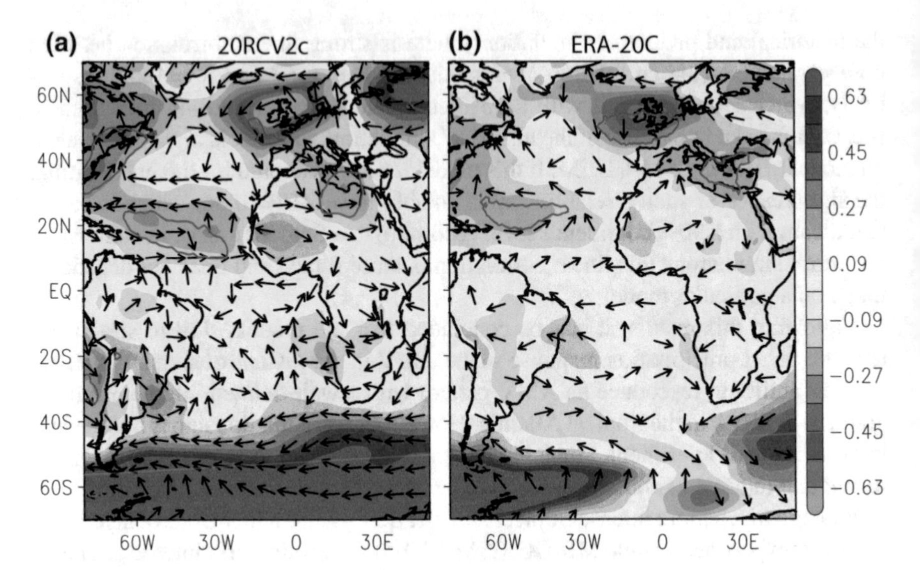

Fig. 4.8 Regression onto the observed AMV index of the unfiltered JAS anomaly of the surface pressure (shaded) (hPa per standard deviation) and the wind direction at 850 hPa (vectors) from the 20CRV2c (**a**) and ERA-20C (**b**) reanalyses. Contours indicate the regions where the surface pressure regression coefficients are significant at the 5% level

tropical band, they show a contrast of surface pressure anomalies between the north and the south of the tropical Atlantic sector that spans West Africa. Such a pressure gradient is consistent with the simulated anomalous equatorial southerly winds over the tropical Atlantic, which is similar to observations, and westerlies around 10°N trough the Atlantic and West Africa. These features are associated with a northward shift of the ITCZ and more eastward intrusion of air from the tropical Atlantic into West Africa that promote a Sahel rainfall increase (Rowell et al. 1992; Knight et al. 2006).

In agreement with the intensity of the tropical Atlantic SSTA gradient and the precipitation anomalies, the surface pressure associated with the AMV that the CMIP5 models simulate is weaker than in observations. Therefore, it can be inferred that the fact that models underestimate the interhemispheric SST and the surface pressure gradients related to the AMV results in changes in the low-level atmospheric circulation of the WAM system that are also weaker than in observations. As a consequence, the simulated Sahel rainfall response to the AMV is also underestimated with respect to observations, in agreement with Martin et al. (2014). Comparing the historical and piControl simulations, although they show similar features, the latter shows more intense and statistically robust atmospheric response to the AMV than the former, consistent with the other model-mean regression patterns.

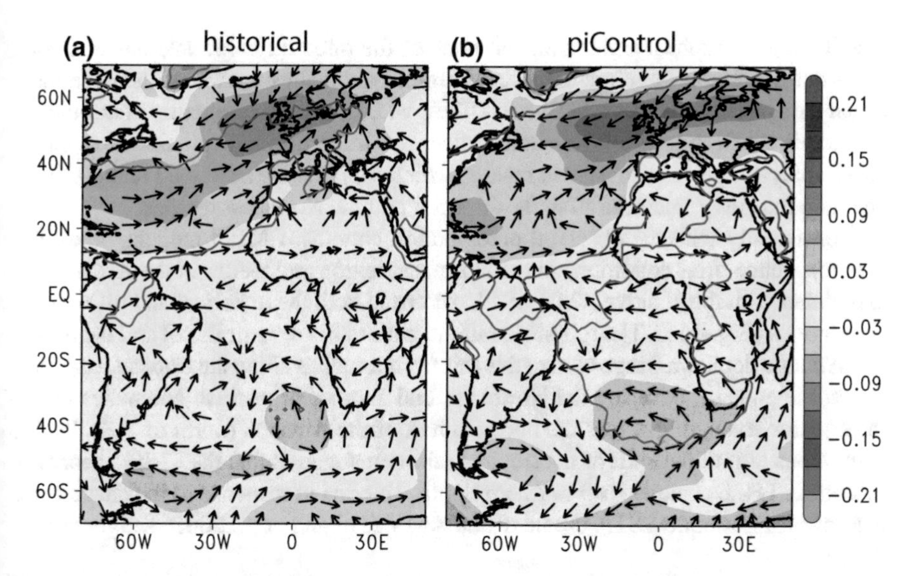

Fig. 4.9 Regression onto the simulated AMV index of the unfiltered JAS anomaly of the surface pressure (shaded) (hPa per standard deviation) and the wind direction at 850 hPa (vectors) from the historical (**a**) and piControl (**b**) simulations averaged among the 17 CMIP5 models. Contours indicate the regions where the surface pressure regression coefficients are significant at the 5% level

4.3 Amazon and Northeast Rainfall Response

Regarding tropical South America, the rainfall response to the AMV during DJF-MAM is anomalously negative over the Northeast of Brazil and positive in most of the vast Amazonia region and further north (Fig. 4.10). The observational regression patterns of precipitation show little statistical significance over most of northern South America. Particularly, the negative rainfall response to the AMV in the Northeast of

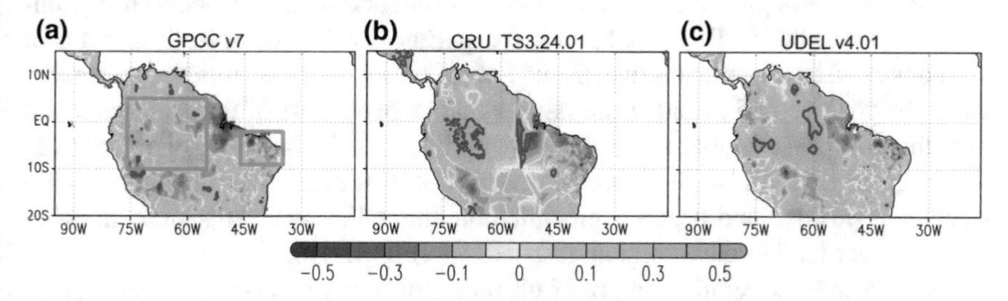

Fig. 4.10 Regression map of the unfiltered DJFMAM precipitation anomaly from the **a** GPCC v7, **b** CRU TS3.24.01 and **c** UDEL v4.01 observational data bases onto the standardized AMV index (units are mm day^{-1} per standard deviation). Contours indicate the regions where the regression is significant at the 5% level from a "random-phase" test. Orange boxes delimit the Amazon and the Northeast regions

Brazil shows no robust significance in none of the three different data bases. Over the Amazon region there are significant rainfall anomalies in two of the observational data bases (CRU TS3.24.01 and UDEL v4.01), though differently distributed. Nevertheless, despite the poor significance, the three different precipitation data bases analyzed provide consistent results, which supports the described relationship between the AMV and rainfall in the Amazon and the Northeast regions.

Consistently with the observed precipitation patterns, CMIP5 models, on average, reproduce drier conditions in the Northeast region and wetter in the Amazonia associated with the positive AMV SSTA pattern in both the historical and piControl simulations (Fig. 4.11). The full-scale rainfall pattern in northern Brazil and the tropical Atlantic depicts a fringe of negative rainfall anomalies along the tropical Atlantic below the equator, covering the Northeast, and another of positive anomalies over the equator, extending across the northern half of the Amazon (north of 7°S). Such anomalous latitudinal shift of the tropical rain-belt suggests that the CMIP5 models reproduce ITCZ changes over northern South America in response to the characteristic tropical Atlantic SSTA dipole of the AMV (Folland et al. 2001; Knight et al. 2006).

Note that the scale used to display the rainfall anomalies of the ensemble-mean regression patterns (ranging from around −0.1 to 0.1 mm day^{-1} per standard deviation) is lower than the one used for observations (from −0.5 to 0.5 mm day^{-1} per standard deviation). The underestimation of the anomalies in the model-mean patterns with respect to observations can be explained, in part, by the fact that precipitation patterns of 17 different models have been averaged. Besides, some CMIP5 models individually underestimate the intensity of rainfall (Yin et al. 2013), especially in the Amazonia (e.g., see IPSL-CM5A-LR in Figs. B.5 and B.6). Such an underestimation has been attributed to unrealistically reproduced moisture transport related with inaccurate representation of surface radiative fluxes or with overestimation of the tropical convective rainfall over the surrounding Pacific and Atlantic oceans (Yin et al. 2013).

Although the precipitation anomalies are similarly distributed in the regression maps of both the forced and unforced experiments, there are some differences between them. Roughly, the most outstanding difference is that the model-mean rainfall response to the AMV of the historical experiment is less statistically significant and consistent among models than the one of piControl. In the historical experiment, there are positive and negative rainfall anomalies over most of the Amazonia and over the Northeast region, respectively, but without high statistical significance in both regions (Fig. 4.11a). In contrast, in the piControl experiment there are highly significant positive and negative precipitation anomalies over Amazonia, north of 5°S, and over the Northeast region, respectively (Fig. 4.11b).

The fact that the rainfall pattern of piControl runs are, on average, more significant and therefore more consistent between the models than the one from historical simulations may be related to the differences between the AMV patterns obtained for both experiments. But which are the features of the AMV pattern that differ from

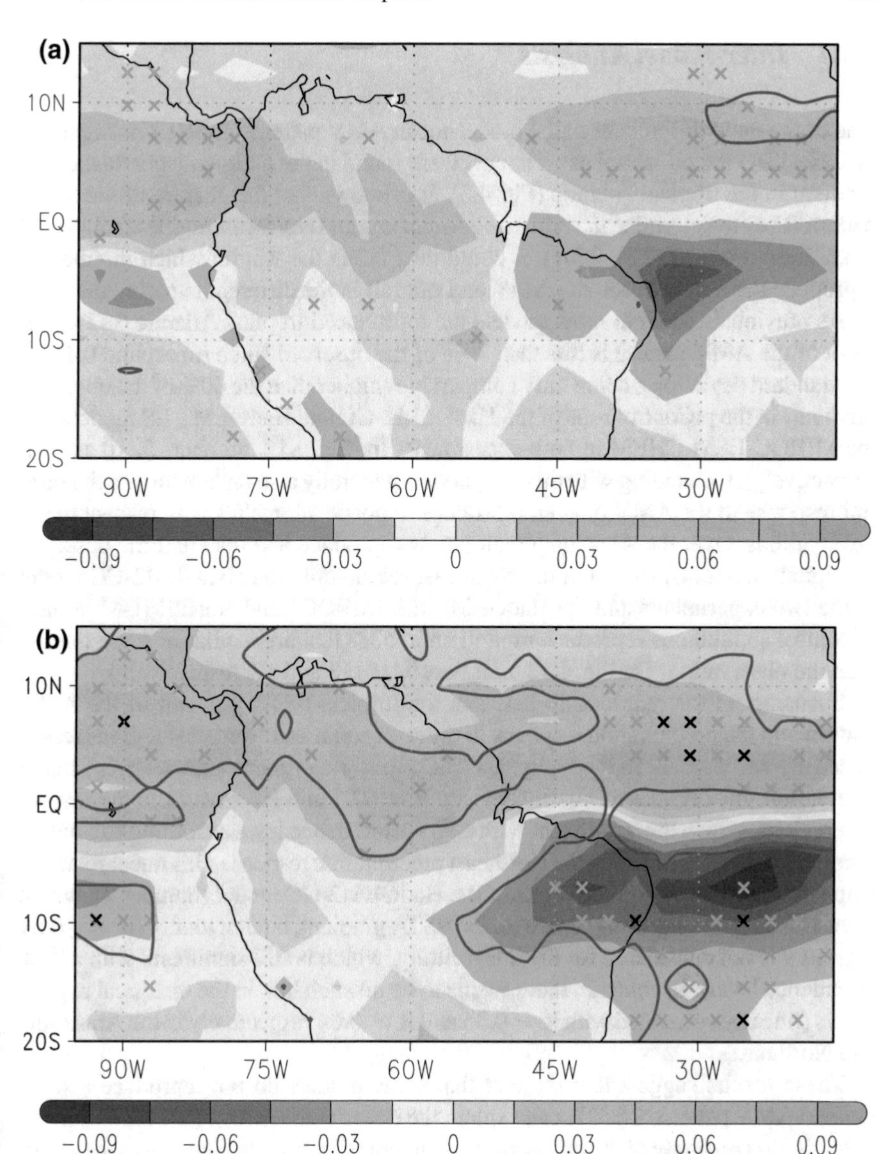

Fig. 4.11 Regression maps of the unfiltered DJFMAM precipitation anomalies (units are mm day^{-1} per standard deviation) on the AMV index averaged among the 17 CMIP5 models in the historical (**a**) and the piControl (**b**) simulations. Black and gray crosses indicate points where the regression coefficient sign coincides in at least 15 and 13 out of the 17 models analyzed, respectively. Contours indicate the regions where the averaged regression is significant at the 5% level from a "random-phase" test. Adapted from Villamayor et al. (2018a)

one experiment to another that induce the differences in the precipitation response? To answer this question, in the following we analyze the AMV patterns simulated by the different models.

4.3.1 Inter-model Analysis

One of the main differences in the model-mean AMV patterns of SSTA between the historical and the piControl experiments were found in the interhemispheric thermal gradient in the tropical Atlantic (Fig. 4.3). It is known that this feature of the SSTA pattern is key to determine the rain in the Amazonia and Northeast regions (e.g., Good et al. 2008; Folland et al. 2001), it could thus affect the way in which the models reproduce the link between the AMV and rainfall in the different experiments.

As previously seen, in most models the reproduced tropical Atlantic SSTA gradient of the AMV pattern is less than 30% of the observed value (of around 0.16 °C per standard deviation), being only comparable (higher than the 60% of the observed gradient) in the piControl runs of the HadGEM2-CC and HadGEM2-ES models and the MIROC-ESM-CHEM in both experiments (in Fig. 4.12, numbers 9, 10 and 14, respectively). Coinciding with this, the models generally also underestimate the rainfall response to the AMV or even reproduce opposite anomalies with respect to the observations. Over the Amazonia region they reproduce less than half the observed precipitation anomalies and in the Northeast region only the HadGEM2-CC model in the two experiments and the HadGEM2-ES, MIROC5 and NorESM1-M in their piControl simulations reproduce rainfall anomalies that are similar or more intense than the observations (in Fig. 4.12, numbers 9, 10, 13 and 17, respectively).

Focusing on the relationship between the tropical SSTA gradient of the AMV pattern and the precipitation response in the Amazonia and Northeast regions reproduced by the models individually, the piControl experiment shows strong linear correlation (the correlation coefficients are $R = 0.81$ and $R = -0.71$ in the respective regions, which are significant with a 95% confidence level according to Student t-test) (Fig. 4.12). Regarding the Northeast precipitation response, this linear relationship strongly depends on the result of the HadGEM2-CC model (number 9), which shows an outstanding strong link with the SSTA gradient, but not totally ($R = -0.51$ if point 9 is not considered for the linear fitting, which is still significant with a 95% confidence level). In contrast, there seems to be no such link in the historical experiments (linearly correlated with $R = 0.35$ and $R = 0.04$ respectively in the Amazonia and Northeast).

These results suggest that the fact that some models do not reproduce a well-defined AMV pattern of SSTA can explain the uncertainties among the models in the precipitation response of the piControl experiment (Fig. B.2). In some cases these patterns show certain relationship between the AMV and the SSTA of other basins, such as the Pacific (Zhang and Delworth 2007; Wu et al. 2011; Levine et al. 2017), which may interfere with the rainfall response to the Atlantic SSTA gradient. According to the results obtained from the AMV influence on Sahel rainfall (Sect. 4.2), in the case of the historical experiment, another source of uncertainty needs to be considered. This question is further discussed in Sect. 4.5.

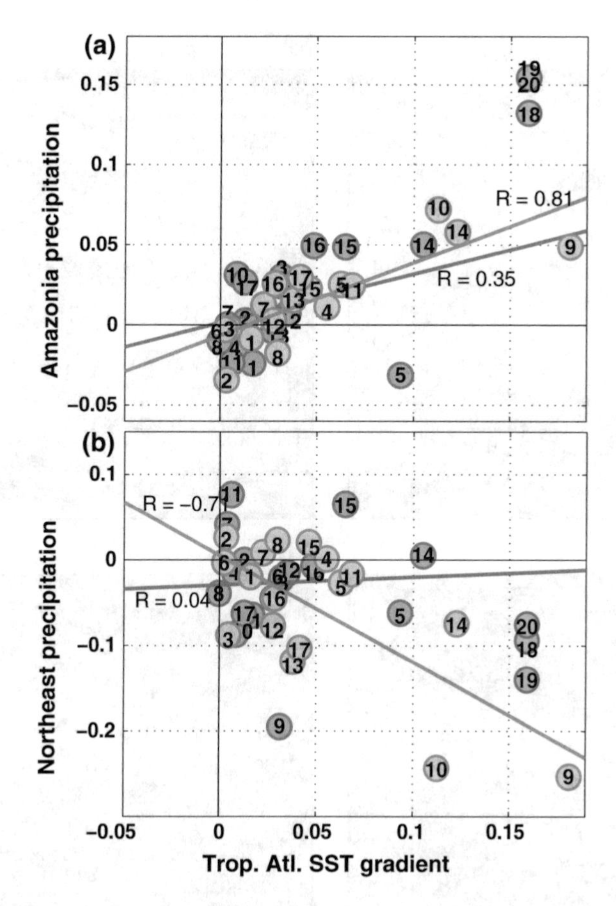

Fig. 4.12 **a** Scatter plot of the regression coefficient of precipitation anomaly over the Amazonia (western delimited region in Fig. 4.10a) and the SSTA tropical Atlantic gradient (difference between the delimited regions in Fig. 4.1b) relative to the AMV of each model in the historical (green) and the piControl (orange) simulations (Figs. B.1, B.2, B.5, and B.6 in Appendix B). The lines indicate the linear regression fitting of the corresponding colored points (R is the correlation coefficient). The numbers from 1 to 17 identify each model individually with the given number in Table 2.1. **b** Same as **a** but using the Northeast region (eastern delimited region in Fig. 4.10a) instead of the Amazonia. Numbers 18, 19 and 20 correspond to CRU TS3.24.01, GPCC v7 and UDEL v4.01 observed data (blue), respectively. Units for the horizontal and vertical axes are mm day^{-1} per standard deviation and K per standard deviation), respectively. From Villamayor et al. (2018a)

4.3.2 Atmospheric Teleconnection Between AMV and Northern Brazil Rainfall

During DJFMAM, the AMV observed signal has stronger effect on the surface pressure over the North Atlantic (Fig. 4.13) than in the boreal summer (JAS, Fig. 4.8). It projects onto a surface low and associated low-level cyclonic circulation, which

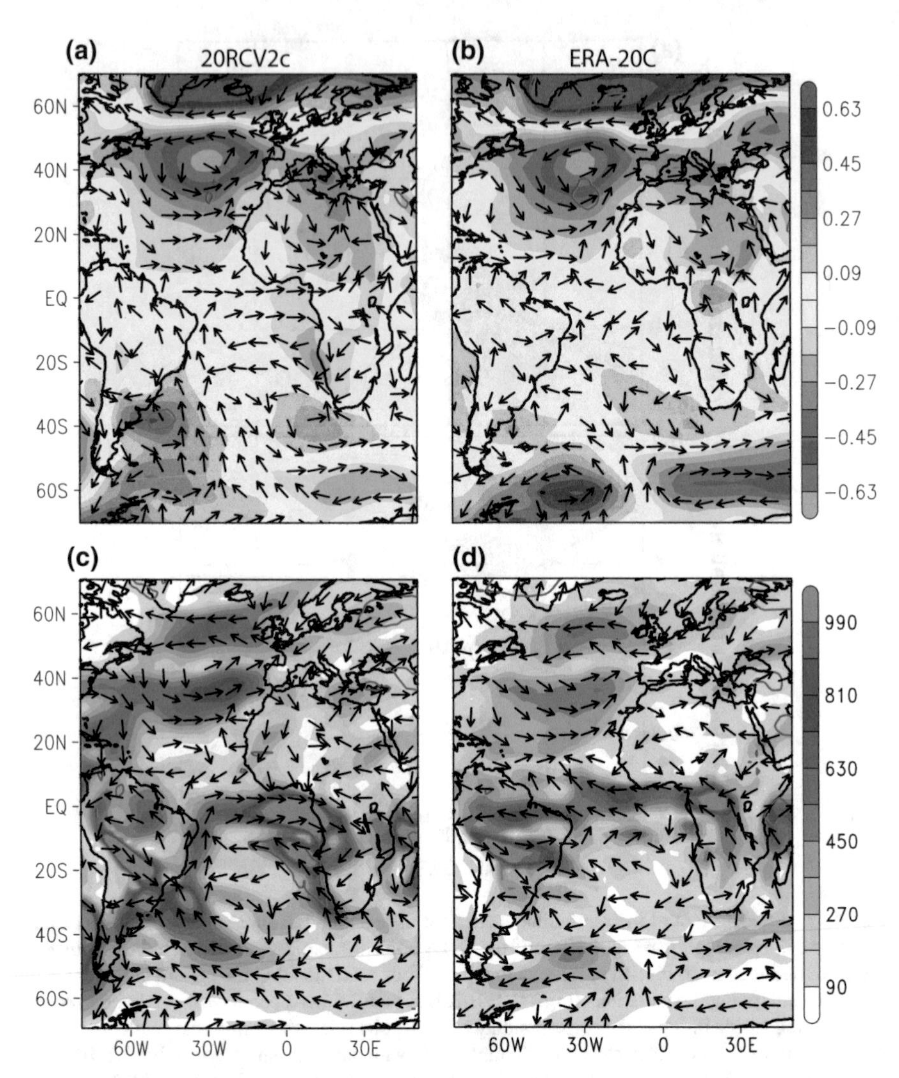

Fig. 4.13 **a**, **b** Regression onto the observed AMV index of the unfiltered DJFMAM anomaly of the surface pressure (shaded) (hPa per standard deviation) and the wind direction at 850 hPa (vectors) from the 20RCV2c and ERA-20C reanalyses. **c**, **d** Homologous regression patters of the magnitude (shaded) and direction (vectors) of the moisture flux (kg/m/day per standard deviation). Contours indicate the regions where the surface pressure and the moisture flux regression coefficients are significant at the 5% level

entails the weakening of the North Atlantic subtropical high (Fig. 4.13a, b). While to the south, it shows positive surface subtropical pressure anomalies. Associated with this interhemipsheric pressure gradient, there are northward anomalous low-level winds over the western part of the tropical Atlantic and northern South America. These winds are, in turn, consistent with the anomalous moisture flux from the tropical Atlantic toward the Amazon River mouth and inland (Fig. 4.13c, d). This low-level circulation also suggest an anomalous meridional circulation with stronger convection over the northern and subsidence over the southern Atlantic basin associated with the decrease and increase of surface pressure, respectively. This anomalous circulation entails the strengthening of the ITCZ north of the equator, reducing the moisture supply in the Northeast (Moura and Shukla 1981; Hastenrath and Greischar 1993; de Albuquerque Cavalcanti 2015). Not only is this mechanism consistent with the anomalous drying of the Northeast (Knight et al. 2006) but also with wetter conditions in the Amazonia region. The northward displacement of the ITCZ provides Amazonia with more humidity advected from the tropical Atlantic toward the Amazon River mouth and inland. With regard to the significance in Fig. 4.13, the observational results do not show high statistical robustness. Nevertheless, both reanalyses reveal the same atmospheric features described, associated with the AMV during DJFMAM. Hence, the coherence between both reanalyses provide confidence to these results.

Consistent with the AMV patterns of SSTA, the model-mean surface pressure response in the historical experiment shows lower statistical significance over the southern Atlantic (between 0° and 40° S) than in the piControl one (Fig. 4.14). CMIP5 models reproduce an anomalous cyclonic circulation over the North Atlantic and a surface pressure contrast with respect to the south (Fig. 4.14a, b), in agreement with observations. However, the North Atlantic cyclone displayed by the surface pressure and low-level wind anomalies is placed more to the northeast than in observations. This is consistent with the distribution of the SSTA in the AMV patterns throughout the tropical North Atlantic. In observations the stronger SSTA are closer to the equator than in the historical and piControl simulations, which are located more to the north (Figs. 4.1 and 4.3a, b, respectively). Despite this difference, models reproduce the observed anomalous northward shift of the cross-equatorial winds and the moisture flux away from the Northeast of Brazil and toward the Amazon basin (Fig. 4.14c, d).

The most remarkable discrepancies between the observed and simulated atmospheric circulation response to the AMV are shown over the South American continent. Observational results form the reanalyses show weak and uncertain surface pressure response over South America (Fig. 4.13a, b). Although there is a strong anomalous northwesterly low-level jet of moisture flux along Bolivia and central Brazil, between 10° and 20° S, that is coherent between both reanalyses (Fig. 4.13c, d). This jet flows southeastward from western Amazonia along the eastern slope of the Peruvian Andes. Such anomalies are related to changes in the low-level winds and the moisture transport over this area that can affect more the climate of subtropical and extratropical regions of the continent (Labraga et al. 2000; Grimm and Zilli 2009; Marengo et al. 2012). On the other hand, models simulate a northward deviation of the anomalous moisture flux from the Amazon and no remarkable signal to the south (Fig. 4.14c, d), in contrast to observations. Such discrepancies between observations and CMIP5 simulations do not affect the rainfall response to the AMV in the Northeast and the northern part of the Amazonia, which are mostly influenced by the easterlies from the tropical Atlantic. In turn, they can substantially affect the

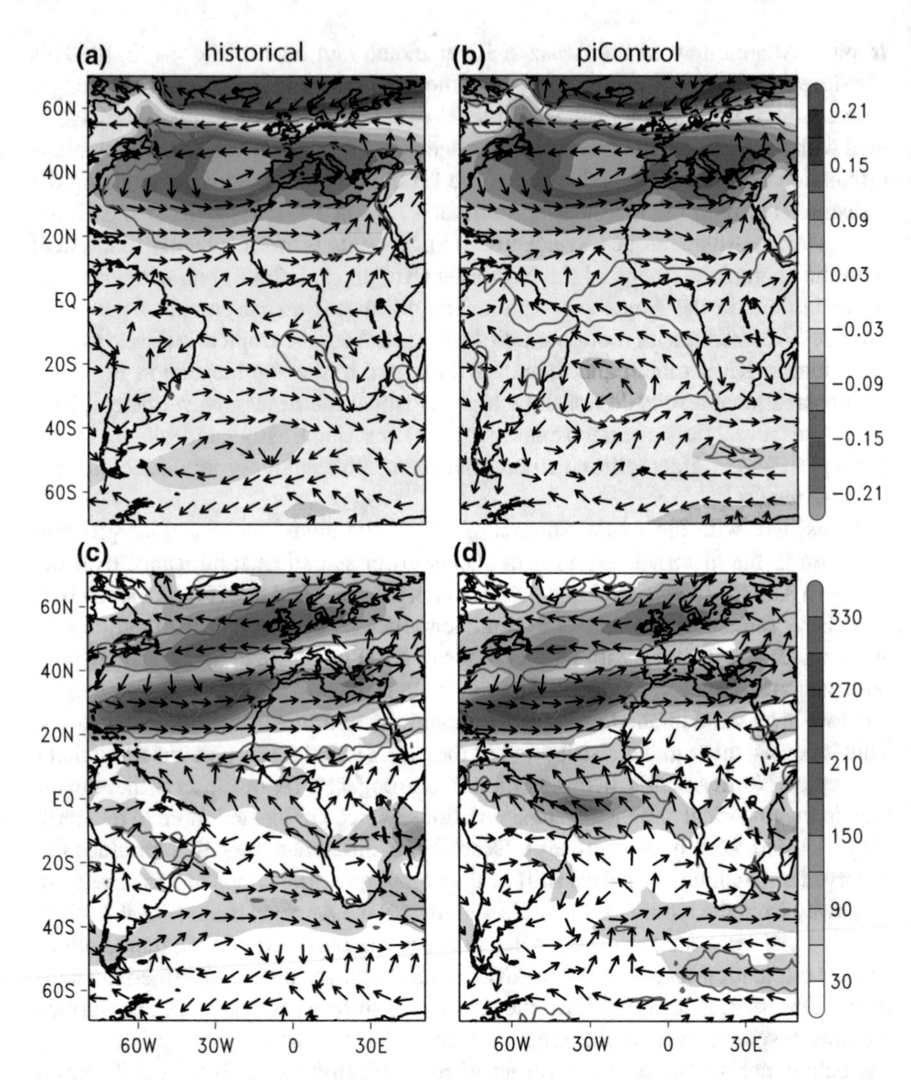

Fig. 4.14 **a**, **b** Regression onto the simulated AMV index of the unfiltered DJFMAM anomaly of the surface pressure (shaded) (hPa per standard deviation) and the wind direction at 850 hPa (vectors) from the historical and piControl simulations. **c**, **d** Homologous regression patters of the magnitude (shaded) and direction (vectors) of the moisture flux (kg/m/day per standard deviation). Contours indicate the regions where the surface pressure and the moisture flux regression coefficients are significant at the 5% level. Adapted from Villamayor et al. (2018a)

way in which CMIP5 models reproduce the relationship between the AMV and rainfall in the south of the Amazonia, as well as other extratropical regions (Marengo et al. 2012), and thus its low-frequency variability.

4.4 RCP8.5 Future Projections

Despite the differences found among some models, the reproduced model-mean AMV pattern and its impact on precipitation over the Sahel, Amazonia and Northeast regions show similar features as in observations in both forced and unforced simulations. On this basis, we can wonder whether the relationship between rainfall and the decadal-to-multidecadal patterns will change or not in a hypothetical future scenario analyzing the model-mean patters of CMIP5 future projections. To this aim, future projections of the RCP8.5 scenario are used.

Regarding the model-mean AMV pattern calculated with the RCP8.5 projection (Fig. 4.15a), in the North Atlantic it depicts a coma-shape SSTA heating similar to the historical and piControl experiments, although slightly less consistent among the models and with lower statistical significance. The global pattern, however, presents colder anomalies than the one given by the historical experiment in the tropical Pacific, the Indian Ocean and the South Atlantic basin. It shows an interhemispheric thermal gradient in the Tropical Atlantic that resembles more to the AMV reproduced by the piControl simulation. Regarding the AMV index, a Fourier analysis reveals

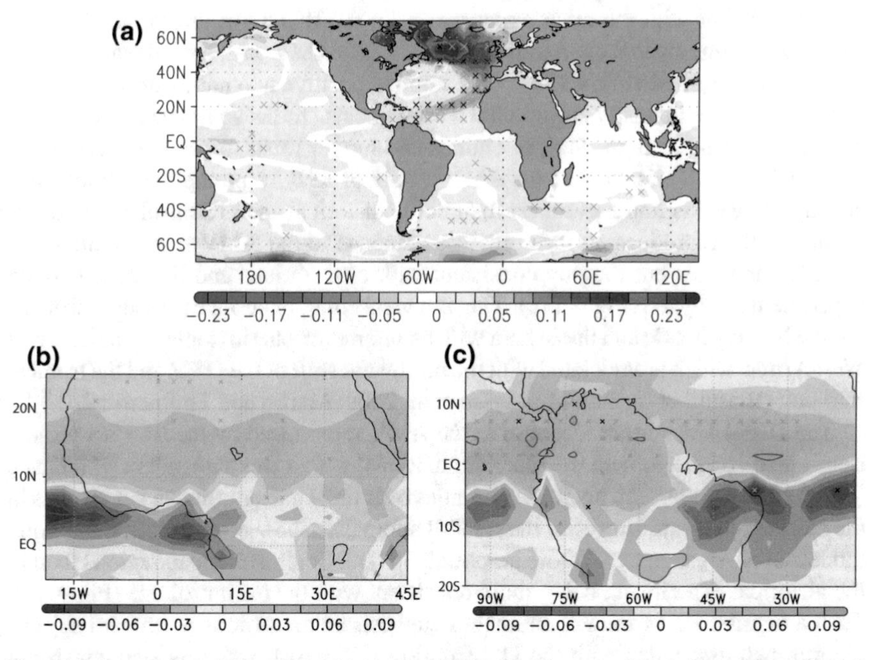

Fig. 4.15 Regression of the unfiltered (**a**) annual SSTA (K per standard deviation) and the (**b**) JAS and (**c**) DJFMAM precipitation anomalies (mm day^{-1} per standard deviation) averaged among the 17 CMIP5 models in the RCP8.5 future projection onto the AMV index. Black and gray crosses indicate points where the regression coefficient sign coincides in at least 15 and 13 out of the 17 models analyzed, respectively. Contours indicate the regions where the regression is significant at the 5% level. Adapted from Villamayor et al. (2018a)

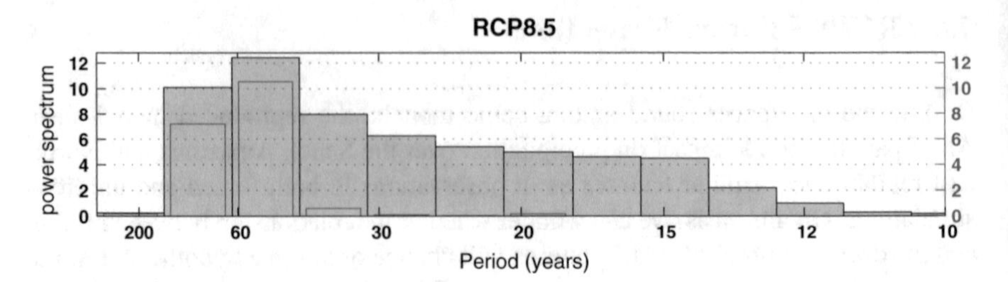

Fig. 4.16 Model-mean power spectra of the Fourier transform of the detrended AMV indices (blue bars) of the RCP8.5 future projection. Red bars indicate the power spectra averaged over the 17 models but taking into account only 95% significant peaks for each of them. The statistical significance for each model's AMV is obtained following a non-parametric hypothesis test in which a probability density function is built from the Fourier spectrum of 13-year low-pass filtered and detrended white noise time series with the same time length as the simulation and averaged across the number of ensemble members in cases where more than one realization is available

that the most characteristic periodicities of the time series obtained from the RCP8.5 future projections of all the models are distributed around 65 years (Fig. 4.16), as in the other simulations analyzed and consistently with observations.

In agreement with the other experiments, in the RCP8.5 future projections the JAS precipitation over West Africa responds to the AMV with a northward shift of the tropical rain-belt (Fig. 4.15b). This implies positive anomalies throughout the Sahel, with more intensity close to the western coast. To the south, negative rainfall anomalies extend from the Gulf of Guinea and further west, spanning the coast line of West Africa. Consistent with the AMV pattern of SST, the precipitation anomalies are slightly weaker and less robust than in the historical and piControl simulations. Similarly, the reproduce surface pressure response to the AMV shows barely significant anomalies but roughly consistent with observations and the other CMIP5 experiments analyzed (Fig. 4.17a). This is a weakening of surface pressure throughout the North Atlantic and the Sahara with an interhemispheric tropical gradient over West Africa, which is associated with the northward shift of the ITCZ and the tropical rain-belt (Biasutti et al. 2009; Lavaysse et al. 2009; Martin and Thorncroft 2014).

The DJFMAM rainfall response to the AMV reproduced by the RCP8.5 projection is also consistent with the other simulations, with lower statistical significance (Fig. 4.15c). This is, with positive anomalies over the Amazonia and negative ones in the Northeast region. However, the rainfall anomalies show more agreement among models over Amazonia and more amplitude in both South American regions than in the historical experiment, being more consistent with the piControl one (Fig. 4.11). Such a distribution of rainfall anomalies suggests an anomalous northward shift of the rain-belt associated with the ITCZ during DJFMAM, which is also consistent with the associated atmospheric dynamic given by RCP8.5 projections: anomalous low pressure at the surface and low-level cyclonic circulation over the north Atlantic basin and northward strengthening of the cross-equatorial winds and of the moisture supply (Fig. 4.17b, c), consistently with the other two experiments.

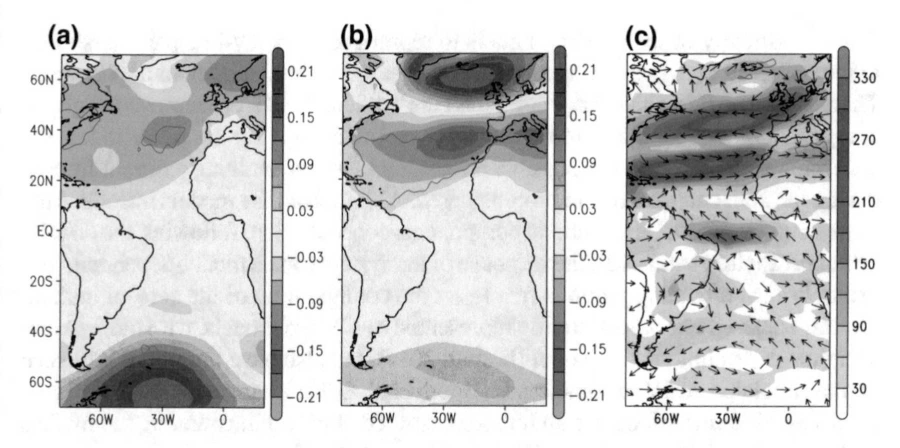

Fig. 4.17 Regression of the unfiltered (**a**) JAS and (**b**) DJFMAM anomaly of the surface pressure (shaded) (hPa per standard deviation) and (**c**) the magnitude (shaded) and direction (vectors) of the DJFMAM moisture flux anomaly integrated from surface to 200 hPa (kg/m/day per standard deviation) onto the AMV index averaged among the 17 CMIP5 models in the RCP8.5 future projection. Contours indicate the regions where the averaged regression is significant at the 5% level

Hence, the RCP8.5 future projection shows that the AMV pattern of SSTA reproduced by the models on average presents the same characteristic features as in other experiments and in observations. Consistently, the rainfall response and the atmospheric mechanism associated with the AMV pattern simulated by RCP8.5 are also similar to the one in the other experiments. Such a result suggests that the AMV mode of SST variability and its impacts are not expected to change in the future, regardless of the concentration of greenhouse gases emitted. On the other hand, it should be pointed out that the low-frequency variability of the AMV may generate a weak signal in the 95-year period run of the RCP8.5 projections. Thus their effect are less robustly captured across the models than in the other experiments. Further long period numerical experiments are still necessary to better understand such variability in the future.

4.5 Discussion on the Possible Role of Aerosols on the AMV

Some important inconsistencies among the models with respect to the rainfall response to the AMV in the Sahel, Amazonia and Northeast regions have been found. From the inter-model analyses presented above, it has been suggested that the models skill to simulate a relationship between the AMV and precipitation similar to the observations is attributed to the accuracy with which they reproduce the interhemispheric tropical Atlantic gradient of SST in the piControl simulations. But such a relationship is arguable in the case of historical simulations. It has been found that the difficulty of the CMIP5 models to reproduce an AMV-like pattern, similar to the observed one, is more recurrent in the historical simulation than in piControl. Therefore, even though we have computed the "residual" SSTA to remove the global component of the external forcing (the GW) before computing the AMV index, the uncertainties induced by the possible effect of an uneven evolving external forcing during the 20th century have to be considered. Although all the models introduce the same configuration of atmospheric components concentration, following the CMIP5 protocol (Taylor et al. 2012), the response of the system can be model-dependent. For example each modeling group is free to set the configuration of the aerosol indirect effects. These effects are differently represented by CMIP5 models, with more or less accuracy depending on the model (Ekman 2014), but also have important influence on the simulated AMV in some cases (Booth et al. 2012; Zhang et al. 2013a). Thus, the aerosol indirect effects are an important source of uncertainty among the models in the way they represent the AMV (Boucher et al. 2013).

One way to identify if the aerosol effects influence the AMV that CMIP5 models simulate is to seek whether the oscillations of the AMV index show any externally forced signal that stands out from the internal variability. For this, we compare the standard deviation of the ensemble-mean AMV index of the historical simulation of each model with a distribution of standard deviation values constructed from the average of a similar sample of AMV indexes randomly chosen from the piControl simulation (Fig. 4.18). From these results, we can conclude that all the ensemble members of the GISS-E2-H and HadGEM2-ES models have a common oscillation

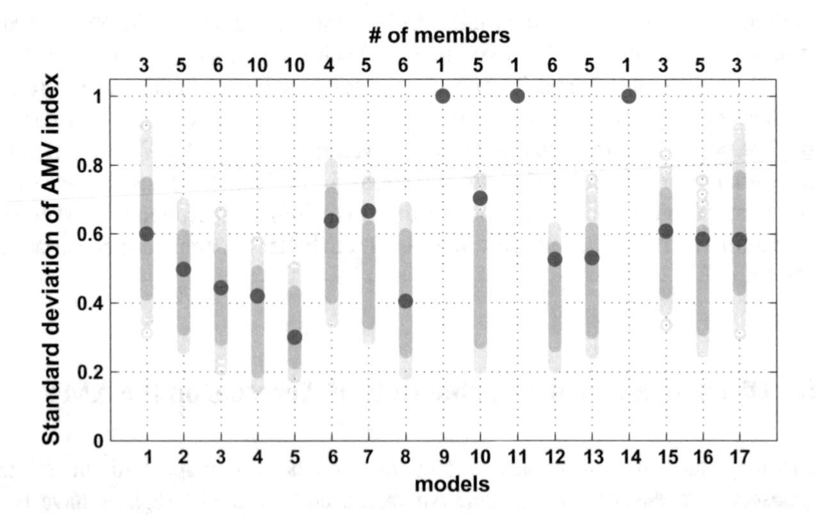

Fig. 4.18 Standard deviation of the ensemble-mean AMV indices of the historical simulations (red dots). The average is done among the indices of the N members available (indicated in the upper axis) for each model (lower axis, same order as in Table 2.1). Grey dots indicate a distribution of 1000 standard deviation values calculated as the average of N sampled AMV indices randomly selected along the piControl index, with the same length as the historical time series. Darker gray dots indicate a 95% of probability distribution

which is significantly independent of the internal variability. Therefore it is reasonable to think that there is a component of the externally forced SST variability in the historical simulations, apart from the GW, that remains in the North Atlantic "residual" SSTA field. This affects the SSTA of the AMV pattern in the historical simulation, setting it apart from the one in piControl of at least the GISS-E2-H and HadGEM2-ES models. But what causes such remaining external forcing in the SST variability? To shed some light on this question, detection and attribution of climate change historicalGHG simulations provided by some CMIP5 models have been analyzed. By computing the AMV patterns (Fig. B.7), with the same method as with historical simulation, and comparing it with its counterpart in the historical and piControl experiments we can infer whether the remaining external forcing is caused by the greenhouse gases or, by default, by the aerosol effects (Taylor et al. 2007).

Regarding the AMV reproduced by the GISS-E2-H model, in the historical simulation it is associated with a spatial pattern of warm SSTA extended almost all across the globe, more intense in the tropics and the North Atlantic (Fig. B.1). In contrast, the piControl (Fig. B.2) and historicalGHG experiments (Fig. B.7) show AMV patterns that, although they are poorly represented, roughly show insignificant or negative anomalies outside the North Atlantic basin. Hence, the difference observed between the AMV pattern in the historical and the unforced piControl experiments is likely produced by external forcing of the aerosol effects that the GISS-E2-H model resolves. The HadGEM2-ES model reproduces a historical AMV pattern with a poor interhemispheric SSTA gradient in the tropical Atlantic in contrast to the piControl and historicalGHG experiments. Therefore we can infer that the aerosol effects in the HadGEM2-ES model also induce discrepancies between the historical and piControl experiments. The differences between the AMV patterns of the historical and piControl experiments of the HadGEM2-CC model could suggest the same, but in this case we cannot test it, as there is only one member available for the historical simulations and the historicalGHG experiment was missing. There are other models that do not provide the historicalGHG experiment either (inmcm4, MIROC5 and MPI-ESM-LR) and hence we cannot infer the causes of the differences between the historical and the unforced simulations. In the historical simulation of the bcc-csm1-1 and GISS-E2-R models there is no noticeable common variability among all the ensemble members (Fig. 4.18). However, the fact that the historical AMV patterns that these two models simulate is globally warmer than the one of the piControl and historicalGHG experiments also suggest that the aerosol effects induce such difference. For the rest of the models (CanESM2, CCSM4, CNRM-CM5, CSIRO-Mk3-6-0, FGOALS-g2, IPSL-CM5A-LR, MIROC-ESM-CHEM, MRI-CGCM3 and NorESM1-M) it is hard to infer any conclusion from these analysis. In some cases this is because the differences between forced and unforced simulated AMV patterns are small; and in other cases, the pattern is not well defined.

4.6 Summary of Main Findings

Regarding the aims of this Thesis, the analysis of the link between the AMV mode of SST variability and precipitation in the Sahel, the Amazonia and the Northeast regions have been presented in this Chapter. The results show that the CMIP5 models can reproduce the relationship between the AMV mode of long-term SST variability and the tropical precipitation in West Africa and northern Brazil rainfall during JAS and DJFMAM, respectively. In its positive phase, the AMV induces intensified rainfall in the Sahel and the Amazonia regions and less precipitation in the Northeast. Though the intensity of rainfall anomalies is notably underestimated (Villamayor et al. 2018a).

In agreement with observations, CMIP5 models reproduce an AMV pattern of SSTA with an interhemispheric thermal gradient in the Atlantic basin with intense warm anomalies north of the equator. Associated with the model-mean AMV pattern, an interhemispheric contrast of surface pressure anomalies over the Atlantic and anomalous latitudinal displacement of the ITCZ over the tropical Atlantic sector are induced. In case of the positive AMV phase, the ITCZ is anomalously displaced toward the Sahel during the rainy season in JAS and hence it rains more. In turn, during DJFMAM the ITCZ experiences a weaker intrusion toward the Northeast of Brazil remaining in latitudes close to the mouth of the Amazon, favoring moisture transport into the Amazonia. This mechanism produces anomalous drying in the Northeast region and wetter conditions in the Amazonia (Villamayor et al. 2018a). The opposite occurs during the cold AMV phases.

To summarize, the main conclusions arisen from this Chapter of results are:

- The CMIP5 models on average can reproduce the main observed features of the AMV pattern, except for the strength of the characteristic tropical SSTA gradient.
- Consequently, they also reproduce the observed AMV influence on rainfall in the Sahel, the Amazon and the Northeast of Brazil regions, though with underestimated intensity.
- The models skill to reproduce the link between rainfall and the AMV is related to the accuracy with which the characteristic interhemispheric gradient of the tropical Atlantic SST pattern of the AMV is represented, particularly in the unforced piControl experiments.
- The aerosols seem to play a relevant role in explaining the differences between the characteristics of the AMV patterns of the historical and piControl experiments, which in turn affects the reproduced link with rainfall.
- The associated atmospheric dynamics with the AMV-precipitation link show that, according to the Atlantic interhemispheric gradient, a surface pressure contrast is induced which entails a latitudinal shift of the ITCZ in CMIP5 simulations similar to observations.
- The analysis of the RCP8.5 future projections suggest that the AMV and its impacts on rainfall will not change under an scenario of large GHGs emission.

References

Alexander, M.A., Halimeda Kilbourne, K., Nye, J.A.: Climate variability during warm and cold phases of the Atlantic Multidecadal Oscillation (AMO) 1871–2008. J. Mar. Syst. **133**, 14–26 (2014). https://doi.org/10.1016/j.jmarsys.2013.07.017

Biasutti, M., Sobel, A.H., Camargo, S.J.: The role of the Sahara low in summertime Sahel rainfall variability and change in the CMIP3 models. J. Clim. **22**, 5755–5771 (2009)

Booth, B.B.B., Dunstone, N.J., Halloran, P.R., Andrews, T., Bellouin, N.: Aerosols implicated as a prime driver of twentieth-century North Atlantic climate variability. Nature **484**, 228–232 (2012). https://doi.org/10.1038/nature10946

Boucher, O., Randall, D., Artaxo, P., Bretherton, C., Feingold, G., Forster, P., Kerminen, V.-M., Kondo, Y., Liao, H., Lohmann, U., et al.: Clouds and aerosols. In: Climate Change 2013: The Physical Science Basis. Contribution of Working Group I to the Fifth Assessment Report of the Intergovernmental Panel on Climate Change, pp. 571–657. Cambridge University Press (2013)

de Albuquerque Cavalcanti, I.F.: The influence of extratropical Atlantic Ocean region on wet and dry years in North-Northeastern Brazil. Front. Environ. Sci. **3**, 1–10 (2015). https://doi.org/10.3389/fenvs.2015.00034

Ekman, A.M.: Do sophisticated parameterizations of aerosol-cloud interactions in CMIP5 models improve the representation of recent observed temperature trends? J. Geophys. Res. Atmos. **119**, 817–832 (2014)

Enfield, D.B., Mestas-Nuñez, A.M., Trimble, P.J.: The Atlantic multidecadal oscillation and its relation to rainfall and river flows in the continental U.S. Geophys. Res. Lett. **28**, 2077–2080 (2001). https://doi.org/10.1029/2000GL012745

Folland, C.K., Colman, A.W., Rowell, D.P., Davey, M.K.: Predictability of northeast Brazil rainfall and real-time forecast skill, 1987–98. J. Clim. **14**, 1937–1958 (2001)

Good, P., Lowe, J.A., Collins, M., Moufouma-Okia, W.: An objective tropical Atlantic sea surface temperature gradient index for studies of south Amazon dry-season climate variability and change. Philos. Trans. R. Soc. Lond. Ser. B Biol. Sci. **363**, 1761–1766 (2008). https://doi.org/10.1098/rstb.2007.0024

Grimm, A.M., Zilli, M.T.: Interannual variability and seasonal evolution of summer monsoon rainfall in South America. J. Clim. **22**, 2257–2275 (2009). https://doi.org/10.1175/2008JCLI2345.1

Hastenrath, S., Greischar, L.: Circulation mechanisms related to northeast Brazil rainfall anomalies. J. Geophys. Res. Atmos. **98**, 5093–5102 (1993)

Kerr, R.A.: A North Atlantic climate pacemaker for the centuries. Science **288**, 1984–1985 (2000). https://doi.org/10.1126/science.288.5473.1984

Knight, J.R.: A signature of persistent natural thermohaline circulation cycles in observed climate. Geophys. Res. Lett. **32**(L20), 708 (2005). https://doi.org/10.1029/2005GL024233

Knight, J.R., Folland, C.K., Scaife, A.A.: Climate impacts of the Atlantic Multidecadal Oscillation. Geophys. Res. Lett. **33**(L17), 706 (2006). https://doi.org/10.1029/2006GL026242

Labraga, J.C., Frumento, O., López, M.: The atmospheric water vapor cycle in South America and the tropospheric circulation. Am. Meteorol. Soc. **13**, 1899–1915 (2000)

Lavaysse, C., Flamant, C., Janicot, S., Parker, D., Lafore, J.-P., Sultan, B., Pelon, J.: Seasonal evolution of the West African heat low: a climatological perspective. Clim. Dyn. **33**, 313–330 (2009)

Levine, A.F., McPhaden, M.J., Frierson, D.M.: The impact of the AMO on multidecadal ENSO variability. Geophys. Res. Lett. **44**, 3877–3886 (2017)

Marengo, J.A., Liebmann, B., Grimm, A.M., Misra, V., Dias, P.L.S., Cavalcanti, I.F.A., Carvalho, L.M.V., Berbery, E.H., Ambrizzi, T., Vera, C.S., Saulo, A.C., Nogues-paegle, J., Zipser, E., Seth, A., Alves, L.M.: Recent developments on the South American monsoon system. Int. J. Climatol. **32**, 1–21 (2012). https://doi.org/10.1002/joc.2254

Martin, E.R., Thorncroft, C.D.: The impact of the AMO on the West African monsoon annual cycle. Q. J. R. Meteorol. Soc. **140**, 31–46 (2014)

Martin, E.R., Thorncroft, C., Booth, B.B.B.: The multidecadal Atlantic SST-Sahel rainfall teleconnection in CMIP5 simulations. J. Clim. **27**, 784–806 (2014). https://doi.org/10.1175/JCLI-D-13-00242.1
Mohino, E., Janicot, S., Bader, J.: Sahel rainfall and decadal to multi-decadal sea surface temperature variability. Clim. Dyn. **37**, 419–440 (2011a). https://doi.org/10.1007/s00382-010-0867-2
Moura, A.D., Shukla, J.: On the dynamics of droughts in Northeast Brazil: Observations, theory, and numerical experiments with a general circulation model (1981)
Rowell, D.P., Folland, C.K., Maskell, K., Owen, J.A., Ward, M.N.: Modelling the influence of global sea surface temperatures on the variability and predictability of seasonal Sahel rainfall. Geophys. Res. Lett. **19**, 905–908 (1992)
Sutton, R.T., Hodson, D.L.R.: North Atlantic forcing of North American and European summer climate. Science **309**, 115–118 (2005). https://doi.org/10.1126/science.1109496
Taylor, K.E., Stouffer, R.J., Meehl, G.A.: A summary of the CMIP5 experiment design. World **4**, 1–33 (2007)
Taylor, K.E., Stouffer, R.J., Meehl, G.A.: An overview of CMIP5 and the experiment design. Bull. Am. Meteorol. Soc. **93**, 485–498 (2012)
Villamayor, J., Ambrizzi, T., Mohino, E.: Influence of decadal sea surface temperature variability on northern Brazil rainfall in CMIP5 simulations. Clim. Dyn. **51**, 563–579 (2018a)
Wu, S., Liu, Z., Zhang, R., Delworth, T.L.: On the observed relationship between the Pacific Decadal Oscillation and the Atlantic Multi-decadal Oscillation. J. Ocean. **67**, 27–35 (2011)
Yin, L., Fu, R., Shevliakova, E., Dickinson, R.E.: How well can CMIP5 simulate precipitation and its controlling processes over tropical South America? Clim. Dyn., 1–17 (2013)
Zhang, R., Delworth, T.L.: Impact of the Atlantic multidecadal oscillation on North Pacific climate variability. Geophys. Res. Lett. **34** (2007)
Zhang, R., Delworth, T.L.: Impact of Atlantic multidecadal oscillations on India/Sahel rainfall and Atlantic hurricanes. Geophys. Res. Lett. **33**(L17), 712 (2006). https://doi.org/10.1029/2006GL026267
Zhang, R., Delworth, T.L., Sutton, R., Hodson, D.L.R., Dixon, K.W., Held, I.M., Kushnir, Y., Marshall, J., Ming, Y., Msadek, R., Robson, J., Rosati, A.J., Ting, M., Vecchi, G.A.: Have aerosols caused the observed Atlantic multidecadal variability? J. Atmos. Sci. **70**, 1135–1144 (2013a). https://doi.org/10.1175/JAS-D-12-0331.1

Chapter 5
Influence of the IPO

Abstract The IPO impact in the regions of the Sahel, the Amazonia and the Northeast of Brazil during their respective rainy seasons is addressed in this Chapter. For that purpose, the IPO mode of SST variability is characterized from CMIP5 simulations and observations, as well as its impact on rainfall in the three regions of interest. Then, the atmospheric mechanisms involved are also shown and discussed. The comparison between externally forced and unforced simulations lets discern whether the radiative forcing contributes to the formation of the IPO pattern of SST or if it has some effects on the link with rainfall. The same analysis is finally applied to the RCP8.5 future projections in order to show how the link between the IPO and rainfall is expected to change in the future or not.

5.1 The IPO Index and Pattern

In accordance with previous studies (Mantua et al. 1997; Mantua and Hare 2002; Deser et al. 2004; Shen et al. 2006; Mohino et al. 2011a; Dai 2013), the IPO index obtained from observations shows cold regimes in the periods of 1909–1925, 1944–1976 and from 1998 onwards, and warm regimes in the periods 1925–1944 and 1976–1998 (Fig. 5.1a). However, such time variability shows no dominant frequency (in accordance with the proxy reconstruction from Shen et al. 2006) but various spectral peaks at decadal and multidecadal periodicities, mainly in the 15–25 and 50–70 year bands (Fig. 5.2), in accordance with previous works (Minobe 1999; Chao et al. 2000; Tourre et al. 2001; Mantua and Hare 2002; MacDonald and Case 2005).

The content of this Chapter has been published in the following articles:

Villamayor, J. & Mohino, E. (2015): Robust Sahel drought due to the Interdecadal Pacific Oscillation in CMIP5 simulations. *Geophys. Res. Lett.*, **42**, 1214–1222.
http://dx.doi.org/10.1002/2014GL062473

Villamayor, J., Ambrizzi, T. & Mohino, E. (2018a): Influence of decadal sea surface temperature variability on northern Brazil rainfall in CMIP5 simulations. *Clim. Dyn.*
https://doi.org/10.1007/s00382-017-3941-1.

J. Villamayor, *Influence of the Sea Surface Temperature Decadal Variability on Tropical Precipitation: West African and South American Monsoon*, Springer Theses, https://doi.org/10.1007/978-3-030-20327-6_5

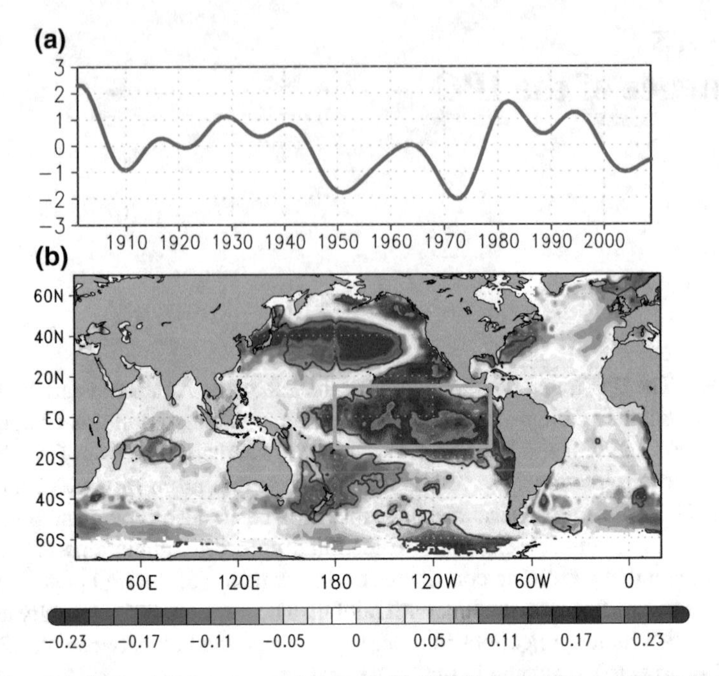

Fig. 5.1 **a** Standardized IPO index obtained with HadISST1 from 1901 to 2009. **b** Regression pattern of the unfiltered HadISST1 SSTA onto the standardized IPO index (units are K per standard deviation). Contours indicate the regions where the regression is significant at the 10% level from a "random-phase" test. Green box delimits the Tropical Pacific area. Adapted from Villamayor et al. (2018a)

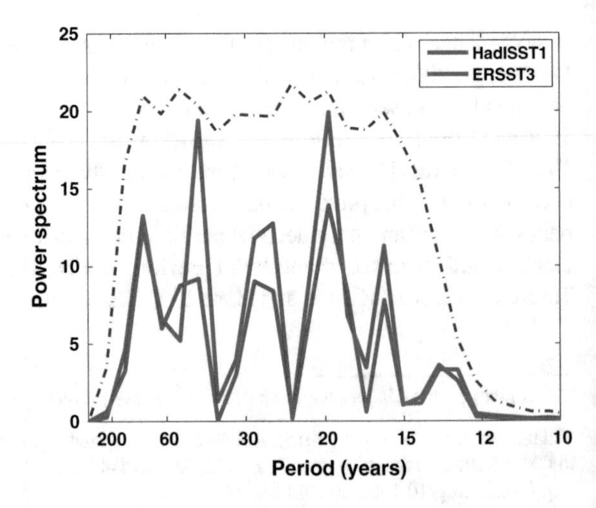

Fig. 5.2 Power spectrum of the Fourier transform of the detrended IPO indices from HadISST1 and ERSST3 data bases. Gray dashed line indicates the 95% confidence level following a non-parametric hypothesis test in which a probability density is built from the Fourier spectrum of 13-year low-pass filtered and detrended white noise time series

The observed IPO pattern is characterized by a statistically significant warming in the tropical Pacific, with an ENSO-like shape, extending to the extratropics along the western coasts of both North and South America (Fig. 5.1b). It also presents two cold tongues of SSTA in mid-latitudes, poleward of around 20°, expanding eastward from the coasts of Asia and Oceania, respectively, and being more prominent in the Northern Hemisphere. In the rest of the global SST there are no remarkable anomalies, except for a significant anomalous warming over the Indian Ocean and cooling along the North American east coast and north of Europe.

The CMIP5 simulations show IPO patterns that are highly consistent with the observed one (Fig. 5.3). In both the historical and piControl simulations, the model-mean SST patterns present strong statistical significance and consistency among models. They show intense warm SSTA throughout the tropical Pacific and poleward along the eastern edge of the basin. In the extratropical Pacific two cold tongues expand from the west, with more intensity in the northern hemisphere. Away from

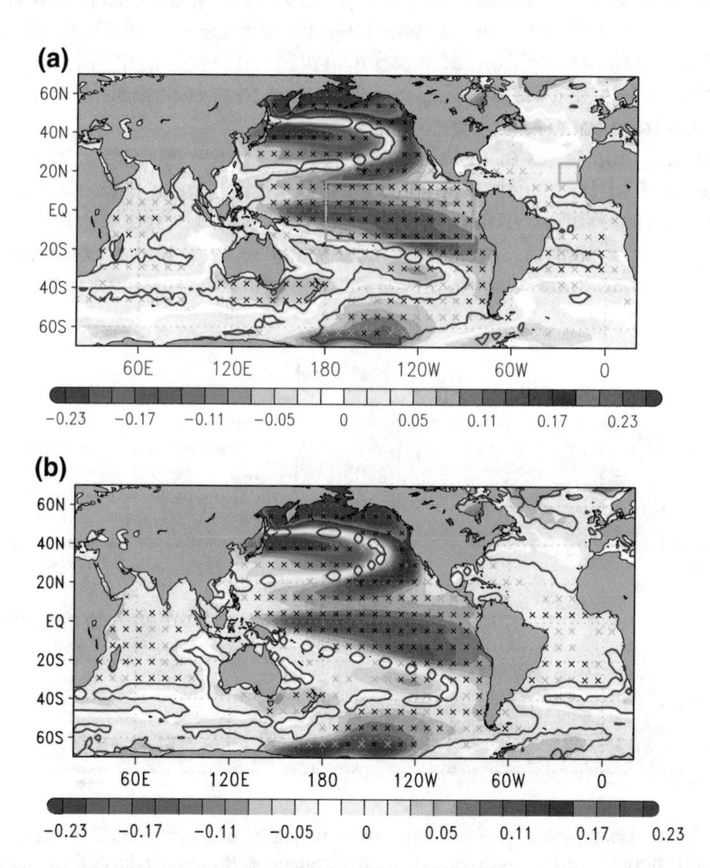

Fig. 5.3 Regression onto the IPO index of the unfiltered SSTA (K per standard deviation) averaged among the 17 CMIP5 models in the historical (**a**) and the piControl (**b**) simulations. Black and gray crosses indicate points where the sign of the regression coefficient coincides in at least 15 and 13 out of the 17 models analyzed, respectively. Contours indicate the regions where the averaged regression is significant at the 5% level from a "random-phase" test. Green boxes delimit the Tropical Pacific region and an area of the Tropical Atlantic. Adapted from Villamayor et al. (2018a)

the Pacific basin the anomalies are also less intense, such as a weak but significant widespread warming of the Indian Ocean surface that is consistent with observations.

Although the anomalies distribution of the IPO pattern reproduced by the models in both simulations is very similar to the observed one, their amplitudes are in general underestimated by the models. Particularly, in the tropical Pacific the observational IPO pattern shows a warming of up to 0.23 K per standard deviation (Fig. 5.1b), while the models on average reproduce a maximum warming of about 0.15 K per standard deviation (Fig. 5.3). Similarly, the anomalous cooling in the Pacific extratropics and the warming of the Ocean Indian are lower than in observations. In the Atlantic basin, the models do not reproduce the observed significant cooling located to the north. Instead, the models reproduce a weak but statistically significant positive warming in the tropical sector associated with the IPO. Regarding the differences between the results obtained from both the forced and the unforced simulations, the model-mean IPO pattern of the piControl experiment is found to be slightly more consistent among the models than the historical one. However, there is little difference between the patterns from both sets of experiments.

As for the time series of the simulated IPO, a Fourier analysis is computed to find whether CMIP5 models reproduce an IPO with characteristic periodicities of oscillation that are similar to the observed ones (Fig. 5.4). The model-mean spectrum resulting from the analysis reveals that models tend to show higher power spectra in two frequency bands, one close to 50–70 years and the other one close to 15–25 years

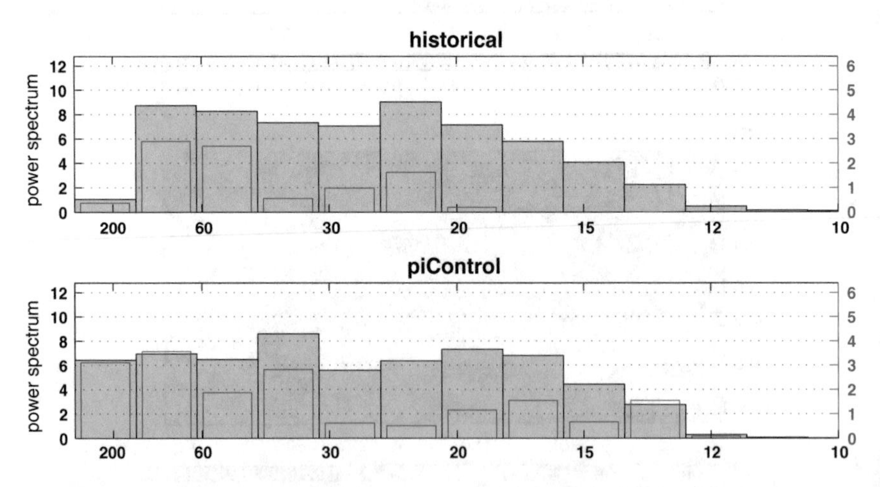

Fig. 5.4 Model-mean power spectra of the Fourier transform of the detrended IPO indices (blue bars) of the historical (top) and piControl (bottom) simulations. Red bars indicate the power spectra averaged over the 17 models but taking into account only 95% significant peaks for each of them. The statistical significance for each model's IPO is obtained following a non-parametric hypothesis test in which a probability density function is built from the Fourier spectrum of 13-year low-pass filtered and detrended white noise time series with the same time length as the simulation and averaged across the number of ensemble members in cases where more than one realization is available

in both the historical and piControl simulations. This suggests that the power spectra in models tends to cluster around two preferred bands, of approximately 15–25 and 50–70 years, which is roughly consistent with observations.

5.2 Sahel Rainfall Response

The observed JAS rainfall anomalies over West Africa associated with the IPO are estimated as the regression pattern onto the IPO index in Fig. 5.5. The result reveals that a weakening of rainfall almost throughout the Sahel region occurs during positive IPO phases. Although the anomalies do not show strong statistical significance, the three data bases analyzed are consistent in most of the Sahel region, which provides confidence to the observational result. Only in the easternmost part of the Sahel, from 5° to 15°W, the anomalies are weak and uncertain among the different data bases. To the east, there are intense negative anomalies over eastern Africa. South of the Sahel, there is enhanced precipitation on the coastal area of Guinea, Sierra Leone and Liberia, as well as along the Gulf of Guinea coast, more intense over Nigeria and Cameroon. Hence, it can be suggested that related to positive IPO phases there is general precipitation decrease in West Africa, with some local exceptions. While rainfall is favored under negative IPO conditions.

Regarding the CMIP5 simulations, associated with a positive IPO, the models simulate a significant pattern of negative precipitation anomalies across the Sahel (Fig. 5.6), in agreement with observations (Fig. 5.5). They also reproduce strong drought conditions over eastern Africa and a local maximum of positive rainfall anomalies in Central Africa, close to the Gulf of Guinea, according to observations. Nevertheless, the observed positive anomalies of precipitation on the coastal region of West Africa are not reproduced by the simulations. In agreement with the reproduced IPO patterns of SSTA, models underestimate the intensity of the precipitation

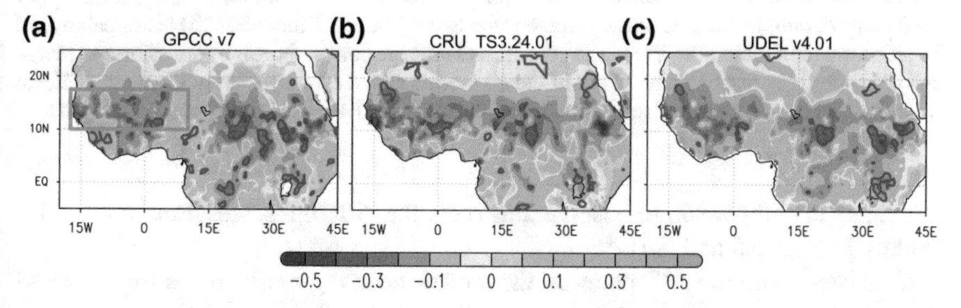

Fig. 5.5 Regression map of the unfiltered JAS precipitation anomaly from the **a** GPCC v7, **b** CRU TS3.24.01 and **c** UDEL v4.01 observational data bases onto the standardized IPO index (units are mm day^{-1} per standard deviation). Contours indicate the regions where the regression is significant at the 5% level from a "random-phase" test. Orange box delimits the Sahel region

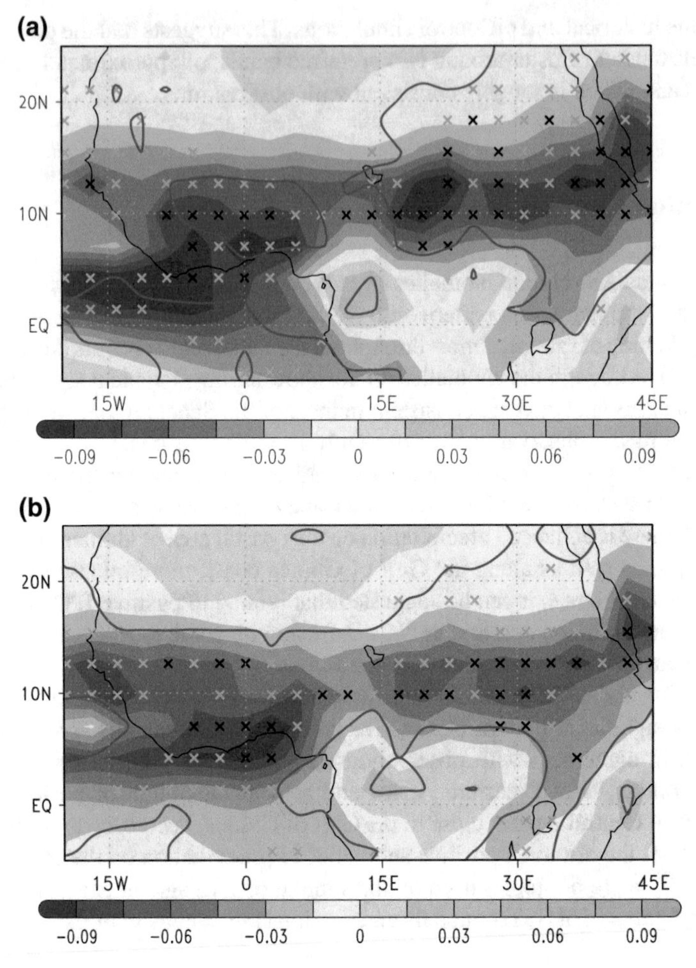

Fig. 5.6 Regression maps of the unfiltered JAS precipitation anomalies (units are mm day^{-1} per standard deviation) on the IPO index averaged among the 17 CMIP5 models in the historical (**a**) and the piControl (**b**) simulations. Black and gray crosses indicate points where the regression coefficient sign coincides in at least 15 and 13 out of the 17 models analyzed, respectively. Contours indicate the regions where the averaged regression is significant at the 5% level from a "random-phase" test

anomalies in comparison to observations (note the different scales used in the color shading in Figs. 5.5 and 5.6).

Consistent with the SST patterns, the model-mean regression maps from the historical experiment show slightly lower statistical significance and less consistency among the models in the precipitation response to the IPO than the piControl simulation. However, the close resemblance between the results from both simulations suggests very low influence of externally forced variability in the recurrent IPO pattern of SST and in its relationship with the Sahel rainfall.

5.2.1 Inter-model Analysis

It has been shown that the CMIP5 models on average reproduce lower rainfall anomalies associated with the IPO than in observations, as well as the pattern of SST. Therefore, a linear relationship between the characteristic IPO pattern and its reproduced rainfall response is sought. According to other studies, Sahel rainfall changes are related to the tropical SST of the Pacific Ocean (Janicot et al. 2001; Joly and Voldoire 2009; Mohino et al. 2011b; Rodríguez-Fonseca et al. 2016). With this in mind, the linear relationship between the anomalous Sahel rainfall intensity and the SSTA averaged in the tropical Pacific, between 15°S–15°N and 180°–95°W (delimited area Fig. 5.1b), associated with the IPO that CMIP5 models reproduce is calculated (Fig. 5.7).

The larger rainfall response obtained for the observations suggest a stronger relationship in the real world. Although the different observational sources show important uncertainty, with the GPCCv7 and CRU TS3.24.01 showing an averaged regression coefficient of −0.09 and −0.14 mm day^{-1} per standard deviation of the IPO index, respectively, and the UDEL v4.01 in between both. Considering this range, the MIROC5 model in the historical experiment, MIROC-ESM-CHEM in both simulations and MPI-ESM-LR in piControl (numbers 13, 14 and 15 in Fig. 5.7, respectively) reproduce rainfall values within it, coinciding with similar tropical Pacific SSTA values to the observed one. The scatter plot in Fig. 5.7 shows strong

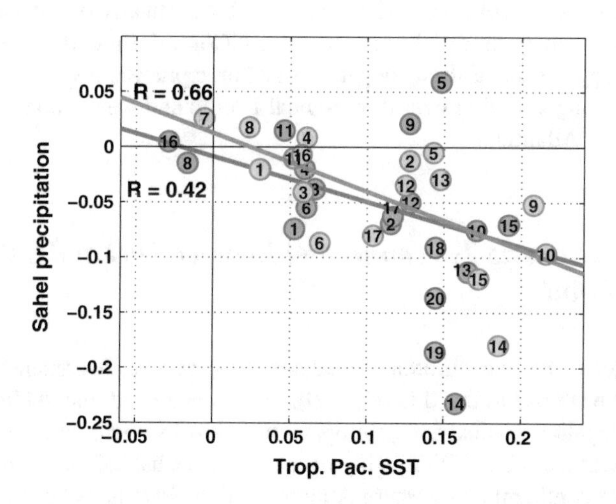

Fig. 5.7 **a** Scatterplot of the regression coefficient of precipitation anomaly over the Sahel (delimited region in Fig. 5.5a) and the SSTA of the tropical Pacific (delimited region in Fig. 5.1b) relative to the IPO of each model in the historical (green) and the piControl (orange) simulations (Figs. C.1, C.2, C.3, and C.4 in Appendix C). The lines indicate the linear regression fitting of the corresponding colored points (R is the correlation coefficient). The numbers from 1 to 17 identify each model individually with the given number in Table 2.1. Numbers 18, 19 and 20 correspond to CRU TS3.24.01, GPCC v7 and UDEL v4.01 observed data (blue), respectively. Units for the horizontal and vertical axes are mm day^{-1} per standard deviation and K per standard deviation), respectively

linear correlation between the Sahel rainfall and the tropical Pacific SSTA associated with the IPO reproduced by the CMIP5 models in both the historical and piControl simulations (with correlation coefficients of $R = 0.42$ and $R = 0.66$ which are statistically significant at the 90% and 95% confidence level, respectively, according to a Student t-test). Hence, the stronger the SSTA warming over the tropical Pacific, the stronger the IPO impact on Sahel drought.

There are, however, a few models (4 and 3 out of 17 models in the historical and piControl experiments, respectively) that produce a positive impact of the IPO on Sahel rainfall, though most of these impacts are not statistically significant. These models tend to show a poorly defined IPO spatial pattern in the Pacific basin with weak positive or even negative SSTA over the Tropical Pacific (inmcm4 and MRI-CGCM3 for the historical simulation and GISS-E2-H and GISS-E2-R for the piControl one; Figs. C.1 and C.2). However, some models with a weak or even positive Sahel rainfall response to IPO show warm SSTA over the Tropical Pacific (for instance the CSIRO-Mk3-6-0 model in the historical simulation; Fig. C.1). The positive SSTA over the northeastern Tropical Atlantic basin, with comparable magnitude to the ones over the Pacific warm tongue could partly explain this behaviour. Such northeastern Tropical Atlantic SSTA has been shown to promote increased rainfall over Sahel (Cook and Vizy 2006; Giannini et al. 2013) and could be overriding the Tropical Pacific influence in some cases. When both regions, Tropical Pacific and northeastern Tropical Atlantic (delimited area in Fig. 5.3a, between 25°–15°N and 30°–18°W), are used to calculate the scatter plot (Fig. 5.8), the linear relationship with the Sahel rainfall increases (with correlation coefficients of $R = 0.52$ and $R = 0.69$, statistically significant at the 95% confidence level, in the case of the historical and piControl simulations, respectively). This result suggests that a strong negative Sahel precipitation response to the IPO is linked to a strong warming over the Tropical Pacific and weak anomalies over the Tropical North Atlantic.

5.2.2 Atmospheric Teleconnection Between IPO and Sahel Rainfall

In observations, the two different reanalysis show certain discrepancies as to the atmospheric response to the IPO (Fig. 5.9). On the one hand, the 20CRV2c reanalysis shows negative tropical and subtropical high-level velocity potential anomalies associated with a positive IPO roughly over the eastern half of the Pacific, spanning the American continent and western Atlantic, and positive ones outside (Fig. 5.9a). This feature indicates high-level anomalous wind divergence over the characteristic warm tongue of the SST IPO pattern and convergence around velocity potential maximums, such as one located over the western coast of the Sahel and another on eastern Africa. At low levels (Fig. 5.9b), there are significant positive velocity potential anomalies over tropical Pacific and divergence over the Maritime Continent. On the rest of the tropical band, the pattern becomes uncertain and difficult to interpret. In

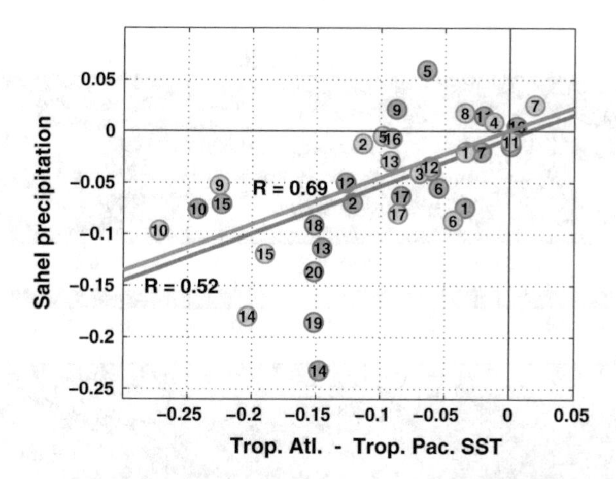

Fig. 5.8 Scatterplot of the regression coefficient of precipitation anomaly over the Sahel (delimited region in Fig. 5.5a) and the SSTA difference between the Tropical Atlantic coastal region next to West Africa (see box over the Tropical Atlantic in Fig. 5.3a) and the SSTA of the tropical Pacific (see box over the Tropical Pacific in Fig. 5.3a) relative to the IPO of each model in the historical (green) and the piControl (orange) simulations (Figs. C.1, C.2, C.3, and C.4 in Appendix C). The lines indicate the linear regression fitting of the corresponding colored points (R is the correlation coefficient). The numbers from 1 to 17 identify each model individually with the given number in Table 2.1. Numbers 18, 19 and 20 correspond to CRU TS3.24.01, GPCC v7 and UDEL v4.01 observed data (blue), respectively. Units for the horizontal and vertical axes are mm day^{-1} per standard deviation and K per standard deviation), respectively

particular over West Africa, although no statistical significance is obtained, low-level divergence is weakly suggested over the central part and anomalous easterlies near the Atlantic coast are shown, which are associated with Sahel drought (Grodsky et al. 2003; Pu and Cook 2010; Nicholson 2013). In association with the pattern at high levels, it can be suggested that the tropical Pacific warm tongue induces enhanced convection which results in anomalous subsidence on the rest of the tropical band.

On the other hand, the patterns of velocity potential and wind direction associated with the IPO from the ERA-20C reanalysis (Fig. 5.9c, d) show two centers of anomalous high-level divergence and low-level convergence deep convection: one over the west and central tropical Pacific and another spanning the African continent. The anomalies are the opposite over northern Indian Ocean and an area that covers the easternmost part of the tropical Pacific, Central America and western Atlantic. This depicts a situation that favors convection over the central Pacific and Africa and subsidence over Central America and the north Indian Ocean. Particularly, focusing on West Africa, the anomalous convective activity suggested by the ERA-20C reanalysis is opposite to the result obtained from 20CRV2c. Therefore, it is difficult to draw any conclusions about the atmospheric teleconnection between the IPO and the Sahel JAS precipitation based on the observational result.

CMIP5 models, on average, reproduce anomalous high-level divergence over central tropical Pacific and convergence over western tropical Atlantic and West Africa,

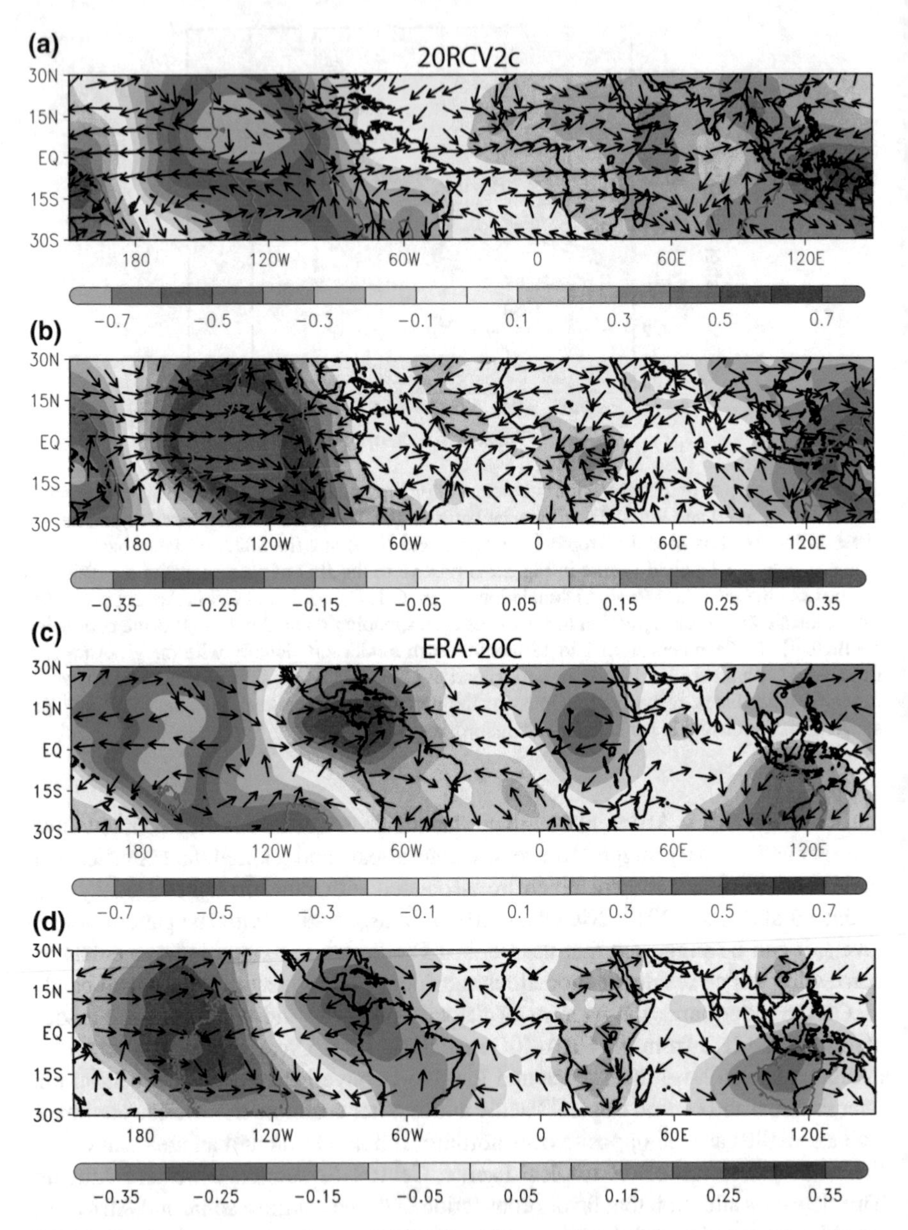

Fig. 5.9 Regression onto the observed IPO index of the unfiltered JAS anomaly of the velocity potential (shaded) (10^6 m^2/s per standard deviation) and the wind direction (vectors) from the 20CRV2c reanalysis at 200 (**a**) and 850 hPa (**b**). **c**, **d** are the same as (**a**) and (**b**), respectively, but for the ERA-20C reanalysis. Contours indicate the regions where the velocity potential regression coefficients are significant at the 5% level

with a maximum on the Indian Ocean in both the historical and piControl simulations (Fig. 5.10). Consistently, at low levels there is anomalous wind convergence and divergence over these regions, respectively, with evident easterly winds in the Sahel. Wile over West Africa the observational result is highly uncertain, due to the discrepancies between reanalyses, the model-mean simulations show an atmospheric response to the IPO with high statistical significance that supports the mechanism suggested by the 20CRV2c reanalysis. Although the amplitude of the velocity potential anomalies is underestimated with respect to observations, consistently with the other variables (note the difference between the color shade scales used in Figs. 5.9 and 5.10).

These results hence suggest an anomalous Walker-type overturning cell that connects upward movements over the central Pacific in response to local warm SSTA there with subsidence over West Africa. This, in turn, reduces convection over the area and the low-level monsoon westerlies, leading to drought conditions over the Sahel. Such mechanism was also proposed for the impact of ENSO events on Sahel rainfall (Janicot et al. 2001; Joly and Voldoire 2009; Mohino et al. 2011b). The similarity between the teleconnection mechanisms at different time scales is consistent with the similar SST patterns over the Tropical Pacific observed for the ENSO and IPO (Deser et al. 2004). This mechanism is shown by the 20CRV2c reanalysis and supported by the CMIP5 simulations, with robust agreement among models, but not by the ERA-20C reanalysis. As a consequence, the atmospheric response to the IPO that the ERA-20C reanalysis reproduces is arguable. Furthermore, the anomalous convective activity associated with the IPO that the ERA-20C reanalysis shows over West Africa is inconsistent with the observed reduced precipitation. Therefore, the ERA-20C reanalysis is considered to be unreliable in terms of the atmospheric response to the IPO pattern related with the Sahel rainfall during JAS.

5.3 Amazon and Northeast Rainfall Response

The observed precipitation anomalies related to the IPO during DJFMAM show significant deficit in the Amazonia and Northeast regions (Fig. 5.11). In the Amazonia, the rainfall anomalies are mostly spread across the entire area, being more intense and statistically significant over the western side. In the Northeast of Brazil the stronger negative anomalies are distributed along the coastal part of the region, especially at the northwest, and decrease in magnitude inland. Although the Northeast precipitation anomalies show low statistical significance, the agreement among the three observational data bases analyzed supports the robustness of this result. The described relationship between precipitation and the IPO coincides with the impact on rainfall in both the Amazonia and Northeast regions produced by the SSTA pattern of ENSO (e.g., Ambrizzi et al. 2004), which is similar to the tropical Pacific component of the IPO and agrees with other works that suggest a similar connection at decadal-to-multidecadal timescales (Dettinger et al. 2001). In addition to this, the observations also show significant positive precipitation anomalies associated with

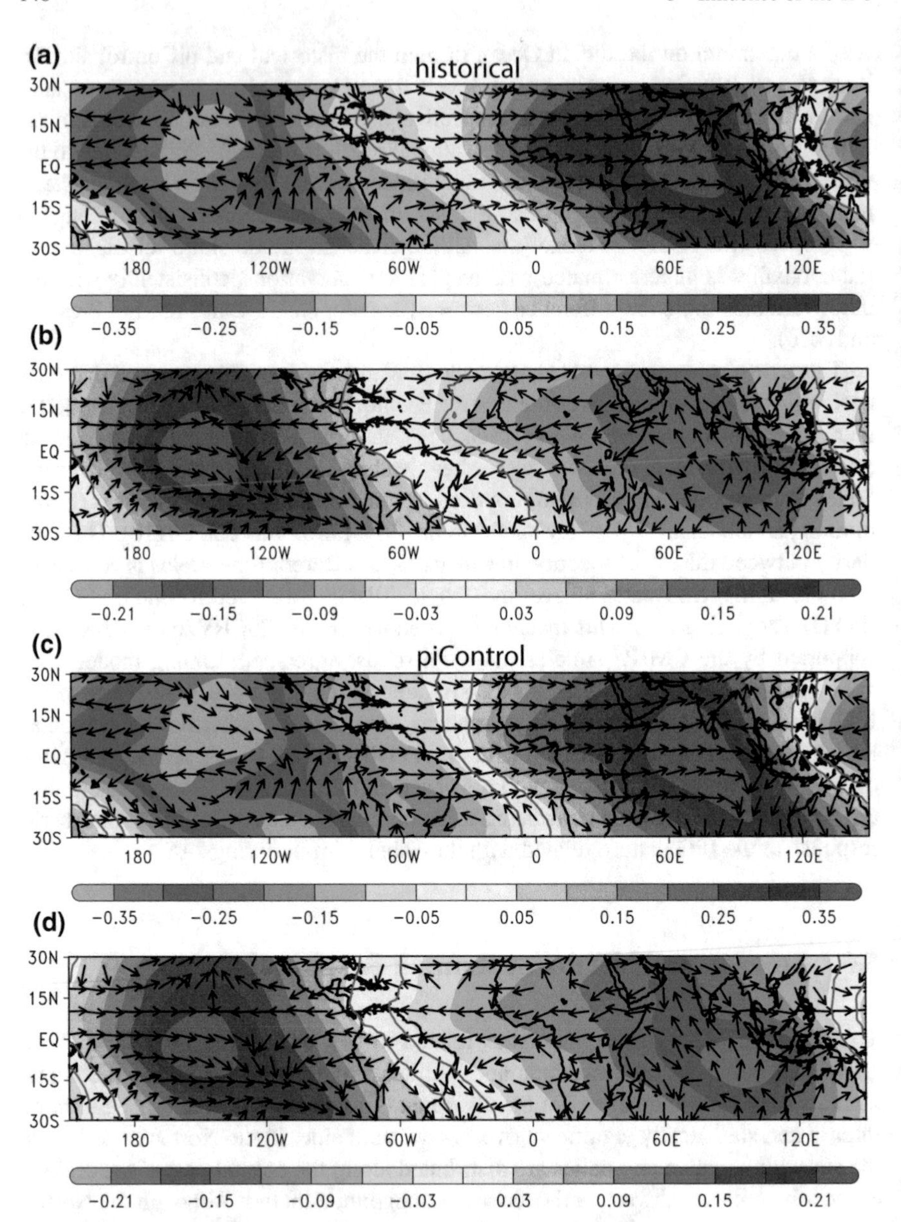

Fig. 5.10 Regression onto the simulated IPO index of the unfiltered JAS anomaly of the velocity potential (shaded) (10^6 m^2/s per standard deviation) and the wind direction (vectors) at 200 (**a**) and 850 hPa (**b**) from the historical simulation averaged among the 17 CMIP5 models. **c**, **d** are the same as (**a**) and (**b**), respectively, but for the piControl simulation. Contours indicate the regions where the velocity potential regression coefficients are significant at the 5% level

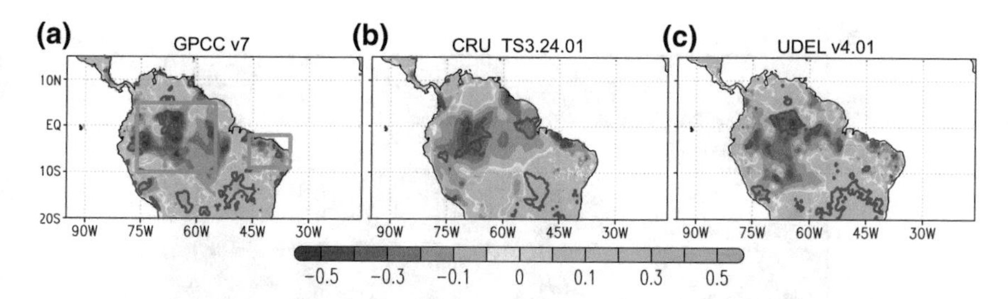

Fig. 5.11 Regression map of the unfiltered DJFMAM precipitation anomaly from the **a** GPCC v7, **b** CRU TS3.24.01 and **c** UDEL v4.01 observational data bases onto the standardized IPO index (units are mm day^{-1} per standard deviation). Contours indicate the regions where the regression is significant at the 5% level from a "random-phase" test. Orange boxes delimit the Amazon and the Northeast regions

the IPO over central Brazil as well as in the north and south of the Amazon Basin: over Venezuela and Bolivia, respectively.

Both historical and piControl experiments present an impact of IPO on rainfall in both the Amazonia and Northeast regions during DJFMAM similar to the observed one, though underestimated in intensity (Fig. 5.12). In the Amazon basin the model-mean rainfall response is negative across most of the region. However, in contrast to the observations, the anomalies are more intense and consistent among the models to the east in both the historical and piControl experiments. In the historical experiment, the precipitation anomalies over the Amazon region associated with IPO are lower than in piControl, especially in the southwestern part where the anomalies are not significant only in the historical simulation. In the Northeast, the models reproduce on average a significant decrease of rainfall across the region with high agreement among the models. In both historical and piControl experiments the precipitation anomalies are similarly distributed, being stronger in the western half of the Northeast region. Nevertheless, these anomalies are more intensely reproduced by the historical simulations which also show stronger negative anomalies over the western edge of the region than the piControl experiment.

Out of northern Brazil, the models reproduce positive rainfall anomalies south of the Amazonia and in central Brazil, as in observations. However the rainfall response given by the models in these areas is much weaker than the observed one, indicating certain disagreement among models. North of the Amazonia region, over Venezuela they reproduce negative anomalies (contrary to observations) and show very robust negative signal over southern Central America, in Costa Rica and Panama, in agreement with observations.

In both the historical and piControl experiments, there is a strong intensification of precipitation along the tropical Pacific between 0° and 10°S, which affects the western slope of the Andes. In contrast, over the Atlantic there is a weakened tropical rain-belt around 5°S that expands westward inland. This suggests that the CMIP5 models reproduce a strengthening of the ITCZ over the warm tropical Pacific and a weakening over the Atlantic and northern Brazil during DJFMAM associated with the IPO pattern.

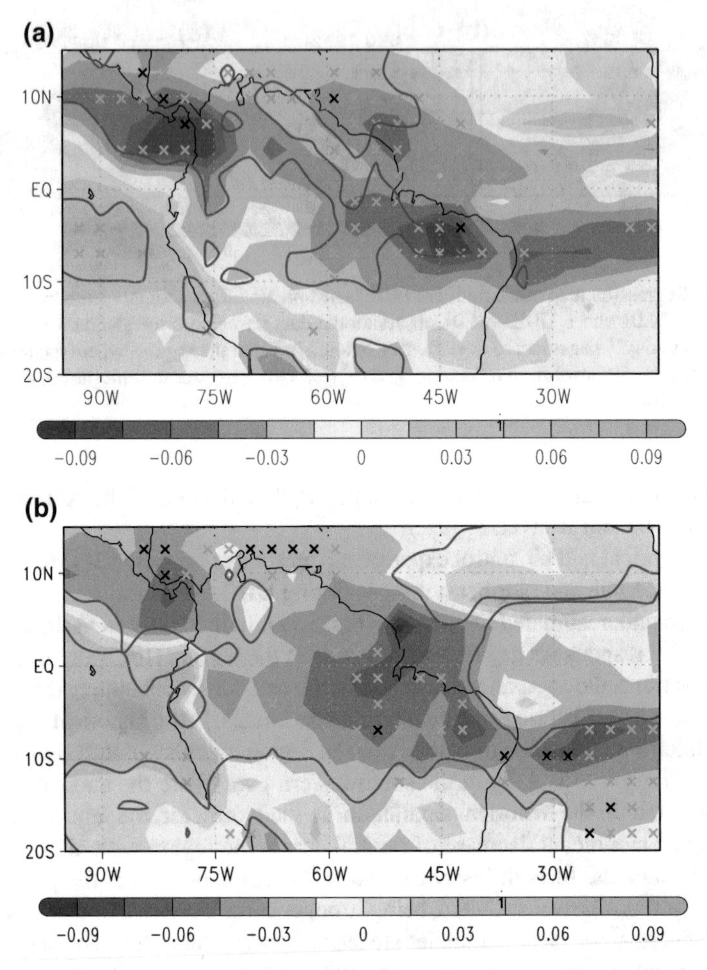

Fig. 5.12 Regression maps of the unfiltered DJFMAM precipitation anomalies (units are mm day^{-1} per standard deviation) on the IPO index averaged among the 17 CMIP5 models in the historical (**a**) and the piControl (**b**) simulations. Black and gray crosses indicate points where the regression coefficient sign coincides in at least 15 and 13 out of the 17 models analyzed, respectively. Contours indicate the regions where the averaged regression is significant at the 5% level from a "random-phase" test. Adapted from Villamayor et al. (2018a)

5.3.1 Inter-model Analysis

Despite the robustness of the IPO SSTA pattern reproduced across all the models (Fig. 5.3), the simulated precipitation response over the Amazonia and Northeast regions is less consistent among the models (Fig. 5.12). Regarding the models individually, roughly half of them (CanESM2, CSIRO-Mk3-6-0, HadGEM2-ES, IPSL-CM5A-LR, MIROC5, MIROC-ESM-CHEM, MPI-ESM-LR) broadly reproduce negative anomalies along the tropical Atlantic sector and inland similarly in both experiments (Figs. C.5 and C.6). The precipitation patterns linked to the IPO of these models roughly show a weakened tropical rain-belt over the north of South America, suggesting an anomalous weakening of the convective rainfall associated with the ITCZ as in the model-mean pattern. However, other models reproduce precipitation patterns that are noisy or present weak anomalies (CNRM-CM5, FGOALS-g2, GISS-E2-H, GISS-E2-R), others display opposite rainfall response to the observed one in some of the two regions of northern Brazil (inmcm4, MRI-CGCM3, CCSM4, NorESM1-M) and there are only two that do not show a consistent rainfall response in the two different experiments (bcc-csm1-1 and HadGEM2-CC).

The differences among the models in the simulated impact of the IPO on the Amazonia and Northeast rainfall can be attributed to the accuracy with which they reproduce the IPO pattern with respect to the observed one. Particularly, focusing on the link between the precipitation response to the IPO and the characteristic tropical Pacific component of the SSTA pattern, it is found that there is a linear relationship (Fig. 5.13). In the Amazonia region, such a relationship is weaker in the historical than in the piControl experiment (linearly correlated with $R = -0.41$ and $R = -0.64$, respectively) (Fig. 5.13a). This is consistent with the fact that the robustness of the SSTA pattern among the models is slightly weaker in the forced than in the unforced simulations (Fig. 5.3). In case of the Northeast, the relationship between precipitation and the IPO pattern is similar in both the historical and piControl experiments (linearly correlated with $R = -0.41$ and $R = -0.40$, respectively) (Fig. 5.13b). The linear fit between the rainfall response to the IPO and the tropical Pacific component of the SSTA pattern in all cases is not highly significant (the correlation coefficients with an absolute value of $R = 0.40$ are barely significant with a 90% confidence level, according to a Student t-test). But it has to be considered that the IPO pattern shows significant loads of SSTA away from the Pacific and hence there are other domains that may also contribute to influence the connection with rainfall.

Therefore, this result suggests that the accuracy with which the models reproduce the precipitation response to the IPO in both the Amazonia and the Northeast regions can be partly related to the magnitude of the SSTA pattern, in particular to its tropical Pacific component. So, the higher the temperature in the tropical Pacific, the lower the precipitation anomalies in both regions and vice versa. In addition, no remarkable discrepancies between the forced and unforced simulations are found.

5.3.2 Atmospheric Teleconnection Between IPO and Northern Brazil Rainfall

The observational patterns of surface pressure and low-level wind direction during DJFMAM associated with the IPO (Fig. 5.14a, c) suggest a zonal and tropical atmospheric mechanism connecting the IPO to rainfall anomalies. Such a mechanism consists of a weakening of pressure and convergent winds at 850 hPa over the Pacific and increased surface pressure across the rest of the tropical regions, spanning the Atlantic sector and eastern South America, consistently shown by both

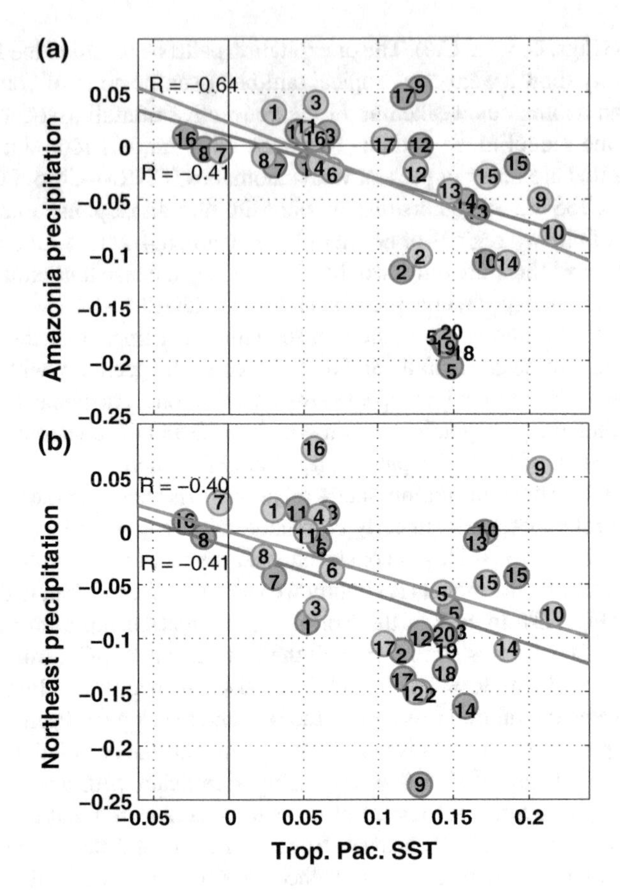

Fig. 5.13 **a** Scatter plot of the regression coefficient of precipitation anomaly over the Amazonia (western delimited region in Fig. 5.11a) and the SSTA of the tropical Pacific (delimited region in Fig. 5.1b) relative to the IPO of each model in the historical (green) and the piControl (orange) simulations (Figures C.1, C.2, C.5, and C.6 in Appendix C). The lines indicate the linear regression fitting of the corresponding colored points (R is the correlation coefficient). The numbers from 1 to 17 identify each model individually with the given number in Table 2.1. **b** Same as **a** but using the Northeast region (eastern delimited region in Fig. 5.11a) instead of the Amazonia. Numbers 18, 19 and 20 correspond to CRU TS3.24.01, GPCC v7 and UDEL v4.01 observed data (blue), respectively. Units for the horizontal and vertical axes are mm day^{-1} per standard deviation and K per standard deviation), respectively. From Villamayor et al. (2018a)

reanalyses. The surface pressure anomalies during DJFMAM suggest an anomalous Walker circulation in DJFMAM, as during JAS. This cell of anomalous circulation has associated increased ascending motion over the warm tropical Pacific, which reduces local surface pressure, and subsidence over the tropical Atlantic and part of northern South America, consistent with the enhanced atmospheric pressure. Such mechanism is similar to the teleconnection between the ENSO and the Amazonia and Northeast, which features subsidence over South America and induces rainfall

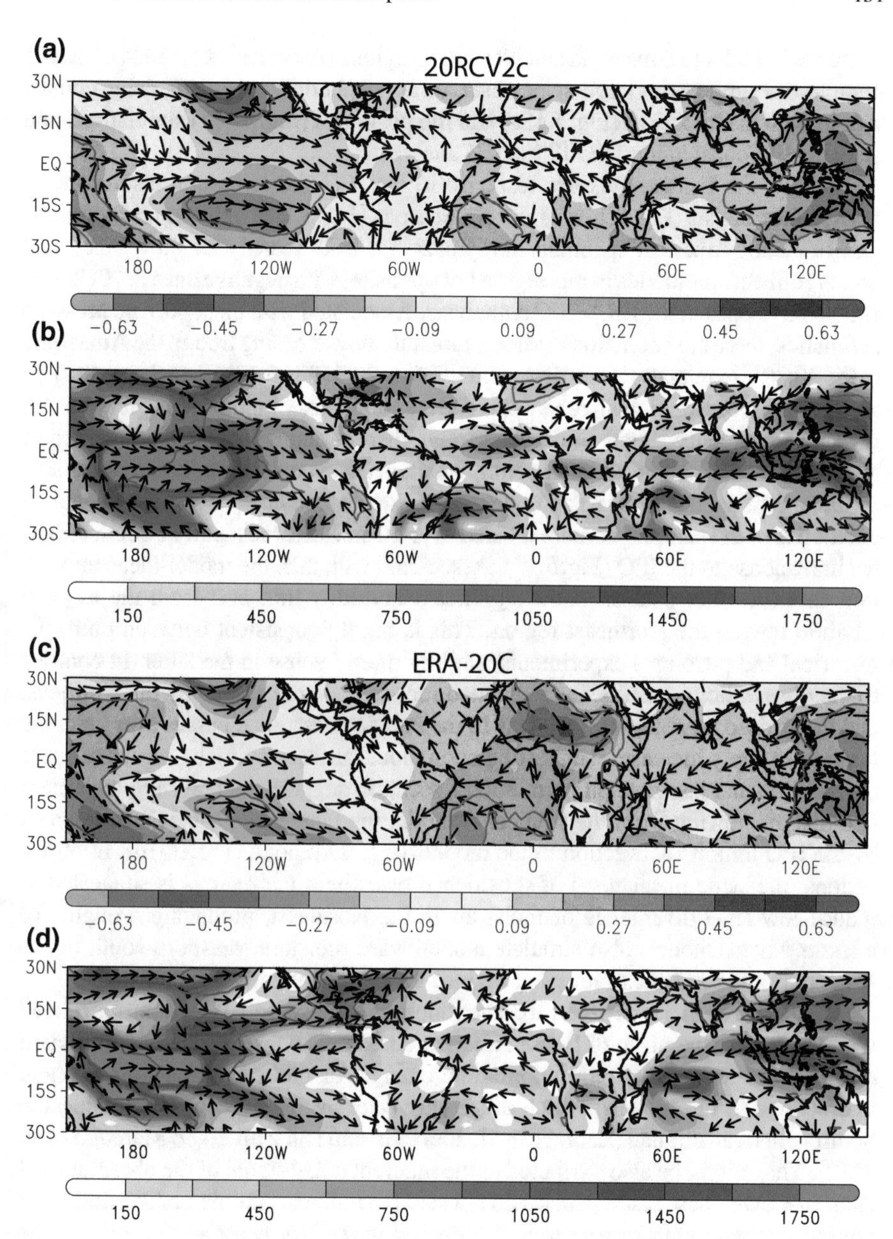

Fig. 5.14 Regression onto the observed IPO index of the unfiltered DJFMAM anomaly of **a** the surface pressure (shaded) (hPa per standard deviation) and the wind direction at 850 hPa (vectors) and **b** the magnitude (shaded) and direction (vectors) of the moisture flux (kg m^{-1} day^{-1} per standard deviation) from the 20CRV2c reanalysis. **c** and **d** are the same as **a** and **b**, respectively, but for the ERA-20C reanalysis. Contours indicate the regions where the surface pressure or the moisture flux magnitude regression coefficients are significant at the 5% level

decrease in both the Amazonia and Northeast regions (Ambrizzi et al. 2004). CMIP5 simulations support this large-scale observed atmospheric mechanism, being consistent in both the historical (Fig. 5.15a) and the piControl (Fig. 5.15c) simulations and statistically robust among the 17 models analyzed.

Locally, the surface pressure response to the IPO shows certain discrepancies between reanalysis (Fig. 5.14a, c) and simulations (Fig. 5.15a, c) in the northernmost part of South America, spanning the Amazonia. Both reanalyses show weak and non-significant anomalies in this region but opposite, with negative ones in 20CRV2c and positive ones in the ERA-20C reanalysis. Associated with these surface pressure anomalies, there are anomalous easterly moisture flow crossing trough the Amazonia in the 20CRV2c reanalysis and divergent in ERA-20C (Fig. 5.14b, d, respectively). In the Northeast of Brazil, both reanalyses show anomalously high surface pressure and moisture divergence. South of the Amazonia, there is anomalous moisture transport toward central Brazil and further south, which is also consistent in both reanalyses.

On the other hand, CMIP5 models reproduce a center of significant negative anomalies of the surface pressure located over the northernmost part of South America in response to the IPO (Fig. 5.15). Associated with this, the model-mean anomalous moisture flux patterns show significant humidity transport from the tropical Atlantic toward the Northeast region. This is highly consistent between both, the historical and piControl experiments, slightly more intense in the latter. In contrast to the observational results, the simulated moisture supply passes by the Northeast and the Amazonia regions all the way to the Pacific coast. As it passes over land, the humidity supply increases, suggesting anomalous surface drying by means of more evaporation and less precipitation. Such a scenario suggests that, even though there is moist air flowing from the tropical Atlantic, the large-scale subsidence induced by the IPO inhibits convection in the models (e.g., Drumond et al. 2010). In observations, the same mechanism of subsidence over the same regions is suggested to induce low-level divergence of moist air in the Northeast, producing drought. To a lesser extent, models also simulate a southward moisture transport, south of the Amazonia and toward extratropical regions in agreement with observations.

These discrepancies between observed and simulated regional mechanisms related with the IPO in the north of South America could be a consequence of insufficient resolution in areas of strong topographic change, or unresolved land-atmosphere interactions in the models, which are highly relevant features in determining the South American climate (Labraga et al. 2000; Grimm and Zilli 2009; Marengo et al. 2012). They might be also attributed to the inherent noisy signal of the observational data. However, these discrepancies do not seem to influence the sign of the simulated rainfall response with respect to that observed in the Northeast and in most of the Amazon region. But they are likely associated with the lack of precipitation anomalies in southwestern Amazon, with respect to observations, and could be highly relevant to resolve the climate variability of other extratropical regions that are also determined by the SAM system (Marengo et al. 2012).

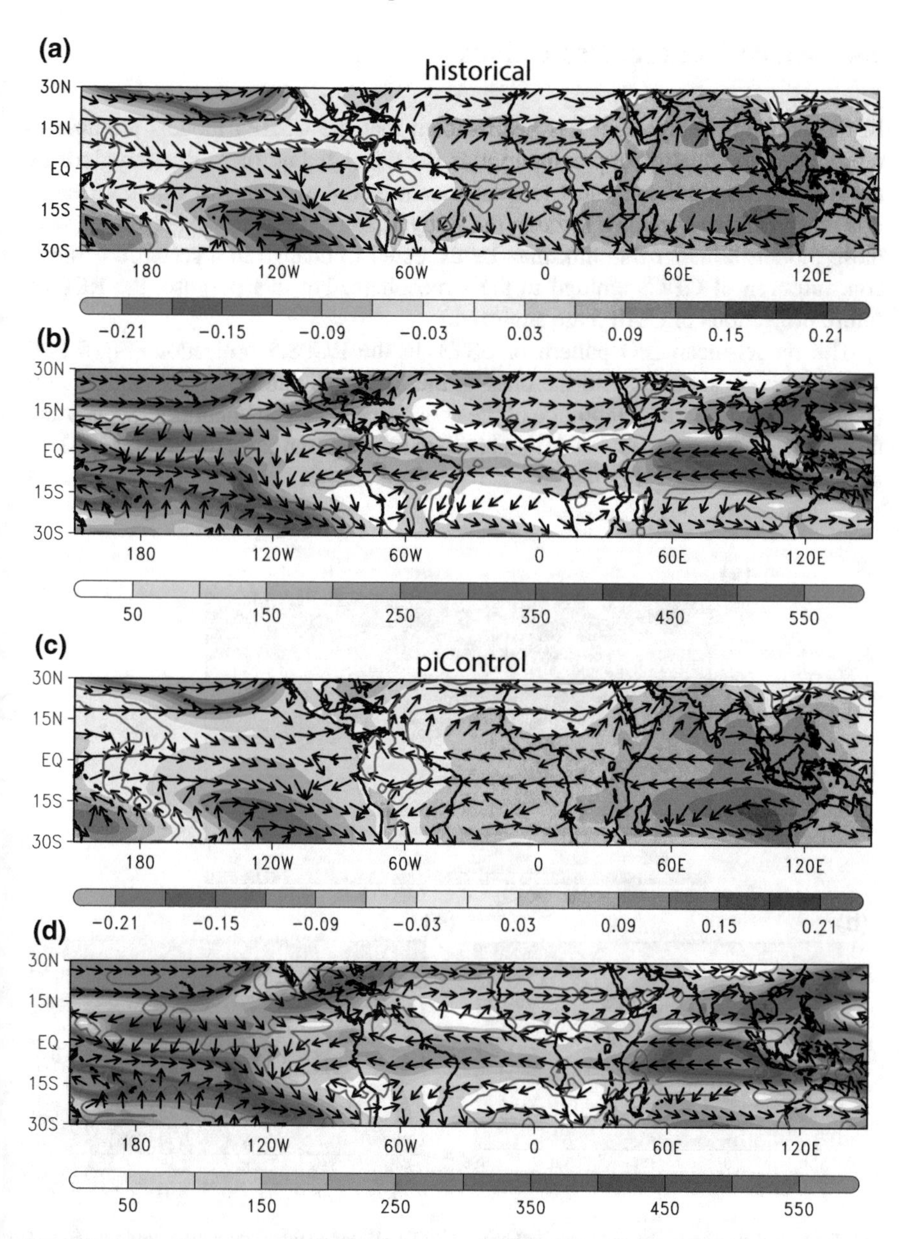

Fig. 5.15 Regression onto the observed IPO index of the unfiltered DJFMAM anomaly of **a** the surface pressure (shaded) (hPa per standard deviation) and the wind direction at 850 hPa (vectors) and **b** the magnitude (shaded) and direction (vectors) of the moisture flux (kg m^{-1} day^{-1} per standard deviation) from the historical simulation averaged among the 17 models. **c** and **d** are the same as **a** and **b**, respectively, but for the piControl simulation. Contours indicate the regions where the surface pressure or the moisture flux magnitude regression coefficients are significant at the 5% level. Adapted from Villamayor et al. (2018a)

5.4 RCP8.5 Future Projections

In the previous analyses, it has been shown that the CMIP5 models are able to reproduce a modulation of precipitation in the Sahel and northern Brazil regions by the IPO robustly and congruent to what was observed over the last century. Hence, it is reasonable to speculate through the study of future projections performed with these models whether this link may be expected to change in a scenario of large concentration of GHGs emitted to the atmosphere. For this purpose, the RCP8.5 future projections of CMIP5 are analyzed.

The model-mean IPO pattern of SSTA in the RCP8.5 projection (Fig. 5.16a) shows high consistency with the ones in the historical and piControl experiments (Fig. 5.3). However, the amplitude of the anomalies, their statistical significance and the consistency among models is notably lower throughout the pattern. This is likely explained by the fact that the RCP8.5 runs are shorter than the others (typically

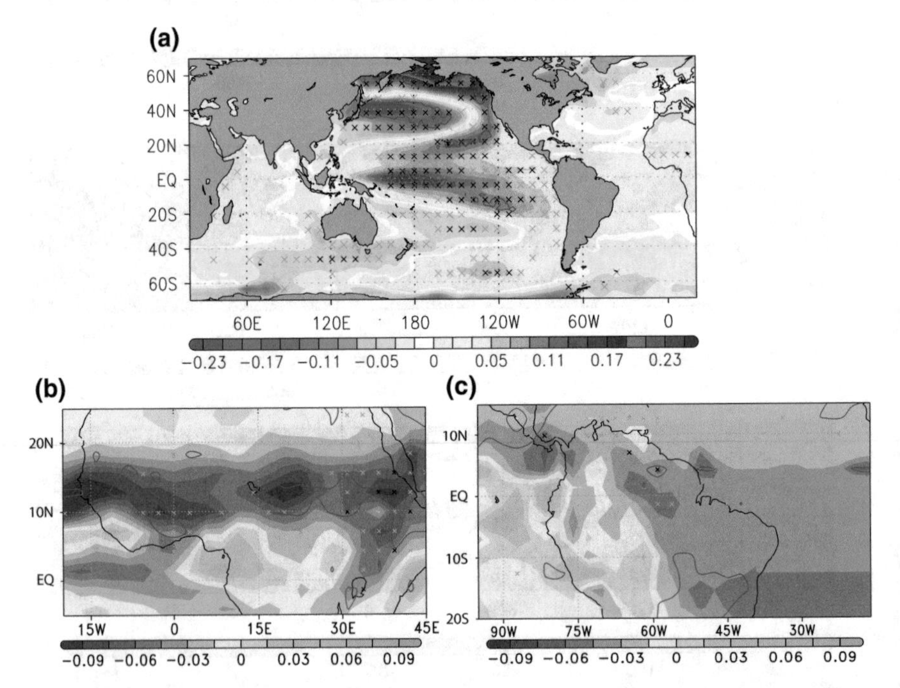

Fig. 5.16 Regression of the unfiltered (**a**) annual SSTA (K per standard deviation) and the (**b**) JAS and (**c**) DJFMAM precipitation anomalies (mm day^{-1} per standard deviation) averaged among the 17 CMIP5 models in the RCP8.5 future projection onto the IPO index. Black and gray crosses indicate points where the regression coefficient sign coincides in at least 15 and 13 out of the 17 models analyzed, respectively. Contours indicate the regions where the regression is significant at the 5% level. Adapted from Villamayor et al. (2018a)

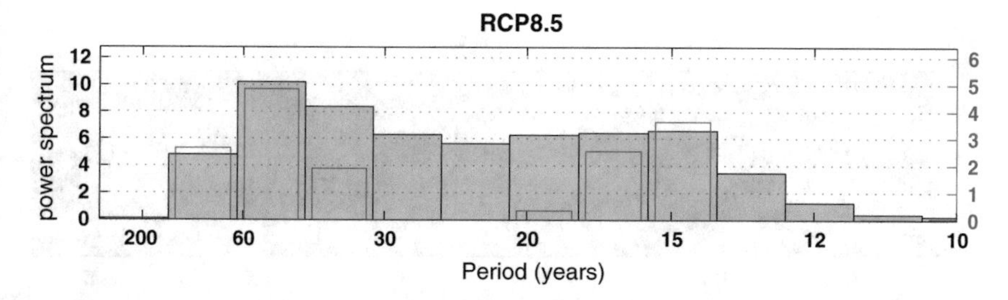

Fig. 5.17 Model-mean power spectra of the Fourier transform of the detrended IPO indices (blue bars) of the RCP8.5 future projection. Red bars indicate the power spectra averaged over the 17 models but taking into account only 95% significant peaks for each of them. The statistical significance for each model's IPO is obtained following a non-parametric hypothesis test in which a probability density function is built from the Fourier spectrum of 13-year low-pass filtered and detrended white noise time series with the same time length as the simulation and averaged across the number of ensemble members in cases where more than one realization is available

from 2006–2100, detailed in Table 2.1). Hence, the representation of the decadal-to-multidecadal SST variability characteristic of the IPO is necessarily less accurate than using simulations providing longer SST time series. The Fourier analysis applied to the IPO indices from the RCP8.5 experiments also reveals a non-unique preferable periodicity but two bands of maximum power spectrum around roughly 15–25 and 50–70 years (Fig. 5.17) as in observations (Minobe 1999; Chao et al. 2000; Tourre et al. 2001; Mantua and Hare 2002; MacDonald and Case 2005) and consistently with the other simulations analyzed.

Consistent with the IPO pattern of SSTA, the rainfall response reproduced in the RCP8.5 experiment is less robust than the others. In the Sahel, the RCP8.5 experiment represents similar JAS rainfall response to the IPO than the other analyzed simulations, regarding both the distribution and intensity of anomalies. The main difference is the reduced consistency among models (Fig. 5.16b). The associated atmospheric mechanisms involved in this link is also similarly reproduced as in the historical and piControl simulations, suggesting anomalous Walker circulation with enhanced convection over the tropical Pacific and subsidence over West Africa (Fig. 5.18a, b).

Over the Amazon region, the negative DJFMAM rainfall response show very low statistical significance and slightly weaker in RCP8.5 than in the other experiments. In the Northeast of Brazil, the rainfall anomalies are not only less intense but also uncertain, mostly negative in the interior but not statistically significant. The associated atmospheric dynamics coincides with the historical and piControl simulations but it is also less consistent among the models (Fig. 5.18c, d). Therefore, these results suggest that, according to the CMIP5 simulations, the SSTA pattern associated with the IPO variability mode and its impacts on Sahel, Amazonia and Northeast rainfall are not expected to change in the future.

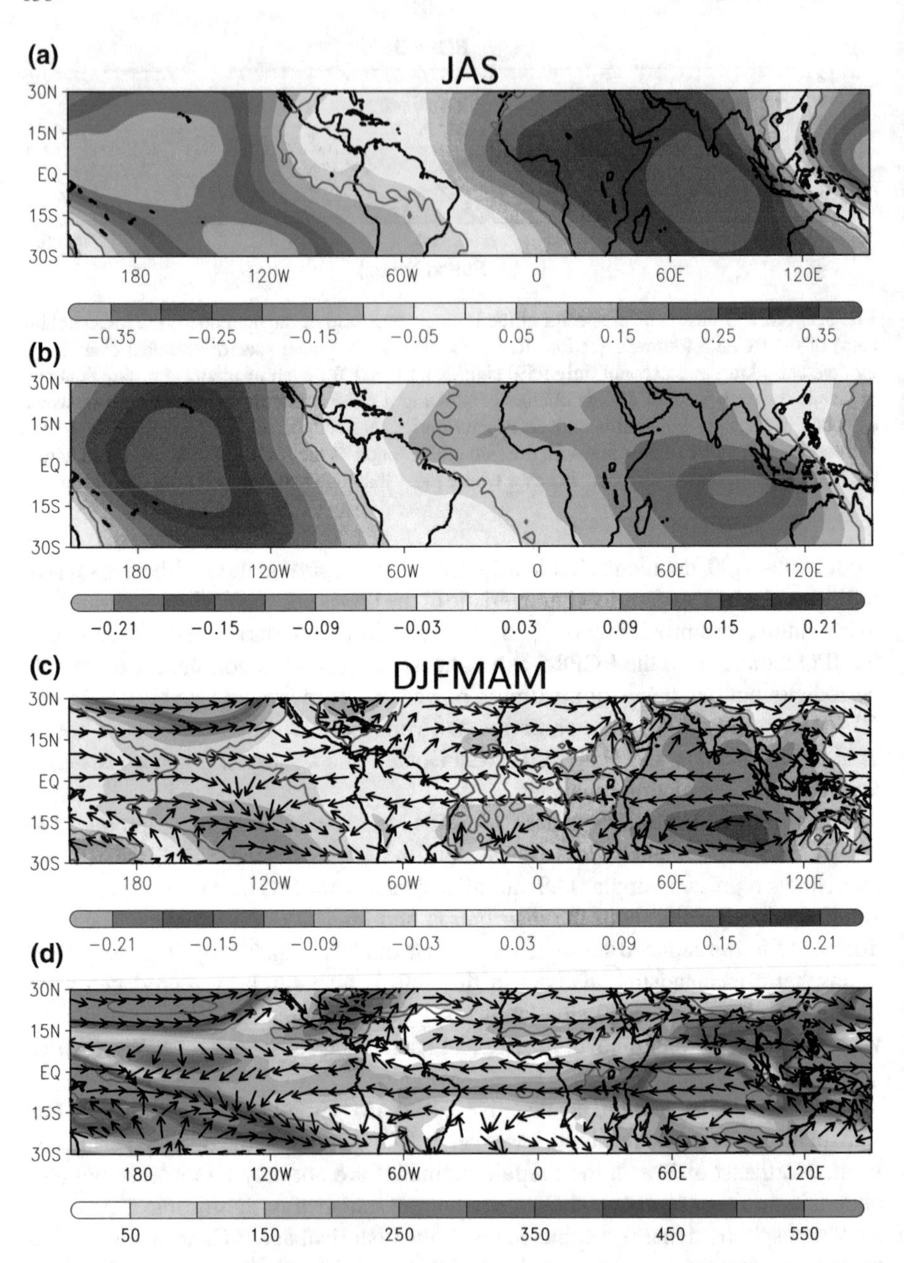

Fig. 5.18 Model-mean regression patterns onto the IPO index from the RCP8.5 future projection of: **a** the unfiltered JAS the velocity potential anomaly (10^6 m^2/s per standard deviation) at 200 and **b** 850 hPa; **c** the DJFMAM anomaly of the surface pressure (shaded) (hPa per standard deviation) and the wind direction at 850 hPa (vectors); and **d** the magnitude (shaded) and direction (vectors) of the DJFMAM moisture flux anomaly integrated from surface to 200 hPa (kg/m/day per standard deviation). Contours indicate the regions where the averaged regression is significant at the 5% level

5.5 Summary of Main Findings

In this Chapter, the aim of the Thesis of determining the influence of the IPO mode of SST variability on the precipitation in the three regions of interest has been addressed. The results show that CMIP5 models can reproduce the main spatial characteristics of the IPO pattern (Villamayor and Mohino 2015). On the Pacific basin, they accurately reproduce a pattern of warm tropical temperatures, more intense to the east, and two extratropical cold tongues, extending from the west in each hemisphere, similar to observations. However, on average they underestimate the intensity of the SSTA.

Regarding the IPO impacts, the model-mean precipitation response in the Sahel during JAS (Villamayor and Mohino 2015), the Amazonia and the Northeast regions during DJFMAM (Villamayor et al. 2018a) is also similar to the observed one. During the positive phase of the IPO, the large-scale atmospheric mechanism observed as well as reproduced by the models is an anomalous Walker circulation, with increased convection over the warm tropical Pacific and subsidence over the Atlantic sector. This atmospheric connection hinders precipitation throughout West Africa and the north of South America, in both observations and CMIP5 simulations. However, it has to be noticed that the ERA-20C reanalysis reproduces a different atmospheric IPO impact during JAS. It depicts anomalous subsidence over Central America and the Maritime Continent and promotes deep convection over central Pacific and the center of Africa, which is inconsistent with reduced Sahel rainfall. This suggests that the ERA-20C reanalysis is unreliable as far as the atmospheric mechanisms during JAS are concerned.

Although on average models succeed in producing the observed IPO features and impacts, the results show that they underestimate the amplitude of the anomalies associated. The origin of such an underestimation may be partially attributed to the weak rainfall response that AGCMs forced with prescribed SSTs show at decadal-to-multidecadal time scales (e.g. Rodríguez-Fonseca et al. 2011; Joly et al. 2007; Kucharski et al. 2013), which could be related to the accuracy with which the vegetation-atmosphere feedbacks are parameterized (Giannini et al. 2003). But it has been shown that the models skill to reproduce intense and negative rainfall anomalies is related to the way in which they reproduce the most characteristic features of the IPO pattern, particularly the amplitude of tropical Pacific SSTA.

It should be noted that, despite the high resemblance between the historical and piControl experiments regarding the IPO pattern and the precipitation response in the Sahel, Amazonia and Northeast regions, the agreement among models is slightly higher in piControl. However, in contrast to the AMV (Fig. 4), these differences are very small. Hence, according to CMIP5 models, it can be suggested that the IPO is an internal mode of SST variability that is hardly affected by external radiative forcing, as well as its impacts on West African and northern South American precipitation.

From the study of the RCP8.5 future projection, similar results to the other simulations are broadly obtained. Therefore, it is inferred that the IPO mode of variability, its impact on rainfall, as well as the atmospheric mechanisms involved, are not expected to change in the future even if there a steep increase of the GHGs emissions. Despite this, the reproduced IPO pattern and its impacts are less significantly represented by the RCP8.5 projection than in the historical and piControl simulations, specially in the northeast of Brazil.

In summary, the major conclusions emerging from the analysis carried out in this Chapter are:

- CMIP5 models robustly reproduce the observed characteristics of the IPO mode.
- The rainfall response to the IPO in West Africa and northern South America is also well captured by the CMIP5 simulations with respect to observations, though they underestimate the amplitude.
- The intenser the tropical Pacific SSTA warming of the reproduced IPO pattern, the more accurate the simulated rainfall response is with respect to observations.
- Little differences between forced and unforced simulations indicate the predominant internal nature of the IPO mode of variability.
- The atmospheric dynamics associated with the IPO, suggested by observations and supported by the CMIP5 simulations, depicts a mechanism of anomalous Walker circulation with subsidence over the studied regions that hampers local convection and, in turn, reduces rainfall.
- RCP8.5 future projection suggests that the IPO mode is not expected to change under an scenario of high GHGs emissions, as well as its impact on rainfall on the regions studied, though with some uncertainty on the Northeast.

References

Ambrizzi, T., Souza, E.B., Pulwarty, R.S.: The Hadley and Walker regional circulations and associated ENSO impacts on South American seasonal rainfall. In: The Hadley Circulation: Present, Past and Future, pp. 203–235 (2004)

Chao, Y., Ghil, M., McWilliams, J.C.: Pacific interdecadal variability in this century's sea surface temperatures. Geophys. Res. Lett. **27**, 2261–2264 (2000). https://doi.org/10.1029/1999GL011324

Cook, K.H., Vizy, E.K.: Coupled model simulations of the West African monsoon system: twentieth- and twenty-first-century simulations. J. Clim. **19**, 3681–3703 (2006)

Dai, A.: The influence of the inter-decadal Pacific oscillation on US precipitation during 1923–2010. Clim. Dyn. **41**, 633–646 (2013). https://doi.org/10.1007/s00382-012-1446-5

Deser, C., Phillips, A.S., Hurrell, J.W.: Pacific interdecadal climate variability: linkages between the tropics and the North Pacific during boreal winter since 1900. J. Clim. **17**, 3109–3124 (2004)

Dettinger, M., Battisti, D., McCabe, G., Bitz, C., Garreaud, R.: Interhemispheric effects of interannual and decadal ENSO-like climate variations on the Americas. In: Interhemispheric Climate Linkages: Present and Past Climates in the Americas and Their Societal Effects, pp. 1–16 (2001)

Drumond, A., Nieto, R., Trigo, R., Ambrizzi, T., Souza, E., Gimeno, L.: A lagrangian identification of the main sources of moisture affecting northeastern Brazil during its pre-rainy and rainy seasons. PLoS One **5** (2010)

Giannini, A., Salack, S., Lodoun, T., Ali, A., Gaye, A.T., Ndiaye, O.: A unifying view of climate change in the Sahel linking intra-seasonal, interannual and longer time scales. Environ. Res. Lett. **8**, 24010 (2013). https://doi.org/10.1088/1748-9326/8/2/024010

Giannini, A., Saravanan, R., Chang, P.: Oceanic forcing of Sahel rainfall on interannual to interdecadal time scales. Science **302**, 1027–1030 (2003). https://doi.org/10.1126/science.1089357

Grimm, A.M., Zilli, M.T.: Interannual variability and seasonal evolution of summer monsoon rainfall in South America. J. Clim. **22**, 2257–2275 (2009). https://doi.org/10.1175/2008JCLI2345.1

Grodsky, S.A., Carton, J.A., Nigam, S.: Near surface westerly wind jet in the Atlantic ITCZ. Geophys. Res. Lett. **30**, 3–6 (2003)

Janicot, S., Trzaska, S., Poccard, I.: Summer Sahel-ENSO teleconnection and decadal time scale SST variations. Clim. Dyn. **18**, 303–320 (2001). https://doi.org/10.1007/s003820100172

Joly, M., Voldoire, A.: Influence of ENSO on the West African monsoon: temporal aspects and atmospheric processes. J. Clim. **22**, 3193–3210 (2009)

Joly, M., Voldoire, A., Douville, H., Terray, P., Royer, J.-F.: African monsoon teleconnections with tropical SSTs: validation and evolution in a set of IPCC4 simulations. Clim. Dyn. **29**, 1–20 (2007)

Kucharski, F., Molteni, F., King, M.P., Farneti, R., Kang, I.-S., Feudale, L.: On the need of intermediate complexity general circulation models: a SPEEDY example. Bull. Am. Meteorol. Soc. **94**, 25–30 (2013)

Labraga, J.C., Frumento, O., López, M.: The atmospheric water vapor cycle in South America and the tropospheric circulation. Am. Meteorol. Soc. **13**, 1899–1915 (2000)

MacDonald, G.M., Case, R.A.: Variations in the Pacific Decadal Oscillation over the past millennium. Geophys. Res. Lett. **32**(L08), 703 (2005). https://doi.org/10.1029/2005GL022478

Mantua, N.J., Hare, S.R.: The Pacific decadal oscillation. J. Ocean. **58**, 35–44 (2002). https://doi.org/10.1023/a:1015820616384

Mantua, N.J., Hare, S.R., Zhang, Y., Wallace, J.M., Francis, R.C.: A Pacific interdecadal climate oscillation with impacts on salmon production. Bull. Am. Meteorol. Soc. **78**, 1069–1079 (1997)

Marengo, J.A., Liebmann, B., Grimm, A.M., Misra, V., Dias, P.L.S., Cavalcanti, I.F.A., Carvalho, L.M.V., Berbery, E.H., Ambrizzi, T., Vera, C.S., Saulo, A.C., Nogues-paegle, J., Zipser, E., Seth, A., Alves, L.M.: Recent developments on the South American monsoon system. Int. J. Climatol. **32**, 1–21 (2012). https://doi.org/10.1002/joc.2254

Minobe, S.: Resonance in bidecadal and pentadecadal climate oscillations over the North Pacific: Role in climatic regime shifts. Geophys. Res. Lett. **26**, 855–858 (1999). https://doi.org/10.1029/1999GL900119

Mohino, E., Janicot, S., Bader, J.: Sahel rainfall and decadal to multi-decadal sea surface temperature variability. Clim. Dyn. **37**, 419–440 (2011a). https://doi.org/10.1007/s00382-010-0867-2

Mohino, E., Rodríguez-Fonseca, B., Mechoso, C.R., Gervois, S., Ruti, P., Chauvin, F.: Impacts of the tropical Pacific/Indian oceans on the seasonal cycle of the west african monsoon. J. Clim. **24**, 3878–3891 (2011b). https://doi.org/10.1175/2011JCLI3988.1

Nicholson, S.E.: The West African Sahel: a review of recent studies on the rainfall regime and its interannual variability. ISRN Meteorol. **2013**, 32 (2013). https://doi.org/10.1155/2013/453521

Pu, B., Cook, K.H.: Dynamics of the West African westerly jet. J. Clim. **23**, 6263–6276 (2010)

Rodríguez-Fonseca, B., Janicot, S., Mohino, E., Losada, T., Bader, J., Caminade, C., Chauvin, F., Fontaine, B., García-Serrano, J., Gervois, S., et al.: Interannual and decadal SST-forced responses of the West African monsoon. Atmos. Sci. Lett. **12**, 67–74 (2011)

Rodríguez-Fonseca, B., Suárez-Moreno, R., Ayarzagüena, B., López-Parages, J., Gómara, I., Villamayor, J., Mohino, E., Losada, T., Castaño-Tierno, A.: A review of ENSO influence on the North Atlantic. A non-stationary signal. Atmosphere **7**, 87 (2016)

Shen, C., Wang, W.-C., Gong, W., Hao, Z.: A Pacific decadal oscillation record since 1470 AD reconstructed from proxy data of summer rainfall over eastern China. Geophys. Res. Lett. **33** (2006)

Tourre, Y.M., Rajagopalan, B., Kushnir, Y., Barlow, M., White, W.B.: Patterns of coherent decadal and interdecadal climate signals in the Pacific Basin during the 20th century (2001). https://doi.org/10.1029/2000GL012780

Villamayor, J., Mohino, E.: Robust Sahel drought due to the Interdecadal Pacific Oscillation in CMIP5 simulations. Geophys. Res. Lett. **42**, 1214–1222 (2015)

Villamayor, J., Ambrizzi, T., Mohino, E.: Influence of decadal sea surface temperature variability on northern Brazil rainfall in CMIP5 simulations. Clim. Dyn. **51**, 563–579 (2018a)

Chapter 6
Contribution of the SST Modes to Rainfall Variability

Abstract In Chaps. 3–5 the impact of the GW, the AMV and the IPO on precipitation in the Sahel, the Amazon and the Northeast regions has been analyzed separately. In this chapter we evaluate how much of the total rainfall variability is modulated by each one of these SST modes at decadal-to-multidecadal time scales arises. To this end we analyze the variance of the low-frequency index of precipitation at each region that the three modes of SST explain through a multi-linear regression model. In this way, we can quantify the contribution of each SST mode to the total rainfall decadal-to-multidecadal variability in observations and find whether the CMIP5 models reproduce a similar behavior. From the RCP8.5 future projections, the same analysis provides an insight on whether the modulating roles of the SST modes on the rainfall regimes are expected to change in the future.

6.1 The Sahel Rainfall

Observational records show a climatological precipitation rate during JAS of 5.9 mm day^{-1} over 1901–2009 (averaged over the three data sets) (Fig. 6.1a). In the historical simulation, the climatological JAS precipitation simulated by CMIP5 models range from 3.1 to 7.3 mm day^{-1}, with an average of 5.1 mm day^{-1} close to the observed one. Conversely, the Sahel rainfall variability is widely underestimated by most models (Fig. 6.1b). Only a few models (CSIRO-Mk3-6-0, MIROC5 and MIROC-ESM-CHEM) match or exceed the observed variance of the Sahel precipitation (0.57 mm day^{-1}, averaged across the three data sets) taking into account all time scales. In turn, approximatively half the observed Sahel precipitation total variance (48%, averaged over the three data sets) takes place at decadal-to-multidecadal time scales. In contrast, in the models the simulated high-frequency variability prevails upon the decadal-to-multidecadal one (which ranges among models from 13 to 36% of the total variance). Therefore, the way in which CMIP5 models simulate the

---

The content of this chapter referring to the contribution of the SST modes on the Sahel rainfall decadal variability is in preparation for publication in the following article: Villamayor and Mohino (2019): SST drivers of Sahel rainfall at decadal timescales in CMIP5 simulations. (In preparation).

J. Villamayor, *Influence of the Sea Surface Temperature Decadal Variability on Tropical Precipitation: West African and South American Monsoon*, Springer Theses, https://doi.org/10.1007/978-3-030-20327-6_6

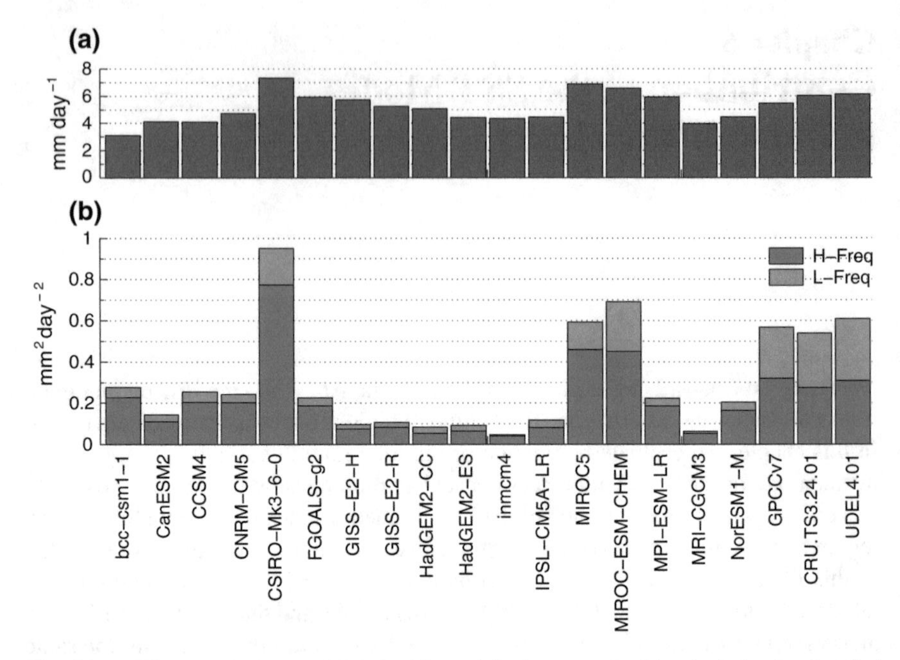

Fig. 6.1 **a** Climatological rate of Sahel JAS precipitation over the historical simulated period (typically 1850–2006, see Table 2.1) of the CMIP5 models and over 1901–2009 in observations (GPCC v7, CRU TS3.24.01 and UDEL v4.01 data sets). **b** Total variance of the Sahel index of JAS precipitation divided into that corresponding to the high-(blue) and low-frequency (orange) variability (periodicities below and above 13 years, respectively). Units are mm day^{-1}

Sahel rainfall variance is model-dependent, although all of them underestimate the observed decadal-to-multidecadal variability.

The multi-linear regression analysis of the 13-year low-pass filtered time series of the Sahel precipitation anomalies in JAS performed with the GW, AMV and IPO indices reveals that most CMIP5 models underestimate the role of the SST modes to modulate the Sahel rainfall variance (decomposed following Eq. 2.29) (Fig. 6.2). In observations, the three modes together account for 73% of the total low-frequency variability of rainfall, on average across the three precipitation data sets analyzed. The AMV is the main SST mode that leads the Sahel precipitation changes at decadal-to-multidecadal time scales with a contribution of 45%. The GW follows with 17% and then the IPO with 8%. Most CMIP5 models, in turn, simulate a combined contribution of the three SST modes that do not exceed 30% of the total variance of the Sahel index. The only exception is the MIROC-ESM-CHEM model, whose reproduced SST modes account for 77% of the Sahel rainfall low-frequency variability, similar to observations. However, the contribution of each SST mode to the total variability that it simulates is different from the observed one. Indeed, all CMIP5 models show different behavior in this sense. In most of them (CCSM4, FGOALS-g2, GISS-E2-H, GISS-E2-R, inmcm4, IPSL-CM5A-LR, MIROC-ESM-CHEM, MPI-ESM-LR,

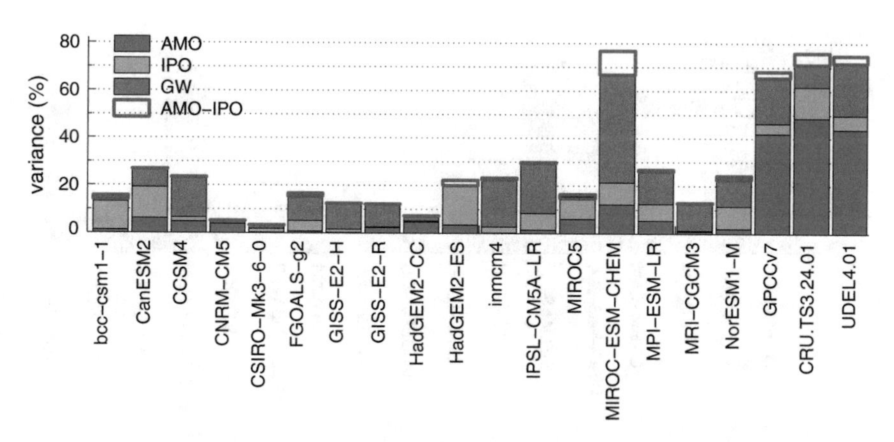

Fig. 6.2 Components of the 13-year low-pass filtered index of Sahel precipitation variance (in %) explained by the multi-linear regression analysis with the GW, AMV and IPO indices from each model in the historical simulation and from observations (using GPCC v7, CRU TS3.24.01 and UDEL v4.01 precipitation data sets and HadISST1 for SST). The components correspond to the GW (brown), the AMV (blue), the IPO (green) and the AMV-IPO covariance (purple contour)

MRI-CGCM3 and NorESM1-M) the GW is the mode that contributes the most to the Sahel precipitation low-frequency variability, in others it is the IPO (bcc-csm1-1, CanESM2, CSIRO-Mk3-6-0, HadGEM2-ES and MIROC5) or the AMV (CNRM-CM5 and HadGEM2-CC).

In the RCP8.5 simulation, a dominant role of the GW mode on the Sahel rainfall low-frequency variability is simulated by almost all the models (Fig. 6.3). In turn, the three modes of SST account for more variance of the Sahel index than in the historical simulation. Only a few models reproduce a dominant driving role of the AMV (HadGEM2-CC) or the IPO (CCSM4 and MPI-ESM-LR). This result suggests that, under the conditions of extremely large GHGs concentrations imposed in the RCP8.5 future projections, CMIP5 models, in general, reproduce Sahel rainfall low-frequency variations that are mainly dominated by the GW rather than the other decadal-to-multidecadal modes of SST. On the other hand, the fact that GW does not show such a remarkable influence on the Sahel precipitation in a few models may be related to the competing effects that GHGs have on Sahel precipitation. Gaetani et al. (2017) show that the direct effect of the GHGs enhances Sahel rainfall through surface heating induced by local net radiation changes, while they also reduce precipitation indirectly via SST warming. Hence, even under very intense GW conditions, the influence of the AMV or the IPO may prevail if the GHGs direct and indirect effects that these models reproduce are balanced.

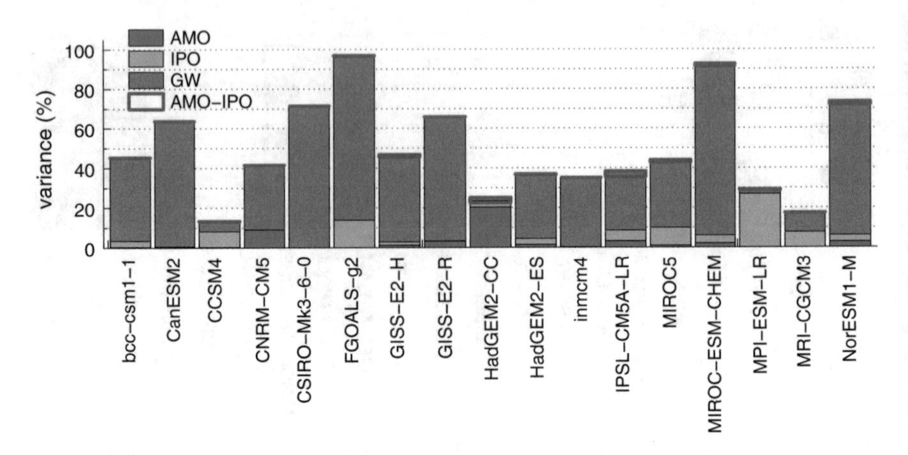

Fig. 6.3 Components of the 13-year low-pass filtered index of Sahel precipitation variance (in %) explained by the multi-linear regression analysis with the GW, AMV and IPO indices from each model in the RCP8.5 future projection. The components correspond to the GW (brown), the AMV (blue), the IPO (green) and the AMV-IPO covariance (purple contour)

6.2 The Amazon Rainfall

The climatology of DJFMAM precipitation in the Amazon regions is 8.6 mm day^{-1} in observations (averaged over the three data sets) (Fig. 6.4a). Models, however, underestimate it with a simulated mean value of 6.1 mm day^{-1}. The simulated rainfall variability at all time scales is, on average, half of the observed one (Fig. 6.4b). However, there is no apparent relationship between the underestimation of the climatological rainfall and the variability. Regarding the low-frequency variability, in observations it represents 32% of the total variance (averaged over the three data sets) and in models, on average, only 17%. The amazon rainfall variability is hence more prominent at longer than decadal-to-multidecadal time scales.

In the Amazon region, the decadal-to-multidecadal modes of SST account for 73% of the total variance of the low-frequency index of precipitation on average across the three observational data sets (Fig. 6.5). Around 40% of the explained variance corresponds to the GW, 17% to the IPO and 14% to the AMV influence. Therefore, following observational data, the GW might be considered as the main modulator of the Amazon rainfall at decadal-to-multidecadal time scales. In the historical simulation, 9 out of the 17 models analyzed also reproduce a major influence of the GW on precipitation compared to the other modes of SST. But they do not agree as to the proportion of the total variance that the SST modes can account for. All CMIP5 models underestimate the observed Amazon rainfall variance that can be explained by the three SST modes. It does not exceed 44% in any of the cases. On the other hand, some models (bcc-csm1-1, CCSM4 and MPI-ESM-LR) reproduce an Amazon rainfall variability that is mostly related with the AMV, though its relationship with the SST decadal-to-multidecadal modes is rather weak. In others

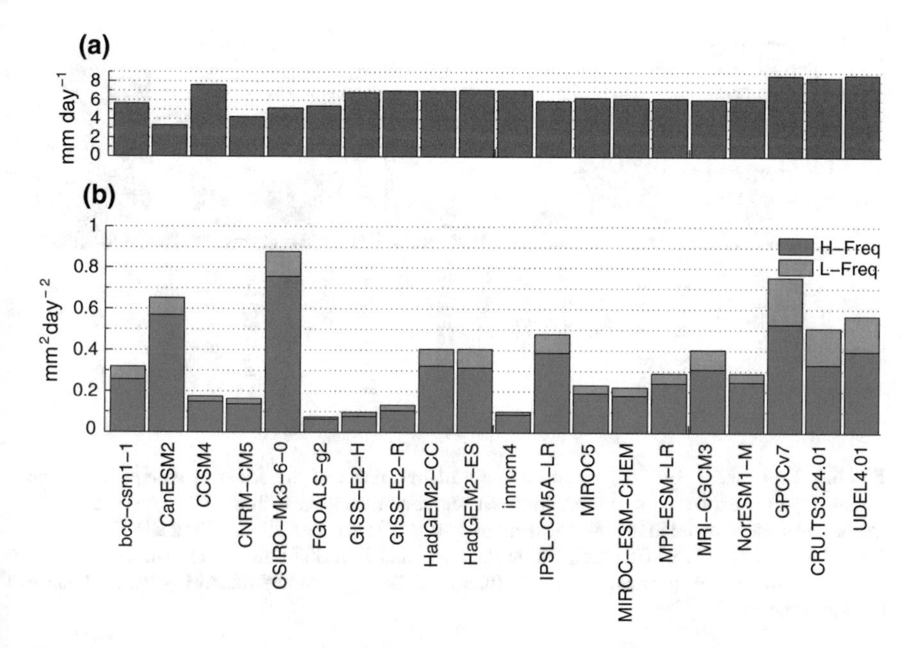

Fig. 6.4 **a** Climatological rate of the Amazon DJFMAM precipitation over the historical simulated period (typically 1850–2006, see Table 2.1) of the CMIP5 models and over 1901–2009 in observations (GPCC v7, CRU TS3.24.01 and UDEL v4.01 data sets). **b** Total variance of the Amazon index of DJFMAM precipitation divided into that corresponding to the high- (blue) and low-frequency (orange) variability (periodicities below and above 13 years, respectively). Units are mm day^{-1}

(CanESM2, CSIRO-Mk3-6-0, HadGEM2-ES and NorESM1-M) the IPO influences more the precipitation variability than the GW and the AMV. In the case of the inmcm4 model, the GW, AMV and IPO modes individually account for around 2%, 17% and 20% of the total variance of the Amazon index, respectively. However, all SST modes together barely explain 17% of it due to the strong correlation between the AMV and the IPO indices resulting from this model ($R = 0.6$, statistically significant at the 99% confidence level). Hence, in this case, it cannot be identified which one of the decadal-to-multidecadal modes of SST drives more the Amazon rainfall low-frequency variability.

Regarding the RCP8.5 future projection, CMIP5 models in general reproduce a stronger modulation of the decadal-to-multidecadal SST modes in the Amazon precipitation than in the historical simulation (Fig. 6.6). As in the case of the Sahel region, the reproduced Amazon rainfall low-frequency variability is mainly led by the GW. Only 4 out of the 17 CMIP5 models analyzed reproduce a stronger relationship with the AMV (bcc-csm1-1) or the IPO (CNRM-CM5, inmcm4 and MRI-CGCM3) than with the GW. Therefore, models suggest with high agreement that, under a hypothetical future scenario of large GHGs concentrations, the GW pattern of SST resulting from the radiative forcing would have a prevailing effect on the Amazon rainfall variability at decadal-to-multidecadal time scales.

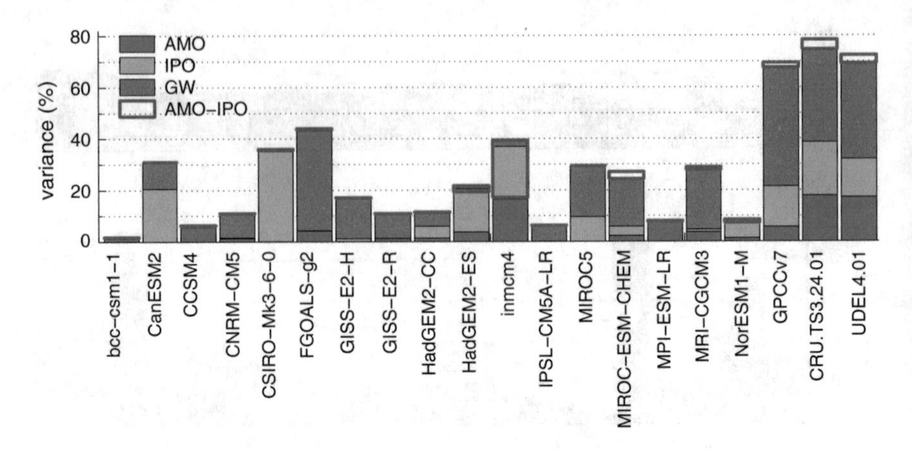

Fig. 6.5 Components of the 13-year low-pass filtered index of the Amazon precipitation total variance (in %) explained by the multi-linear regression analysis with the GW, AMV and IPO indices from each model in the historical simulation and from observations (using GPCC v7, CRU TS3.24.01 and UDEL v4.01 precipitation data sets and HadISST1 for SST). The components correspond to the GW (brown), the AMV (blue), the IPO (green) and the AMV-IPO covariance (purple contour)

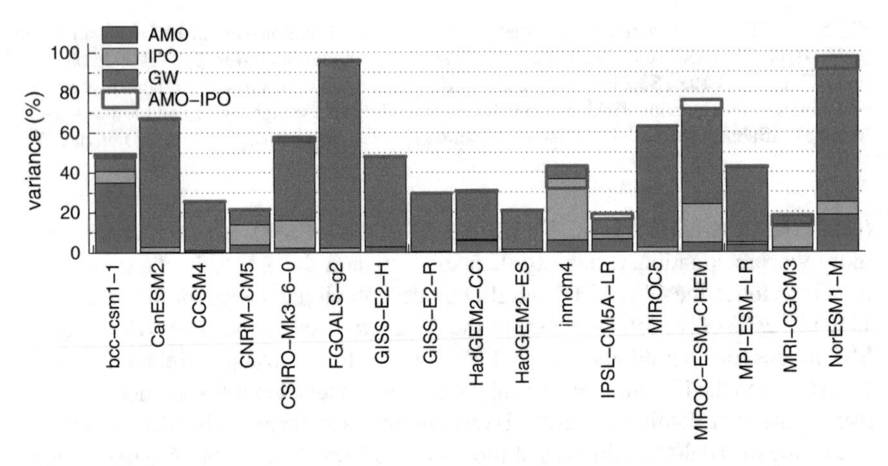

Fig. 6.6 Components of the 13-year low-pass filtered index of the Amazon precipitation total variance (in %) explained by the multi-linear regression analysis with the GW, AMV and IPO indices from each model in the RCP8.5 future projection. The components correspond to the GW (brown), the AMV (blue), the IPO (green) and the AMV-IPO covariance (purple contour)

6.3 The Northeast Rainfall

In the Northeast of Brazil, the observed climatological precipitation in DJFMAM is 5.0 mm day^{-1} (averaged across the three databases) (Fig. 6.7a). On average, CMIP5 models reproduce a comparable climatological precipitation rate of 5.2 mm day^{-1}.

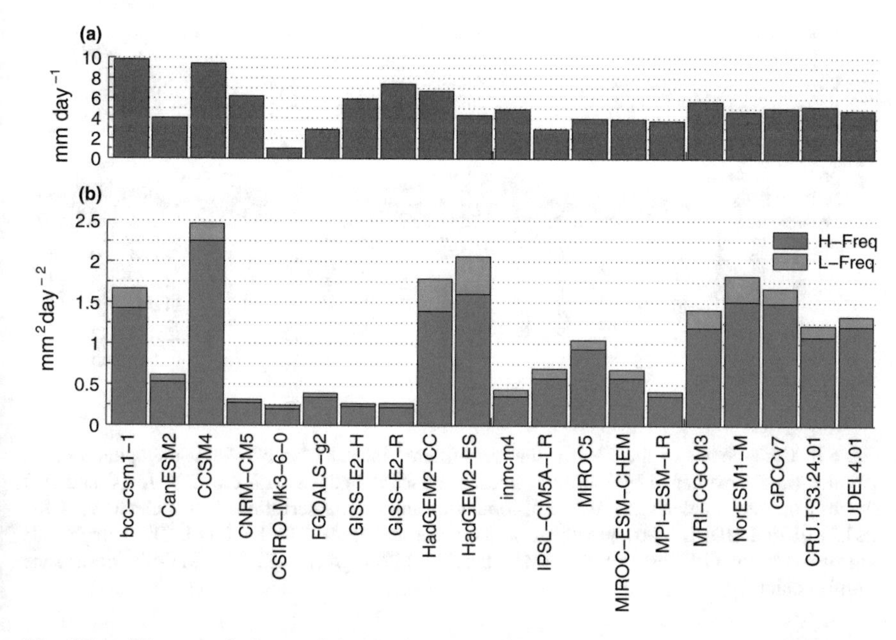

Fig. 6.7 **a** Climatological rate of the Northeast DJFMAM precipitation over the historical simulated period (typically 1850–2006, see Table 2.1) of the CMIP5 models and over 1901–2009 in observations (GPCC v7, CRU TS3.24.01 and UDEL v4.01 data sets). **b** Total variance of the Northeast index of DJFMAM precipitation divided into that corresponding to the high- (blue) and low-frequency (orange) variability (periodicities below and above 13 years, respectively). Units are mm day^{-1}

Although this value is highly variable among the different models, ranging from 0.1 to 9.8 mm day^{-1}. The index of the Northeast precipitation in observations shows a high variance of 1.4 mm day^{-1} (averaged among the three data sets) at all time scales (Fig. 6.7b). Models reproduce a similar total variance of 1.0 mm day^{-1}, on average. But individually, they present diverse behaviors (ranging from 0.2 to 2.5 mm day^{-1}). Similarly, observations and CMIP5 models, on average, show similar low-frequency variability representing 11% and 16% of the total variance, respectively. Compared to the Sahel and the Amazon precipitation, whose decadal-to-multidecadal variations account for half and one third of the total variance, respectively, the Northeast rainfall has an outstanding variability that predominately takes place at rather interannual time scales.

The multi-linear analysis of the low-frequency index of rainfall anomalies in the Northeast reveals that little variance is explained by the three modes of SST in observations, ranging from 13 to 26% depending on the data base of precipitation used (Fig. 6.8). In addition, it is not possible to identify the SST mode that accounts for more variance due to the discrepancies between the different rainfall data sets used. Depending on whether the GPCC v7, CRU TS3.24.01 or UDEL v4.01 data set is considered, the SST mode that is shown to explain more variability is the GW,

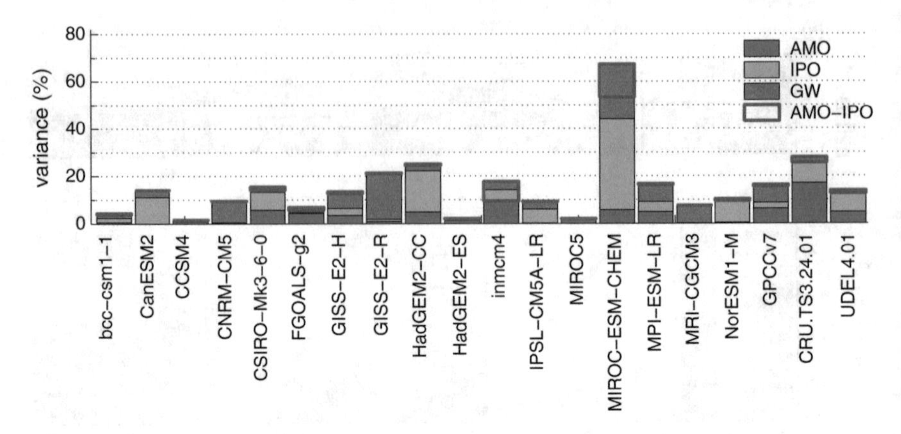

Fig. 6.8 Components of the 13-year low-pass filtered index of the Northeast precipitation total variance (in %) explained by the multi-linear regression analysis with the GW, AMV and IPO indices from each model in the historical simulation and from observations (using GPCC v7, CRU TS3.24.01 and UDEL v4.01 precipitation data sets and HadISST1 for SST). The components correspond to the GW (brown), the AMV (blue), the IPO (green) and the AMV-IPO covariance (purple contour)

the AMV or the IPO, respectively. In the historical simulation of CMIP5 models, in general, the three modes of SST do not account for more than 25% of the total low-frequency variance of the Northeast rainfall. Only the MIROC-ESM-CHEM model reproduces a dependence of more than half of the total rainfall low-frequency variability on the three main decadal-to-multidecadal SST modes. Models show large discrepancies concerning the contribution of each SST mode on the variability of the Northeast precipitation. 6 out of 17 models reproduce a larger dependency of the low-frequency variability of rainfall on the AMV (CCSM4, FGOALS-g2, HadGEM2-ES, inmcm4 and MIROC5), 7 on the IPO (bcc-csm1-1, CanESM1, CSIRO-Mk3-6-0, HadGEM2-CC, ISPSL-CM5A-LR, MIROC-ESM-CHEM and NorESM1-M) and 5 on the GW (CNRM-CM5, GISS-E2-H, GISS-E2-R, MPI-ESM-LR and MRI-CGCM3).

In Chaps. 3–5, it has been shown that the influence of the main decadal-to-multidecadal modes of SST individually on the Northeast rainfall is not statistically significant in observations, which may be attributed to the inherent uncertainty of gridded data sets. Though the connection between them is supported by different sources of precipitation data. In addition, the impact of SST modes on the Northeast precipitation is robustly reproduced by CMIP5 models and coincide with observations. Hence, CMIP5 models provide confidence to the observational result and it can be concluded that the link between the Northeast rainfall and the decadal-to-multidecadal SST modes is robust. However, CMIP5 models in their historical simulation do not clarify the role of each SST mode in modulating the Northeast rainfall, which is also uncertain in observations. Therefore, it is not possible to determine

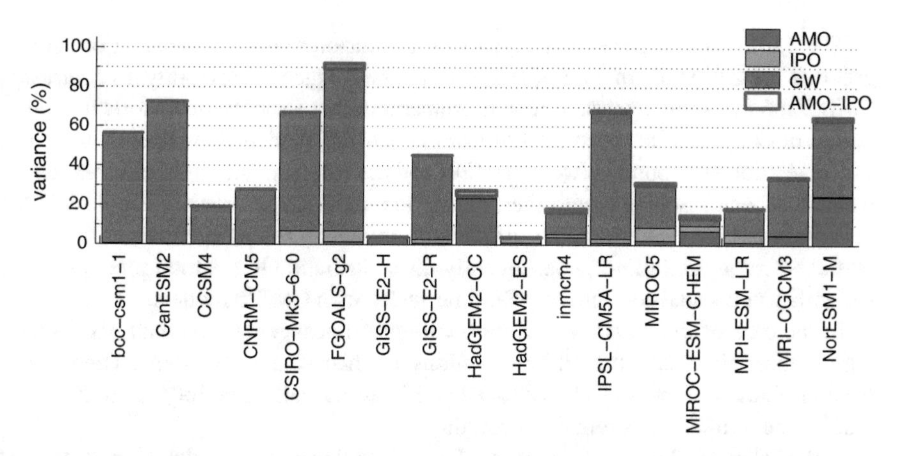

Fig. 6.9 Components of the 13-year low-pass filtered index of the Northeast precipitation total variance (in %) explained by the multi-linear regression analysis with the GW, AMV and IPO indices from each model in the RCP8.5 future projection. The components correspond to the GW (brown), the AMV (blue), the IPO (green) and the AMV-IPO covariance (purple contour)

which mode of SST contributes the most to the Northeast precipitation variability at decadal-to-multidecadal time scales.

In the RCP8.5 future projections, CMIP5 models, in general, reproduce stronger influence of the decadal-to-multidecadal SST modes on the Northeast rainfall low-frequency variability than in the historical simulation (Fig. 6.9). It can be also seen that the low-frequency variability of precipitation reproduced by most models is mainly dominated by the GW. Only a few models reproduce a leading role in modulating the Northeast rainfall of the AMV (GISS-E2-H, HadGEM2-CC and MIROC-ESM-CHEM) or the IPO (HadGEM2-ES). Hence, this result suggests that the intense GW forced by large concentrations of GHGs is expected to lead the precipitation changes in the Northeast of Brazil at decadal-to-multidecadal time scales, prevailing over the AMV and the IPO, in a hypothetical future under the RCP8.5 scenario.

6.4 Discussion

In the Sahel region, observations show that the main modulator of rainfall at decadal-to-multidecadal time scales is the AMV, according to other studies (Mohino et al. 2011a). However, models do not reproduce this result. In contrast, most of the CMIP5 models suggest a dominant role of the GW. The simulated precipitation is more or less responsive to each SST mode depending on the model analyzed. Therefore, although in Chaps. 3–5 it has been shown that CMIP5 models robustly support the observed influence of the main decadal-to-multidecadal modes of SST on Sahel rainfall, they cannot reproduce the observed relative contribution of each SST mode to the total rainfall variability.

The multi-linear regression analysis of the Amazon rainfall in observations suggest that the GW is the main modulator of its low-frequency variability. In contrast, CMIP5 models show uncertain results concerning this issue in the historical simulation. In Chap. 3, it has been also shown that CMIP5 models do not reproduce the strong relationship found between the GW and the Amazon rainfall in observations, which has been associated with their failure to reproduce the tropical Pacific SSTA of the GW pattern. Therefore, it is difficult to confidently asses the impact of the GW on the Amazon rainfall nor, consequently, its contribution to the total precipitation variability at decadal-to-multidecadal time scales with CMIP5 models.

In the case of the Northeast rainfall, the multi-linear regression analysis shows highly uncertain results both in observations and historical simulations. Therefore, the contribution of the SST modes to the Northeast rainfall variability at decadal-to-multidecadal time scales remain uncertain.

In the Sahel and the Amazon regions, the contribution of the decadal-to-multidecadal SST modes to rainfall low-frequency variability that CMIP5 models reproduce in the historical simulation is, in most cases, notably weaker than in observations. Besides, the simulated rainfall in the three regions studied presents less variance than observations. Following this, one may wonder if the reproduced rainfall low-frequency variability is weaker than in observations as a result of an underestimated modulation of the SST modes. The fact that CMIP5 models do not achieve to reproduce the observed rainfall response to the SST at decadal-to-multidecadal time scales may be attributed to weak simulated modes of SST variability and/or to inaccurate ability to transmit their modulating effect on rainfall.

Regarding the simulated SST modes, they present a variance similar to the observed one (Table 6.1). On average, models reproduce a GW index with a variance of 0.029 K, similar to the observed one (0.027 K). The total variance of the simulated SST in the North Atlantic at decadal-to-multidecadal time scales that the AMV accounts for is, on average, (44%) similar to the observed one (37%). Similarly, the simulated model-mean IPO explains a fraction of the Pacific SST variance that is close to the observations (ranging from 40% to 31%, respectively). Therefore we can rule out the possibility that the models underestimate the decadal variability of these SST modes. In Chaps. 3–5, we have also seen that the amplitude of the SSTA patterns associated with the GW, AMV and IPO are comparable to the observed ones. Nevertheless, the models skill to reproduce the observed relationship between SST modes and precipitation has been related to the correct distribution of anomalies in the spatial patterns.

From the previous results, we have also suggested that the underestimation of the link between the SST modes and rainfall that models simulate may be related with the accuracy with which they reproduce the atmospheric teleconnections. There are several works addressing this issue and showing that the AGCM component of GCMs present weak response to SST forcing (Joly et al. 2007; Rodríguez-Fonseca et al. 2011; Kucharski et al. 2013; Vellinga et al. 2016). In this line, we have further analyzed the changes in the intensity of the atmospheric response to the AMV and the IPO modes of SST associated with Sahel rainfall. In this Thesis, we have related the AMV effects on Sahel rainfall with low-level zonal wind anomalies close to the coast

Table 6.1 var_{GW} is the variance of the GW index (units are degree K) for observations (HadISST1) and the CMIP5 historical simulations calculated over the common period 1870–2004. $fvar_{AMV}$ and $fvar_{IPO}$ are the explained variance of the North Atlantic and Pacific SSTA at decadal-to-multidecadal time scales (in %) by the AMV and IPO modes (defined as the PC1, see Chap. 2) in a 140-year period (to match the observed period)

	var_{GW}	$fvar_{AMV}(\%)$	$fvar_{IPO}(\%)$
Observations	0,027	37	40
Model-mean	0,029	44	31
1. bcc-csm1-1	0,061	43	34
2. CanESM2	0,033	38	40
3. CCSM4	0,058	49	29
4. CNRM-CM5	0,018	38	34
5. CSIRO-Mk3-6-0	0,014	60	40
6. FGOALS-g2	0,023	30	26
7. GISS-E2-H	0,027	32	19
8. GISS-E2-R	0,022	65	23
9. HadGEM2-CC	0,011	59	30
10. HadGEM2-ES	0,012	48	27
11. inmcm4	0,021	43	32
12. IPSL-CM5A-LR	0,078	58	33
13. MIROC5	0,017	38	30
14. MIROC-ESM-CHEM	0,023	44	35
15. MPI-ESM-LR	0,051	26	24
16. MRI-CGCM3	0,013	34	32
17. NorESM1-M	0,019	36	34

of West Africa and an interhemispheric surface pressure gradient over the Atlantic sector in agreement with other works (e.g. Martin and Thorncroft 2014; Nicholson 2013). We have also related the link with the IPO to the low-level winds and to anomalous Walker circulation induced by the tropical Pacific SSTA, according to other studies (e.g. Janicot et al. 2001; Mohino et al. 2011a), which has a remarkable imprint on the high-level velocity potential.

Figures 6.10 and 6.11 present scatterplots of coefficients that estimate the AMV and the IPO impacts on low-level westerlies (LLW), a surface pressure gradient (SPG) and high-level velocity potential (HLVP) over West Africa versus the impact on Sahel precipitation (SAH). The LLW coefficient is estimated by averaging the regression of the zonal wind anomalies at 850 hPa over 35°–15°W and 8°–11°N on the AMV or the IPO index obtained for the historical simulation of the CMIP5 models individually. The SPG coefficient is calculated as the averaged regression of the surface pressure anomalies onto the AMV index over 40°–20°W and 10°S–5°N. The HLVP is estimated as the average of the regression of the velocity potential at

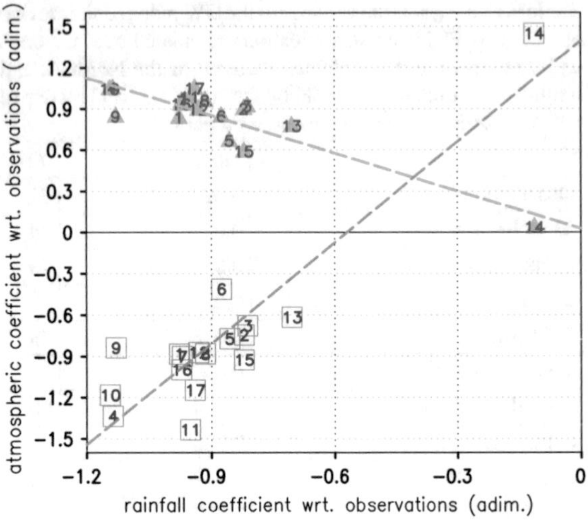

Fig. 6.10 Scatter plot of the a dimensional coefficients ΔSPG_i and ΔLLW_i versus ΔSAH_i (see details in the text) for each of the 17 models in the historical simulation (triangles and squares, respectively) associated with the AMV effects. Linear regressions for each scatterplot are depicted as the green and orange lines, respectively. Each number in the plot corresponds to a model (see Table 2.1)

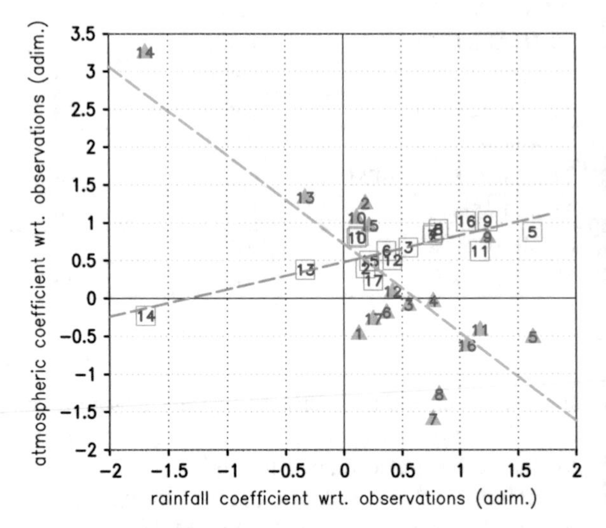

Fig. 6.11 Scatter plot of the adimensional coefficients $\Delta HLVP_i$ and ΔLLW_i versus ΔSAH_i (see details in the text) for each of the 17 models in the historical simulation (triangles and squares, respectively) associated with the IPO effects. Linear regressions for each scatterplot are depicted as the green and orange lines, respectively. Each number in the plot corresponds to a model (see Table 2.1)

200 hPa on the IPO index averaged over 20°W–20°E and 0°–25°N. For SAH we use the area 17.5°W–10°E and 10°N–17.5°N in the regression patterns in Figs. B.3 and C.3. As we are interested in the response of the models with respect to observations, the observed coefficient is subtracted to the simulated one and the result is scaled to the absolute value of the observed one:

$$\Delta SAH_i = \frac{SAH_i - SAH_o}{| SAH_o |}, \tag{6.1}$$

$$\Delta LLW_i = \frac{LLW_i - LLW_o}{\mid LLW_o \mid}, \tag{6.2}$$

$$\Delta SPG_i = \frac{SPG_i - SPG_o}{\mid SPG_o \mid}, \tag{6.3}$$

$$\Delta HLVP_i = \frac{HLVP_i - HLVP_o}{\mid HLVP_o \mid}, \tag{6.4}$$

where, subscripts i and o refer to models ($i = 1, ..., 17$) and observations, respectively. The new coefficients ΔSAH_i, ΔLLW_i, ΔSPG_i and $\Delta HLVP_i$) are the model's i coefficients with respect to observations. The observational coefficients are estimated from the GPCC v7 data set (for SAH) and the 20CRV2c reanalysis (for LLW, SPG and HLVP).

Figure 6.10 shows the scatterplot of ΔSPG_i and ΔLLW_i coefficients versus ΔSAH_i. As all models underestimate the rainfall response, all points of the scatterplot are located to the left of the origin (their coefficients are less positive than the one from observations, which is 0.30 mm day^{-1} per standard deviation). Regarding the response of the low-level zonal winds and the interhemispheric gradient of surface pressure, most models also underestimate them, so most squares are below zero (all but MIROC-ESM-CHEM) (the intensification of westerlies at 850 hPa results in positive values of the LLW coefficient, which in observations is 0.15 ms^{-1} per standard deviation) and all triangles are above zero (negative surface pressure gradient, with negative anomalies to the north and positive ones to the south, result in negative SPG values, which in observations is -19.75 Pa per standard deviation). The AMV response of the analyzed atmospheric variables (with respect to the observations) shows a linear relationship with the associated rainfall anomalies. The linear fitting of the ΔSPG_i and ΔSAH_i coefficients shows highly significant correlation coefficient (R $= -0.88$, significant at the 99% confidence level according to a Student t-test). Similarly, the ΔLLW_i coefficient linearly fits with ΔSAH_i (R $= 0.91$, significant at the 99% confidence level). Although these results seem very dependent on the MIROC-ES-CHEM model, they are not so much. If we eliminate MIROC-ESM-CHEM from the linear adjustment of ΔSPG_i and ΔLLW_i, we still get highly significant correlation coefficients (R $= -0.55$ and R $= 0.60$, respectively, significant at the 99% confidence level). Therefore, there is an evident linear relationship between the strength of the atmospheric and the precipitation response to the AMV pattern of SST: the stronger the underestimation in one, the stronger in the other too. The trends of the linear fitting of the ΔSPG_i and ΔLLW_i coefficients with ΔSAH_i are close to minus one and one (with values of m $= -0.91$ and m $= 0.88$), respectively. This indicates that the effects on atmospheric and rainfall are underestimated by CMIP5 models in roughly same proportion.

As far as the IPO influence on Sahel rainfall is concerned, Fig. 6.11 shows the scatterplot of the corresponding $\Delta HLVP_i$ and ΔLLW_i coefficients versus ΔSAH_i. Most models (all but MIROC5 and MIROC-ESM-CHEM) underestimate the rainfall response to the IPO, therefore most points of the scatterplot are located to the right of the origin (the SAH_i coefficients are less negative than the one from observations, which in observations is 0.09 mm day^{-1} per standard deviation). Regarding the response of the velocity potential at 200 hPa over West Africa, it is also underestimated by models in the majority of the cases (10 out of 17), so most triangles are below zero (low velocity potential anomalies with respect to observations, which is 1.9 10^5 m^2 s^{-1} per standard deviation, result in negative $\Delta HLVP_i$ coefficients). Models also underestimate the response of the westerlies at low levels (all but MIROC-ESM-CHEM), so most squares are above zero (the reduction of westerlies results in negative values of the ΔLLW_i coefficient, and the models coefficients are less negative than the observed one, which is –0.28 m s^{-1} per standard deviation). Consistently with the previous result, the underestimated response of the atmosphere and Sahel rainfall (with respect to the observations) to the IPO seem to be related. A linear fitting of the $\Delta HLVP_i$ and ΔSAH_i coefficients reveals a high and significant correlation coefficient (R = −0.75, significant at the 99% confidence level). The ΔLLW_i coefficients are also well linearly related with ΔSAH_i (R = 0.83, significant at the 99% confidence level). Though this result may look strongly conditioned to the MIROC-ESM-CHEM outlier model, if it is not considered for the linear fitting the correlation coefficients remain significantly high (R = –0.50 and R = 0.65, significant at the 95% and 99% confidence levels, respectively). The slope of the linear fitting between the $\Delta HLVP_i$ and the ΔSAH_i coefficients is close to minus one (m = −1.17), indicating that the high-level atmosphere response is underestimated approximately in the same proportion as precipitation. On the other hand, the trend of the ΔLLW_i ans ΔSAH_i linear adjustment is lower than one (m = 0.36). Such result suggests that the underestimation of the simulated Sahel rainfall in response to the IPO is even higher than the one in the winds at low levels. Therefore, there is an evident linear relationship in the extent to which models underestimate the atmospheric and rainfall response to the IPO.

The conclusions drawn form these results reinforce the idea that the atmospheric component of the CMIP5 GCMs show low responsiveness to the SST as suggested by other works (Joly et al. 2007; Rodríguez-Fonseca et al. 2011; Kucharski et al. 2013; Vellinga et al. 2016) and, as a consequence, the modulation on precipitation is lower than in observations.

6.5 Summary of Main Findings

This chapter has addressed the objective of assessing the contribution of the main modes of SST to the total rainfall variance in the Sahel, Amazonia and Northeast of Brazil at decadal-to-multidecadal time scales. The main conclusions drawn from the results obtained are:

- The observed decadal-to-multidecadal rainfall variability in the Sahel is induced by the AMV to a greater extent. In contrast, most CMIP5 models suggest a dominant role of the GW (this result will be published in Villamayor and Mohino 2019).
- The GW is the main modulator of the Amazon precipitation decadal-to-multidecadal variability in observations and also in some CMIP5 models, although there are important uncertainties among themselves.
- Neither observations nor the CMIP5 models show a clear dominant contribution of one of the three SST modes on the Northeast precipitation decadal-to-multidecadal variability.
- The underestimation of the rainfall response to the SST modes is associated with the distribution of the SSTA and also with the low atmospheric sensitivity to SST changes that CMIP5 models simulate.
- The RCP8.5 future projection suggest that, under strong radiative forcing related with the GHGs, the GW could be expected to prominently dominate the rainfall low-frequency variability in the three regions studied.

References

Gaetani, M., Flamant, C., Bastin, S., Janicot, S., Lavaysse, C., Hourdin, F., Braconnot, P., Bony, S.: West African monsoon dynamics and precipitation: the competition between global SST warming and CO_2 increase in CMIP5 idealized simulations. Clim. Dyn. **48**, 1353–1373 (2017)

Janicot, S., Trzaska, S., Poccard, I.: Summer Sahel-ENSO teleconnection and decadal time scale SST variations. Clim. Dyn. **18**, 303–320 (2001). https://doi.org/10.1007/s003820100172

Joly, M., Voldoire, A., Douville, H., Terray, P., Royer, J.-F.: African monsoon teleconnections with tropical SSTs: validation and evolution in a set of IPCC4 simulations. Clim. Dyn. **29**, 1–20 (2007)

Kucharski, F., Molteni, F., King, M.P., Farneti, R., Kang, I.-S., Feudale, L.: On the need of intermediate complexity general circulation models: a SPEEDY example. Bull. Am. Meteorol. Soc. **94**, 25–30 (2013)

Martin, E.R., Thorncroft, C.D.: The impact of the AMO on the West African monsoon annual cycle. Quart. J. R. Meteorol. Soc. **140**, 31–46 (2014)

Mohino, E., Janicot, S., Bader, J.: Sahel rainfall and decadal to multi-decadal sea surface temperature variability. Clim. Dyn. **37**, 419–440 (2011a). https://doi.org/10.1007/s00382-010-0867-2

Nicholson, S.E.: The West African Sahel: a review of recent studies on the rainfall regime and its interannual variability. ISRN Meteorol. **2013**, 32 (2013). https://doi.org/10.1155/2013/453521

Rodríguez-Fonseca, B., Janicot, S., Mohino, E., Losada, T., Bader, J., Caminade, C., Chauvin, F., Fontaine, B., García-Serrano, J., Gervois, S., et al.: Interannual and decadal SST-forced responses of the West African monsoon. Atmos. Sci. Lett. **12**, 67–74 (2011)

Vellinga, M., Roberts, M., Vidale, P.L., Mizielinski, M.S., Demory, M.-E., Schiemann, R., Strachan, J., Bain, C.: Sahel decadal rainfall variability and the role of model horizontal resolution. Geophys. Res. Lett. **43**, 326–333 (2016)

Villamayor, J., Mohino, E.: SST drivers of Sahel rainfall at decadal timescales in CMIP5 simulations (2019). (In preparation)

Part III
Results II: Case Study

This part of the Thesis addresses the study of the SST influence on a particular decadal rainy period in the Sahel occurring during the late-19th century. This anomalous rainy period is reproduced, for the first time, by AGCM simulations forced with realistic boundary conditions available since year 1854. From these results, the atmospheric mechanisms involved in the abundance of precipitation during this early period in connection with SST are identified and compared with the well-documented decadal rainy period of the mid-20th century. Finally, attention is focused on the role of the Atlantic SST in controlling the late-19th century rainy period through the performance of a set of sensitivity experiments.

Chapter 7
Atlantic Control of the Late 19th Century Sahel Humid Period

Abstract In Chaps. 3–5, it has been shown that the variability of precipitation in the Sahel region is linked to that of the SST at decadal-to-multidecadal time scales. The shifts of the Sahel precipitation regime during the late 20th century have been extensively studied due to its dramatic humanitarian and economic consequences such as during the 1970s and 1980s severe droughts. However, little is known about the decadal variability prior to the 20th century. Some evidences suggest that during the second half of the 19th century the Sahel was as much or even more rainy than during the 1950s and 1960s. The aim of this Chapter is to provide further evidences of such anomalous Sahel humid period in the late-19th century reproducing it by means of climate simulations, as well as to give an explanation to its origin.

7.1 Model Validation

In order to study the Sahel rainfall during the second half of the 19th century, an ensemble of 19 simulations with realistic boundary conditions since year 1854 have been performed with the LMDZ model (more details in Sect. 2.2.1). In this section we evaluate the performance of the model in reproducing the climatology and the robustness of the precipitation response to the boundary conditions imposed in the ensemble of simulations performed.

First of all, the ability of the LMDZ model to reproduce the main features of the West African rainfall that affect the Sahel are evaluated by comparing the climatological Sahel precipitation during JAS simulated by the ensemble mean with respect to observations over the 1901–2000 common period (Fig. 7.1). The model successfully reproduces a maximum value of the climatological rainfall in JAS toward the north of West Africa of around 10 mm day^{-1}, averaged between 17.5°W and 10°E, which is similar to observations. However, while in observations this maximum reaches latitudes between 10° and 12°N, the one of the simulations remains at lower

The content of this chapter is published in the following article: Villamayor et al. (2018b): Atlantic Control of the Late Nineteenth-Century Sahel Humid Period. *J. Climate* https://doi.org/10.1175/JCLI-D-18--0148.1.

J. Villamayor, *Influence of the Sea Surface Temperature Decadal Variability on Tropical Precipitation: West African and South American Monsoon*, Springer Theses, https://doi.org/10.1007/978-3-030-20327-6_7

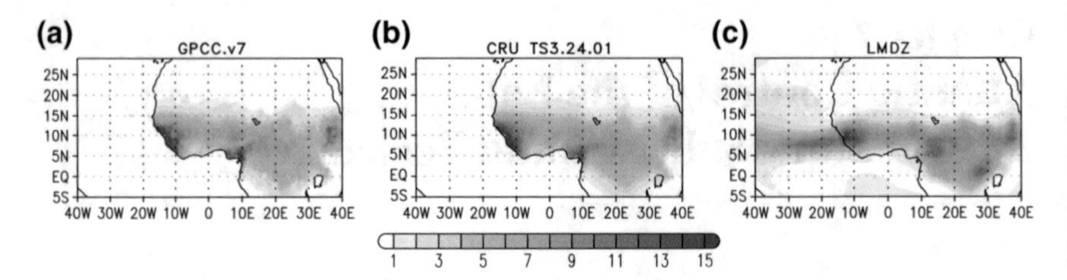

Fig. 7.1 Climatology over 1901–2000 of JAS precipitation (mm^{-1} day) for the GPCC v7 (**a**) and CRU TS3.24.01 (**b**) observational data bases and for the ensemble-mean of the LMDZ simulations (**c**). From Villamayor et al. (2018b)

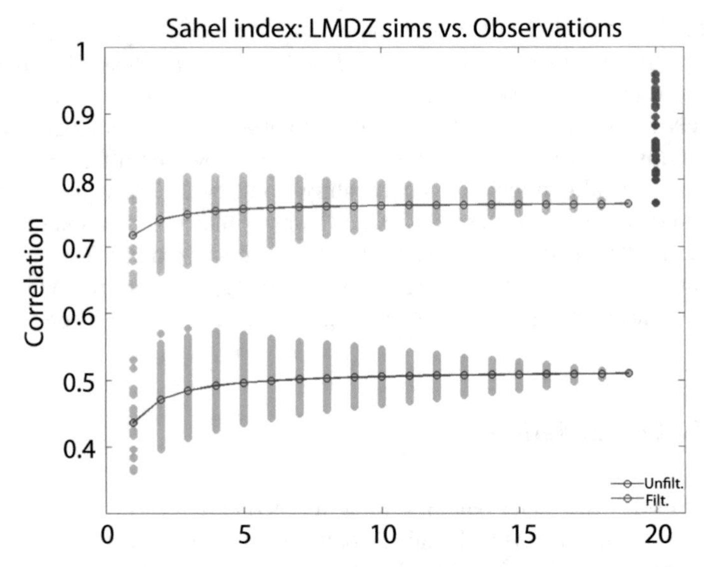

Fig. 7.2 Correlation coefficients between the 8-years low-pass filtered (orange dots) and unfiltered (light blue dots) Sahel JAS rainfall index of the GPCC v7 database and the ensemble mean over all possible combinations of 1–19 reproduced members in the LMDZ simulations. The horizontal axis indicates the number of members averaged. In circles, there are the mean values among all the correlation coefficients obtained for each ensemble. The dark dots to the right denote the correlation coefficients between the 8-years low-pass filtered (red) and unfiltered (blue) Sahel index of each individual member and the ensemble mean of the remaining 18. From Villamayor et al. (2018b). © American Meteorological Society. Used with permission

latitudes, slightly south of 10°N. As a consequence, the reproduced climatological JAS precipitation amount over the Sahel region is underestimated by 23% with respect to observations (from both the GPCC v7 and CRU TS3.24.01 data bases).

The robustness of the Sahel rainfall response to the forcings imposed that is obtained from the ensemble-mean LMDZ simulations with respect to the uncertainty coming from the intrinsic weather noise of the model is also evaluated. To do this, a plot of the correlation between the index of the observed Sahel rainfall (from GPCC

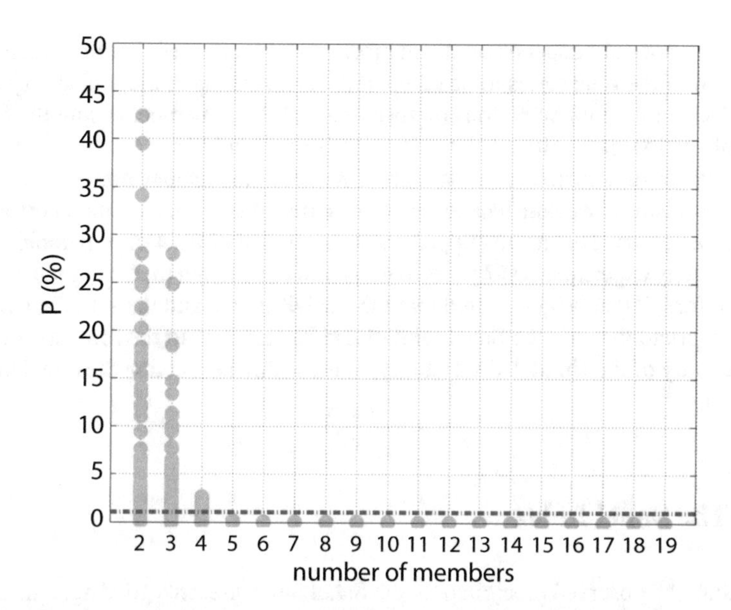

Fig. 7.3 Probability values (%) from the ANOVA analysis applied to the first 12 years of the Sahel indices for all possible combinations of ensembles of 2–19 members from the LMDZ simulations. The red dashed line indicate the threshold of the 99% confidence level

v7 data between 1901 and 2000) and the ensemble-mean index using from 1 to 19 members in all the possible combinations is done, following Caminade and Terray (2010) (Fig. 7.2). The mean of the correlation coefficients of the unfiltered indices rapidly converges to around R = 0.50. Regarding the 8-year low-pass filtered indices, the mean value of the correlation coefficients also quickly converges to R = 0.75. This means that the simulated Sahel precipitation can explain around 30% and 60% of the observed variability at all and at decadal time scales, respectively. The rapid convergence of the correlation mean coefficients to a stable value suggest that the correlation with observations will not increase significantly by adding more members. Therefore, for the sensitivity experiments ATLVAR and INPVAR (see Sect. 2.2.1) we chose 5 members as a compromise between computational effort and robustness of the results. The correlation between the Sahel index of each one of the 19 members separately with the ensemble mean of the remaining 18 simulations is also shown (filled dark red and blue dots in Fig. 7.2). Their high values of about 0.85 and 0.95, respectively at all-time scales and at decadal ones, also suggest the strong influence of the boundary conditions (observed SST, sea ice cover, GHGs concentrations and stratospheric aerosols) in the Sahel rainfall variability with respect to the random noise of the model.

The ANOVA technique is also used to calculate the probability that the variance among the Sahel indices of the different realizations is significantly lower than the common variance (i.e., that the variance led by the imposed forcings is significantly larger than the one associated with the internal weather noise of the model). This probability not only depends on the number of members used, but also strongly on the length of the time series to be compared. As an example, the probability values for all the possible combinations of ensembles of 2–19 members using the first 12 years of the low-pass filtered Sahel indices are shown in Fig. 7.3. In this case, we see that a set of 5 members in enough to neglect the weather noise effect on the ensemble mean. If we consider longer time series, the number of members needed rapidly decreases. This result supports that the choice of a set of 5 members to run the sensitivity experiments (57 years long) is prudent enough to obtain robust results.

Therefore, it can be considered that the LMDZ model and the set of simulations are appropriate to study the Sahel rainfall and its response to the forcings imposed. For the study of the simulated variability, in the following section the Sahel index is analyzed.

7.2 The Sahel Index

Regarding the observed precipitation, the Sahel index presents good agreement with the ASWI index of Gallego et al. (2015) (Fig. 7.4), which is based on the persistence of southwesterlies close to the coast of West Africa (more details in Sect. 1.4). They show marked decadal variability, with an anomalously rainy period over 1945–1963 followed by a dry one lasting to the end of the 20th century. Over the late-19th century the ASWI index suggest a long-lasting wet period over 1845–1890 and more intense than the one of the 20th century. In turn, the Sahel index based on the semi-

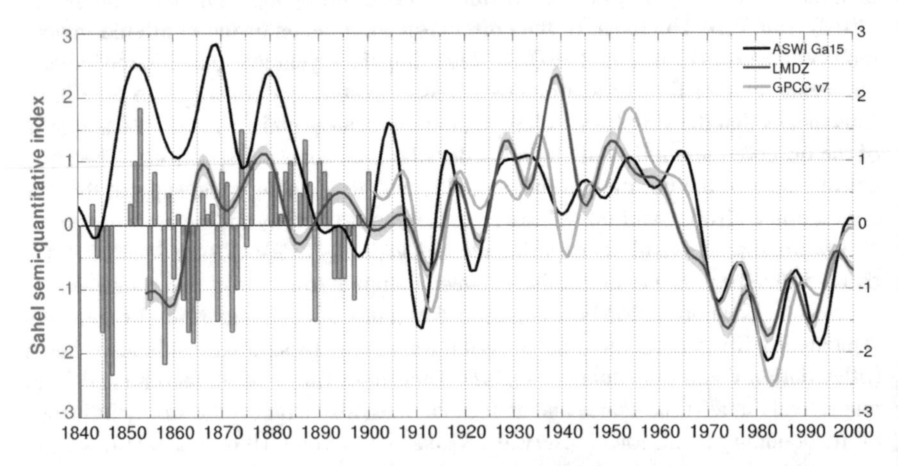

Fig. 7.4 In bars, the semi-quantitative index of Sahel precipitation of the reconstruction of Nicholson et al. (2012). The lines represent the seasonal ASWI in JAS of Gallego et al. (2015) (denoted with the subindex Ga15) (black) and the indices of the JAS seasonal precipitation in the Sahel (averaged in 17.5°W–10°E, 10°–17.5°N) in observations (green) and in the ensemble-mean simulation (blue). The last three indices have been low pass-filtered with an 8-years cut-off period and standardized with respect to the observed period (1901–2000). The blue shade is the standard deviation among the 19 members simulated. From Villamayor et al. (2018b)

quantitative reconstruction of Nicholson et al. (2012) supports the occurrence of such an early long rainy period but shorter (roughly from 1866 to 1892) and less intense than the one in the 20th century (see Fig. 1.27).

The Sahel index of precipitation that results from the ensemble-mean of the LMDZ simulations performed with realistic boundary conditions describes an evolution along the 20th century that is similar to the observed one (Fig. 7.4). Regarding its variability, the model underestimates 50% of the observed Sahel index standard deviation over the period 1901–2000 at all-time scales (using unfiltered indices) and 60% of it at decadal time scales (using 8-year low-pass filtered indices). This is consistent with the models underestimation of the atmospheric and precipitation response to SSTA shown in Chap. 6 and with other studies also showing that, in general, AGCMs have a weak response of the West African rainfall to SST forcing (e.g. Rodríguez-Fonseca et al. 2011; Joly et al. 2007; Kucharski et al. 2013; Vellinga et al. 2016). This is typically attributed to the coarseness of the AGCMs spatial resolution and to the accuracy with which the land-atmosphere feedbacks are parameterized (Giannini et al. 2003; Cook and Vizy 2006; Vizy et al. 2013; Vellinga et al. 2016). However, although the LMDZ simulations underestimate the climatological Sahel precipitation and its variability in comparison with observations, the low-pass filtered simulated and observed Sahel indices are strongly correlated over the 20th century (R = 0.77, significant at the 99% confidence level according to a "random-phase" test). The simulations successfully reproduce the observed low-frequency evolution between peaks of maximum or minimum precipitation in the Sahel, as well as the anomalous humid period of the mid-20th century and the subsequent dry one. The main discrepancy is around year 1940, when the simulations reproduce an intense increase of precipitation in contrast to observations. This inconsistency may be attributed to the high uncertainty of the SST data during the Second World War rather than to the model skills (Thompson et al. 2010; Kennedy 2014; Huang et al. 2016). In addition, the shift from positive to negative anomalies occurs during 1964, 5 years earlier than in observations and the ASWI of Gallego et al. (2015).

Regarding the Sahel rainfall over the second half of the 19th century, the simulations reproduce an anomalously rainy period from 1863 to 1883 with respect to the 20th century climatology (Fig. 7.4). As far as the simulated rainfall intensity and the duration are concerned, the late-19th century rainy period is comparable to the mid-20th century (1945–1963) one. The spatial pattern of precipitation shows significant positive anomalies over the Sahel (Fig. 7.5a). South of 10°N, there is a strong deficit of precipitation in the tropical Atlantic and non-significant anomalies inland.

At decadal time scales, the reproduced Sahel index is positively correlated with the observed ASWI of Gallego et al. (2015) in the 1854–2000 period (R = 0.53, significant at the 87% confidence level). There is, however, disagreement on the duration of the late-19th century wet period (Fig. 7.4): Conversely to Gallego et al. (2015), before 1863 the simulations reproduce anomalous drought conditions in the Sahel, in agreement, however, with Nicholson et al. (2012). The simulated rainy period ends around 1883, 6 years before than suggested by Gallego et al. (2015).

Therefore, these first results show that, albeit small differences in terms of timing and intensity, our simulations succeed in reproducing an unprecedented decadal wet period in the late-19th century, supporting the evidences of its existence (Gallego et al. 2015; Nicholson et al. 2012), and which has similar features to that of the mid-20th century. Such favorable comparison between simulated and observed Sahel rainfall variability since the 19th century suggests that the LMDZ experiments can help shed some light on the physical mechanisms at play.

7.3 Mechanisms Involved in the Late-19th Century Wet Period

During the Sahel humid period of the late-19th century, the LMDZ simulations show a reinforced monsoonal rainfall in West Africa (Fig. 7.5a). There is significant abundance of precipitation throughout a zonal band that covers the Sahel region

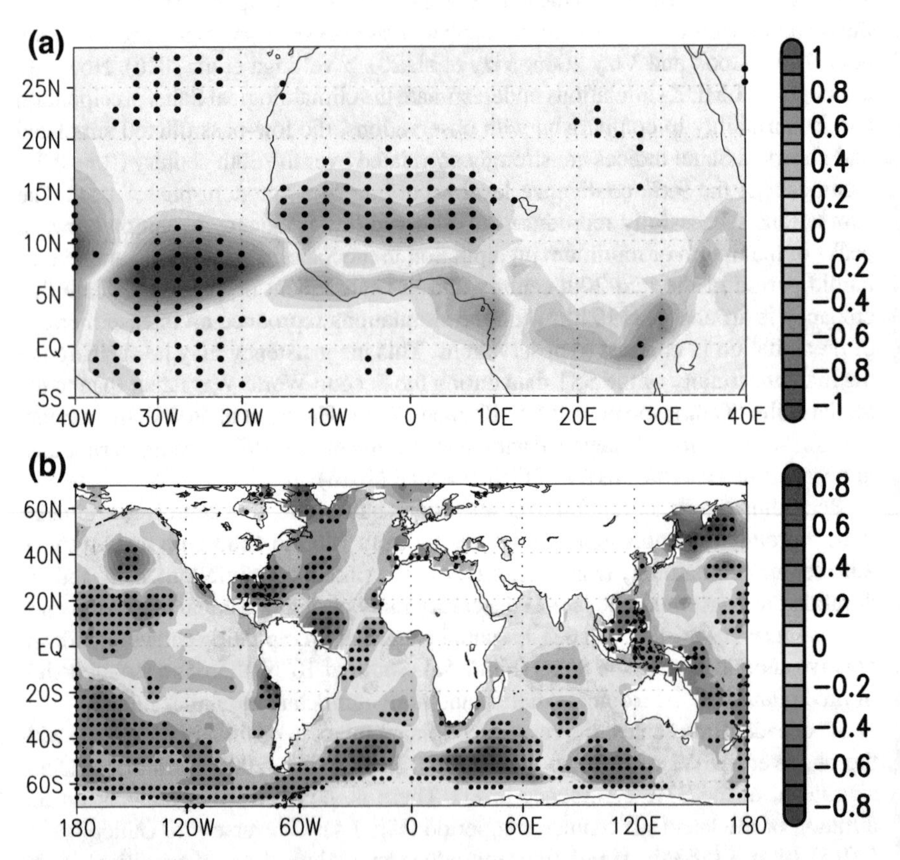

Fig. 7.5 1863–1883 mean JAS seasonal anomalies with respect to the 1854–2000 climatology of **a** the precipitation (mm^{-}1 day) of the ensemble-mean LMDZ simulation and **b** the SSTA (K) of the ERSST.v4 data base. Black dots indicate the regions where the averaged anomalies over the period 1863–1883 are significantly different from zero with 95% confidence level (from a t-test). From Villamayor et al. (2018b). © American Meteorological Society. Used with permission

(from the western coast to 10°E and between 10°–17.5°N), while the surrounding continental areas do not present significant rainfall anomalies. Over the tropical Atlantic, roughly within 3°–12°N, there is a marked deficit of rainfall. Such pattern is suggestive of a northward shift of the tropical rain-belt, typically associated with anomalous latitudinal displacements of the ITCZ over West Africa (Rowell et al. 1992; Knight et al. 2006).

This distribution of precipitation anomalies is associated with SST anomalies that broadly present an inter-hemispheric gradient in the Atlantic basin, with mostly warm SST anomalies (SSTA) in the northern half and cold ones to the south (Fig. 7.5b). An intense cold band of SSTA extends from southwestern to northeastern tropical North Atlantic. This cooling, along with the warm coastal SSTA, creates a strong thermal contrast around 10°N in the vicinity of West Africa. In more remote basins, there are remarkable negative SSTA in the central Pacific basin and positive ones in the extratropics, the maritime continent and the southern tropical Indian Ocean.

In order to understand the mechanisms driving the rainfall increase in the Sahel during the late-19th century, we analyze the mean simulated atmospheric state over the same period. In the Pacific, two subtropical cores of high surface pressure indicate anomalous cyclonic formations reminiscent of the Matsuno-Gill response to the tropical SST cooling (Matsuno 1966; Gill 1980) (Fig. 7.6a). Consistent with this pattern, the velocity potential at the 200 hPa level shows anomalous convergence over the tropical Pacific and divergence over the Maritime Continent and Central America. Such pattern suggests a weakening of the Walker Circulation associated with the anomalous tropical Pacific cooling (Gill 1980). However, this mechanism does not affect significantly the region of West Africa. On the other hand, there is a widespread decrease of surface pressure over the warm North Atlantic. These anomalies create an oblique pressure gradient from the negative ones in the northwest to the positive ones in the southeast tropical North Atlantic, which is statistically significant and coherent with the SSTA pattern. Associated with this, there is anomalous upward vertical wind at the 500 hPa level right over the Sahel region and downward to the south, being more significant over the tropical Atlantic (Fig. 7.6b). This indicates an anomalous northward shift of the ITCZ that favors rainy conditions in the Sahel (Rowell et al. 1992; Knight et al. 2006). Near the surface (at the 925 hPa level), there is anomalous westerly moist advection from the tropical Atlantic inland between 5°–10°N and northeasterly over the Sahel land surface. These moist advection anomalies are primarily due to anomalous horizontal winds (Pu and Cook 2010) rather than to changes in the air humidity content.

In the vertical profile (Fig. 7.7), we can see how the near-surface cross-equatorial monsoon flow presents a strengthening as well as convergence slightly south of 10°N. In turn, the vertical moisture transport at medium and low levels is reinforced south of the Sahel, with a maximum of anomalous humidity around 850 hPa. Then, the anomalous winds carry part of this humidity upwards to middle levels (700–500 hPa) and towards the north, over the Sahel. At mid-levels, the zonal moist advection shows mostly positive anomalies. This suggest a weakening of the mid-level climatological easterly winds, associated with the African Easterly Jet, also contributes to the middle troposphere moistening above the Sahel (Cook 1999). To the north, there is low-level

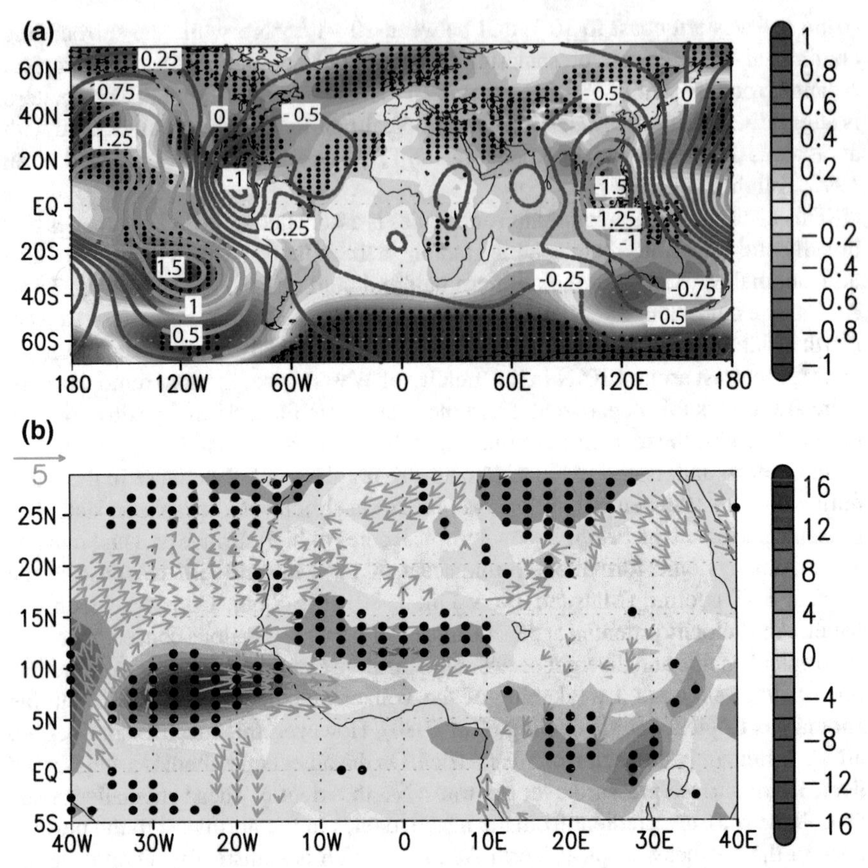

Fig. 7.6 1863–1883 mean of the JAS seasonal anomalies with respect to the entire simulated period (1854–2000) from the ensemble mean of different variables: **a** In colors, the surface pressure (hPa) (black dots indicate statistical significance) and, in contours, the velocity potential at the 200 hPa level (10^6 m^2 s^{-1}) (green contours where it is significant). **b** In colors, the omega vertical velocity (10^{-3} Pa m^{-1}) (black dots indicate significance) and, in vectors, the vertical moist advection (g kg^{-1} m s^{-1}) (plotted where the anomalies of any of the two wind components are significant). The significance of all the anomalies is calculated for a 95% confidence level following a t-test applied to the average over the period 1863–1883. Adapted from (Villamayor et al. 2018b). © American Meteorological Society. Used with permission

moisture transport associated with wind anomalies from the Atlantic between 15°–20°N that penetrates north of the Sahel. Such anomalous moisture inflow, together with a weaker one coming from the Mediterranean Sea, results in large accumulation of low-level humidity in the northernmost Sahel. The humid air, in turn, rises contributing to moisten even more the mid-level air over the Sahel. The air rise is favored by strengthened convergence at around 18°N, which climatologically produces dry convection in the boundary between the Sahel and the Sahara desert (Hall and Peyrillé 2006). Above 600 hPa, there is an intensification of rising air between 10°–20°N and

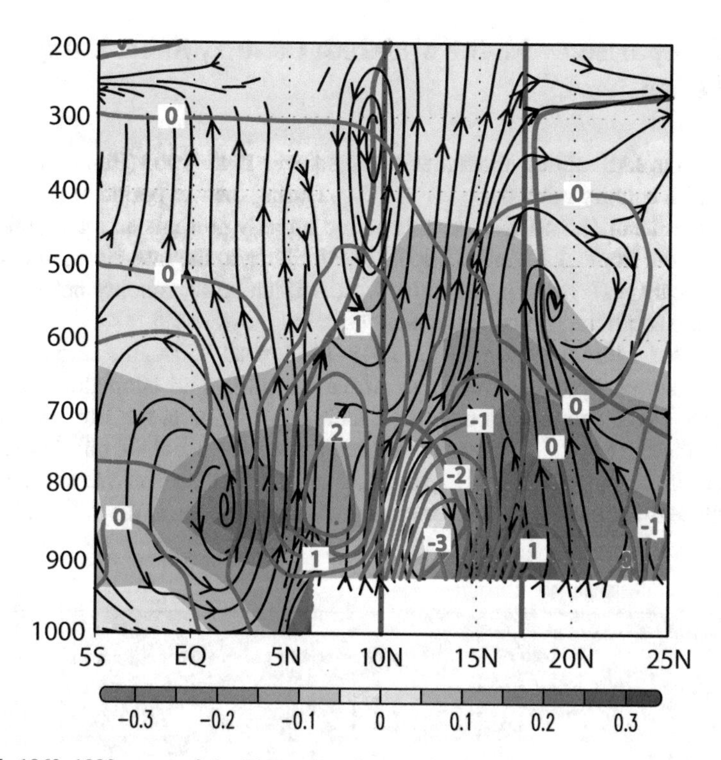

Fig. 7.7 1863–1883 mean of the JAS seasonal anomalies with respect to the entire simulated period (1854–2000) from the ensemble mean of the vertical profile (averaged between 10°W–10°E) of the specific humidity (g kg^{-1}) in colors (gray contours indicate significance), the zonal moist advection (g kg^{-1} m s^{-1}) in red contours and in black streamlines its vertical and latitudinal components (unrealistic units). The significance of the anomalies is calculated for a 95% confidence level following a t-test applied to the average over the period 1863–1883. Purple lines indicate the latitudinal limits of the Sahel. Adapted from Villamayor et al. (2018b). © American Meteorological Society. Used with permission

a weakening between 5°–10°N. This means stronger deep convection over the Sahel associated with the already noted anomalous northward shift of the ITCZ. Therefore, the anomalously abundant intrusion of humidity at mid-levels, combined with the intensification of deep convection, explains the Sahel rainfall enhancement. However, anomalous northeasterly low-level winds over the Sahel, likely associated with the aforementioned anomalous convergence slightly south of 10°N, advect humidity out of the region drying the air in the southern half where, on the other hand, it rains more. Summarizing, this result shows that the reproduced late-19th century period of abundant Sahel precipitation is associated with anomalous humidity supply and deep convection in the middle and high troposphere related to wind anomalies which, most likely, are induced by the Atlantic SSTA.

7.4 Comparison with the Mid-20th Century Sahel Rainy Period

The pattern of JAS rainfall anomalies averaged over 1945–1963 (Fig. 7.8a) is akin to the late-19th century rainy period (Fig. 7.5a). During the wet period of the mid-20th century, the mean JAS SSTA over this period broadly presents an interhemispheric contrast in the tropical Atlantic (Fig. 7.8b). Similarly to the late-19th century, there is a significant SST warming near the West African coast, roughly north of 10°N, while to the south the anomalies remain weak but mostly cold.

Associated with the pattern of the 1945–1963 mean SSTA, the surface pressure presents a strong gradient over West Africa that favors an anomalous northward shift of the ITCZ and rainfall over the Sahel (Fig. 7.9), as in the late-19th century. However there are some differences between the late-19th and the mid-20th century rainy periods, in particular regarding the moisture distribution over West Africa. In the mid-20th century, there is a stronger moisture transport towards the Sahel (Fig. 7.9b–

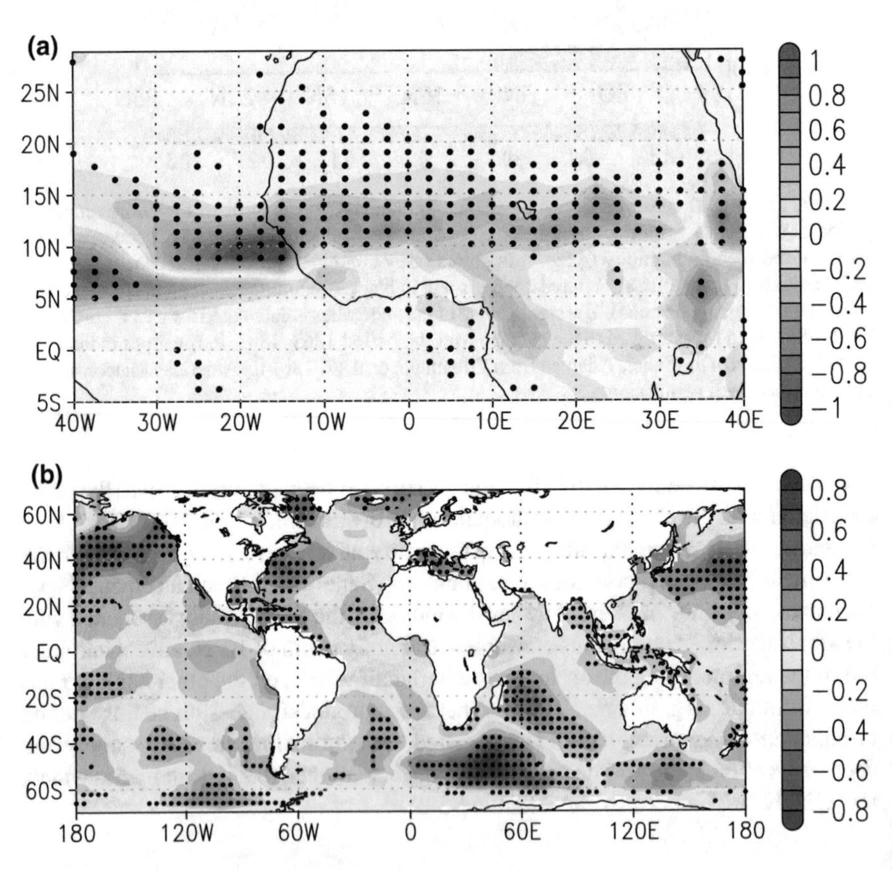

Fig. 7.8 Same as Fig. 7.5 but for the mean of the 1945–1963 period

c), probably related to the stronger latitudinal surface pressure gradient (Fig. 7.9a). The vertical profile of moisture advection shows low-level moist air inflow from the Gulf of Guinea to the Sahel region where it accumulates and raises by convection (Fig. 7.9c), in contrast to the late-19th century. However, despite this difference, the amounts of the reproduced rainfall averaged over each wet period are practically the same (0.26 mm day^{-1} above the climatology).

Therefore, these results suggest that the late-19th century rainy period was produced under similar conditions to the mid-20th century one.

7.5 The Role of the Atlantic and Indo-Pacific

To evaluate the role of the Atlantic and Indo-Pacific Oceans in driving the unprecedented Sahel rainy period of the late-19th century, two additional SST-sensitivity experiments for the period 1854–1910 are performed in which all the SST variability is left, respectively, in the Atlantic (ATLVAR) or Indo-Pacific (INPVAR) basins while the remaining global oceanic surface is fixed to the monthly climatology relative to 1854–1910 (more details in Section 2.2.1). The other boundary conditions remain unchanged with respect to the previous simulations of reference (hereinafter REF).

Figure 7.10 highlights the role of the Atlantic SST in driving the long term Sahel precipitation variability along 1854–1910. The Sahel index of the ATLVAR experiments shows a late-19th century rainy period similar to the one of the REF simulations, even with higher peak values. While, in contrast, the INPVAR experiment reproduces less precipitation. After 1883, ATLVAR and INPVAR experiments, respectively, reproduce less and more Sahel precipitation than the REF simulations. Therefore, the SST variability of the Atlantic and Indo-Pacific basins apparently have contrary effects on the Sahel rainfall in the 1854–1910 period.

Regarding the atmospheric response to the SST imposed in the sensitivity experiments, the average over 1863–1883 in the ATLVAR experiment shows higher surface pressure over the Indian Ocean than the REF simulations (Fig. 7.11a). Consistently, there is more high-level convergence locally and more high-level divergence over West Africa in the former than in the latter, consistently with more Sahel rainfall in ATLVAR than in REF (Fig. 7.10). In addition, over the North Atlantic, the surface pressure is also lower in ATLVAR than in REF, which is associated with a wider ITCZ shift that favors Sahel rainfall. On the other hand, the resulting surface pressure from the INPVAR experiment is high with respect to REF simulations all over the Atlantic basin and the surrounding continents, particularly over West Africa, coinciding with high-level convergence (Fig. 7.11b). While the surface pressure is lower than in REF in the rest of the globe and there is more high-level divergence over the central Pacific and the northern Indian Ocean. So, whereas the Atlantic seems to drive changes on the Sahel rainfall regime, shifting from dry to rainy in 1863 and back to dry in 1883, the role of the Indo-Pacific SST is rather to attenuate these effects through local subsidence anomalies.

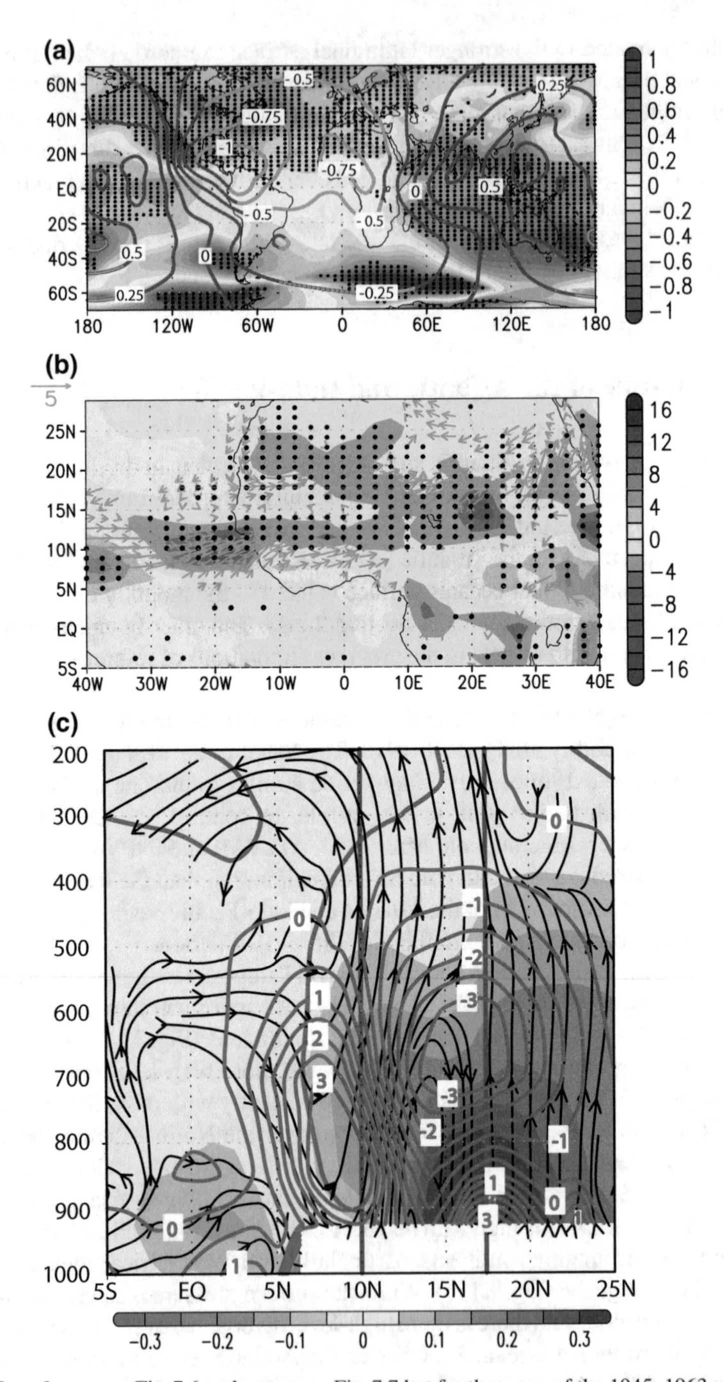

Fig. 7.9 **a–b** same as Fig. 7.6 and **c** same as Fig. 7.7 but for the mean of the 1945–1963 period

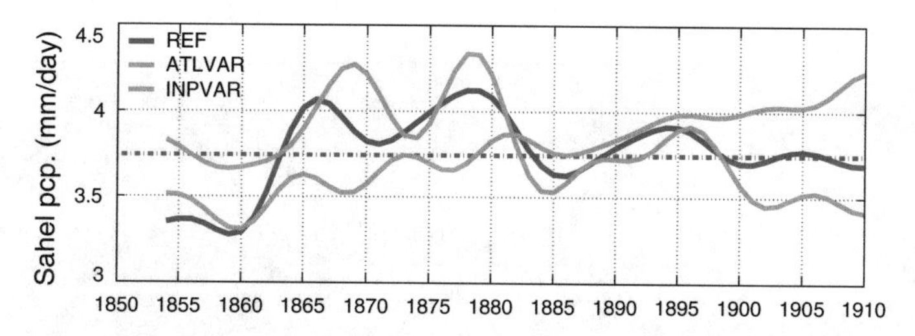

Fig. 7.10 8-years low pass-filtered Sahel indices (mm day^{-1}) of the ensemble-mean REF simulation (dark blue) and the ATLVAR and INPVAR sensitivity experiments. The hatched blue line indicates the 1854–2000 climatological value of the REF simulation Sahel index. The bar chart represents the components (in %) of the Sahel index total variance at decadal timescales reproduced by REF explained by a multi-linear regression analysis with the Sahel indices of both sensitivity experiments. The components correspond to the Sahel index of ATLVAR (orange), INPVAR (green), the covariance between both (light blue contour) and the residual of the multi-linear regression fitting (black contour). From Villamayor et al. (2018b). © American Meteorological Society. Used with permission

A multi-linear regression analysis reveals that the Sahel index of the ATLVAR experiment accounts for most (76%) of the total Sahel index variance of the REF simulations (Fig. 7.12a). This result evidences that the Atlantic SST variability itself is responsible for the vast majority of the reproduced low-frequency changes in the Sahel precipitation over the period 1854–1910. In addition, the compensation between the variance explained by the INPVAR and the co-variance term suggests that the variability of the Atlantic and Indo-Pacific SST is not completely decoupled.

In contrast, at inter-annual time scales these two components are notably decoupled (Fig. 7.12b). In this case, the multi-linear regression model shows that the Sahel index of the INPVAR experiment is the component that accounts for more of the Sahel index variance (40%) reproduced by REF. This is consistent with the important link between the inter-annual tropical Pacific SST variability and the Sahel rainfall (Janicot et al. 1996, 1998, 2001; Mohino et al. 2011b). Instead, the Atlantic SST influence on the inter-annual Sahel index variability (30%) plays a secondary role.

The similar multi-linear regression analysis applied to the ASWI indices shows that the one obtained from the ATLVAR experiment accounts for most of both the low- and the high-frequency ASWI variability (77% and 80%, respectively) (Fig. 7.12c–d). This suggests that the simulated ASWI responds almost exclusively to the Atlantic SST variability at any time scales. Furthermore, the low-pass filtered ASWI and Sahel indices simulated by REF are strongly correlated ($R = 0.89$, significant with a 99% confidence level) along the entire simulated period. While at short time scales (8-year high-pass filtering both indices) this relationship is weaker ($R = 0.59$, significant with a 99% confidence level). Such tight relationship at decadal time scales between both indices is in agreement with observations, as shown by Gallego et al. (2015). Therefore, the ASWI can be considered as a good indicator of the Sahel rainfall variability (as hypothesized by Gallego et al. (2015)) as long as it is modulated by the Atlantic SST, which occurred at decadal time scales in the late-19th century.

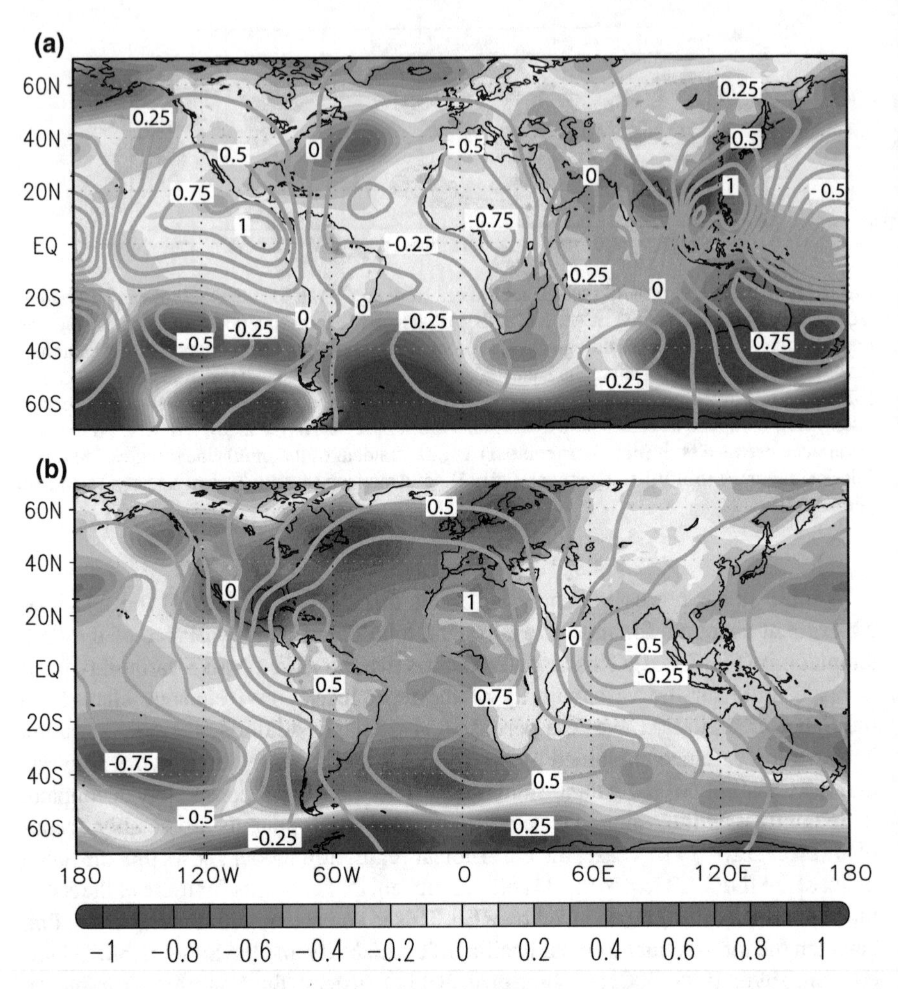

Fig. 7.11 Difference of the 1863–1883 mean JAS surface pressure (hPa), in colors, and the velocity potential at the 200 hPa level (10^6 m^2 s^{-1}), in contours, resulting from the **a** ATLVAR and **b** INPVAR experiments minus the REF simulations. The variables are obtained from the ensemble-mean of all the simulations. From Villamayor et al. (2018b). © American Meteorological Society. Used with permission

7.6 Discussion

In this chapter we have addressed the objective of achieving more evidence of a decadal rainy period in the Sahel during the late-19th century. Our results show that a decadal period of abundant Sahel precipitation in the late-19th century can be reproduced by forcing an AGCM with observational boundary conditions since 1854. This result presents further evidence for the existence of such a rainy period. For the first time, we add information about the mechanisms involved in this long sequence

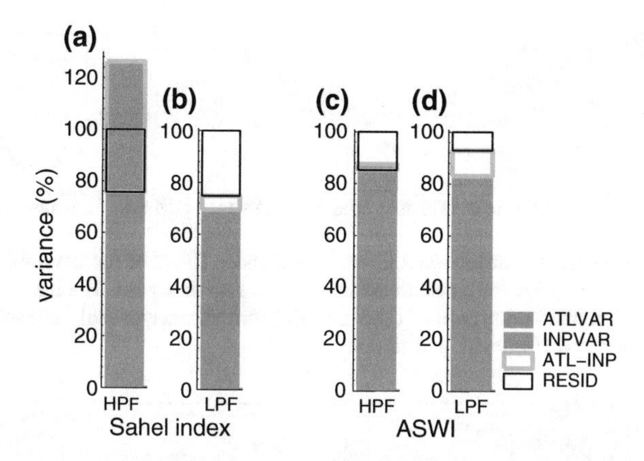

Fig. 7.12 Bar charts of the components (in %) of the total variance of the Sahel index reproduced by REF at **a** decadal (8-year low-pass filtered, shown in Fig. 7.10) and **b** inter-annual timescales (8-year high-pass filtered) explained by the two indices of the sensitivity experiments using a multi-linear regression analysis (Eq. 2.28). **c** and **d** represent the same as (**a**) and (**b**), respectively, but for the ASWI. The components correspond to the indices of ATLVAR (orange) and INPVAR (green), the covariance between both (light blue contour) and the residual of the multi-linear regression fitting (black contour)

of rainy years. The simulations show that, associated with a thermal gradient in the North Atlantic produced by a contrast of the SSTA near the coast of West Africa, there was a strong shift of the ITCZ over the Sahel that favored abundant precipitation with respect to climatology. A similar mechanism is reproduced for the mid-20th century rainy period. Therefore, it can be suggested that both decadal periods may have occurred under similar circumstances (these results are published in Villamayor 2018b).

A set of sensitivity experiments reveals that the Atlantic SST led the Sahel rainfall variability at decadal time scales throughout 1854–1910. So the origin of the late-19th century rainy period may be attributed to the decadal variability of the Atlantic SST. More specifically, it could be related to the AMV since it is the main mode of the Atlantic SST variability at decadal-to-multidecadal time scales. Furthermore, the positive AMV pattern of SSTA is reminiscent of the mean SSTA obtained for the late-19th century (Fig. 7.5b). The standardized Sahel index and the AMV time series computed with the ERSST.v4 data are strongly correlated along the 1854–2000 period ($R = 0.72$, significant at a confidence level of 97%) and the periods of positive (negative) rainfall anomalies in the Sahel roughly coincide with those of positive (negative) AMV phases (Fig. 7.13). This result is in agreement with other works addressing the Sahel rainfall low-frequency variability during the 20th century in observations and general circulation coupled models (e.g. Mohino et al. 2011a; Zhang and Delworth 2006; Hoerling et al. 2006; Knight et al. 2006; Ting et al. 2009; Martin et al. 2014; Martin and Thorncroft 2014). Focusing on the late-19th century, the wet period (1863–1883) coincides with a positive AMV phase of large amplitude.

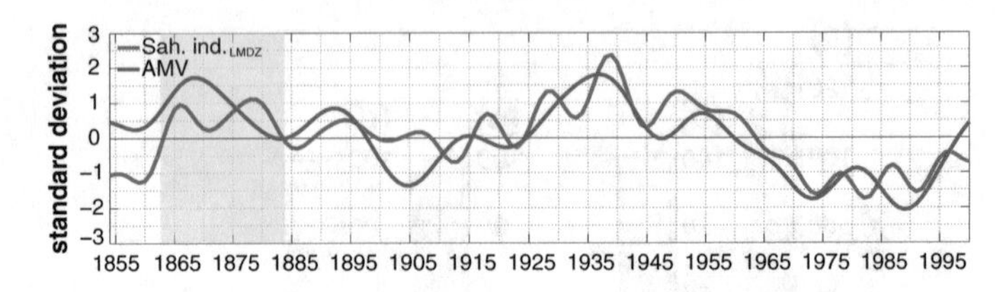

Fig. 7.13 Standardized 8-year low-pass filtered Sahel index (blue) of the ensemble-mean LMDZ simulations and AMV index (red) calculated from ERSST.v4 data base. The light blue band highlights the late-19th century rainy period (1863–1883). From Villamayor et al. (2018b). © American Meteorological Society. Used with permission

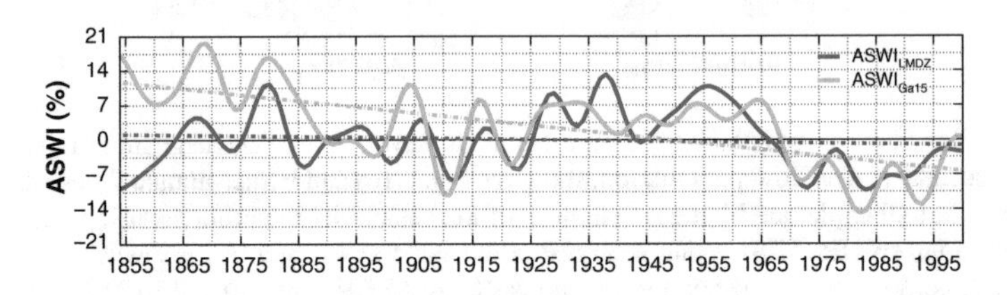

Fig. 7.14 8-year low-pass filtered ASWIs, normalized with respect to their climatology over 1854–2000, of both the ensemble-mean LMDZ simulations (red) and Gallego et al. (2015) (green). Hatched lines indicate the linear trend of the ASWIs over 1854–2000. The slopes of both indices are −0.016 and −0.124 ASWI units (%) per year, respectively. From Villamayor et al. (2018b). © American Meteorological Society. Used with permission

Considering all this, it is reasonable to believe that during the late-19th century there was a connection between the AMV mode of variability and the Atlantic SSTA responsible for the forcing of the Sahel humid period.

In spite of the above results, there are still some unanswered questions regarding Sahel rainfall. On the one hand, our simulations show that this period was as rainy as the mid-20th century one. Nicholson et al. (2012) rather suggests that there was moderate rainfall during the late-19th century and Gallego et al. (2015) that the precipitation was even more abundant than during the observed mid-20th century. To accurately determine the precipitation amounts during the late-19th century, a greater effort to reconstructed or collect reliable observational precipitation data from this period is required. On the other, the trends along the 1854–2000 period of the ASWI long-time series of Gallego et al. (2015) and our simulations do not agree (Fig. 7.14). Both indices show negative trends, however they disagree on the strength: 12.4 ASWI units (%) per century (significant at a 99% confidence level) for the ASWI from Gallego et al. (2015) and 1.6 ASWI units (%) per century (not statistically significant at the 80% confidence level) for the LMDZ simulation. Such disagreement could be due to an underestimation of the drying trend by the model. According to Gaetani et al. (2017), this underestimation could be related to the sensitivity of the model to the competing effects of the global SST warming and the atmospheric carbon dioxide concentration increase, which dampen and favor the Sahel rainfall, respectively. Moreover, the results from Nicholson et al. (2012) also disagree with Gallego et al. (2015) on the strength of the 19th century wet period, suggesting the results from the latter could be overestimating the negative trend. Therefore, this result hinders our knowledge of the evolutions of the WAM strength at longer than decadal time scales with our simulations and, by extension, the possible implications on the future projections of Sahel rainfall (Biasutti and Giannini 2006; Biasutti et al. 2008; Biasutti 2013; Monerie et al. 2012; Vizy et al. 2013).

References

Biasutti, M., Giannini, A.: Robust Sahel drying in response to late 20th century forcings. Geophys. Res. Lett. **33**, L11 706 (2006). https://doi.org/10.1029/2006GL026067

Biasutti, M.: Forced Sahel rainfall trends in the CMIP5 archive. J. Geophys. Res. Atmos. **118**, 1613–1623 (2013)

Biasutti, M., Held, I.M., Sobel, A.H., Giannini, A.: SST forcings and sahel rainfall variability in simulations of the twentieth and twenty-first centuries. J. Clim. **21**, 3471–3486 (2008). https://doi.org/10.1175/2007JCLI1896.1

Caminade, C., Terray, L.: Twentieth century Sahel rainfall variability as simulated by the ARPEGE AGCM, and future changes. Clim. Dyn. **35**, 75–94 (2010)

Cook, K.H.: Generation of the African easterly jet and its role in determining West African precipitation. J. Clim. **12**, 1165–1184 (1999)

Cook, K.H., Vizy, E.K.: Coupled model simulations of the West African monsoon system: twentieth- and twenty-first-century simulations. J. Clim. **19**, 3681–3703 (2006)

Gaetani, M., Flamant, C., Bastin, S., Janicot, S., Lavaysse, C., Hourdin, F., Braconnot, P., Bony, S.: West African monsoon dynamics and precipitation: the competition between global SST warming and CO_2 increase in CMIP5 idealized simulations. Clim. Dyn. **48**, 1353–1373 (2017)

Gallego, D., Ordóñez, P., Ribera, P., Peña-Ortiz, C., García-Herrera, R.: An instrumental index of the West African Monsoon back to the nineteenth century. Quart. J. R. Meteorol. Soc. **141**, 3166–3176 (2015)

Giannini, A., Saravanan, R., Chang, P.: Oceanic forcing of Sahel rainfall on interannual to interdecadal time scales. Science **302**, 1027–1030 (2003). https://doi.org/10.1126/science.1089357

Gill, A.E.: Some simple solutions for heat-induced tropical circulation. Quart. J. R. Meteorol. Soc. **106**, 447–462 (1980). https://doi.org/10.1002/qj.49710644905

Hall, N.M., Peyrillé, P.: Dynamics of the West African monsoon, vol. 139, pp. 81–99, EDP Sciences (2006)

Hoerling, M., Hurrell, J., Eischeid, J., Phillips, A.: Detection and attribution of twentieth-century Northern and Southern African rainfall change. J. Clim. **19**, 3989–4008 (2006). https://doi.org/10.1175/jcli3842.1

Huang, B., Thorne, P.W., Smith, T.M., Liu, W., Lawrimore, J., Banzon, V.F., Zhang, H.-M., Peterson, T.C., Menne, M.: Further exploring and quantifying uncertainties for extended reconstructed sea surface temperature (ERSST) version 4 (v4). J. Clim. **29**, 3119–3142 (2016)

Janicot, S., Harzallah, A., Fontaine, B., Moron, V.: West African monsoon dynamics and eastern equatorial Atlantic and Pacific SST anomalies (1970–88). J. Clim. **11**, 1874–1882 (1998). https://doi.org/10.1175/1520-0442(1998)011<1874:WAMDAE>2.0.CO;2

Janicot, S., Moron, V., Fontaine, B.: Sahel droughts and ENSO dynamics. Geophys. Res. Lett. **23**, 515–518 (1996)

Janicot, S., Trzaska, S., Poccard, I.: Summer Sahel-ENSO teleconnection and decadal time scale SST variations. Clim. Dyn. **18**, 303–320 (2001). https://doi.org/10.1007/s003820100172

Joly, M., Voldoire, A., Douville, H., Terray, P., Royer, J.-F.: African monsoon teleconnections with tropical SSTs: validation and evolution in a set of IPCC4 simulations. Clim. Dyn. **29**, 1–20 (2007)

Kennedy, J.J.: A review of uncertainty in in situ measurements and data sets of sea surface temperature. Rev. Geophys. **52**, 1–32 (2014)

Knight, J.R., Folland, C.K., Scaife, A.A.: Climate impacts of the Atlantic multidecadal oscillation. Geophys. Res. Lett. **33**, L17 706 (2006). https://doi.org/10.1029/2006GL026242

Kucharski, F., Molteni, F., King, M.P., Farneti, R., Kang, I.-S., Feudale, L.: On the need of intermediate complexity general circulation models: a SPEEDY example. Bull. Am. Meteorol. Soc. **94**, 25–30 (2013)

Martin, E.R., Thorncroft, C.D.: The impact of the AMO on the West African monsoon annual cycle. Quart. J. R. Meteorol. Soc. **140**, 31–46 (2014)

Martin, E.R., Thorncroft, C., Booth, B.B.B.: The multidecadal Atlantic SST-Sahel rainfall teleconnection in CMIP5 simulations. J. Clim. **27**, 784–806 (2014). https://doi.org/10.1175/JCLI-D-13-00242.1

Matsuno, T.: Quasi-geostrophic motions in the equatorial area. J. Meteorol. Soc. Jpn Ser. **II**(44), 25–43 (1966)

Mohino, E., Janicot, S., Bader, J.: Sahel rainfall and decadal to multi-decadal sea surface temperature variability. Clim. Dyn. **37**, 419–440 (2011a). https://doi.org/10.1007/s00382-010-0867-2

Mohino, E., Rodríguez-Fonseca, B., Mechoso, C.R., Gervois, S., Ruti, P., Chauvin, F.: Impacts of the tropical Pacific/Indian oceans on the seasonal cycle of the West African monsoon. J. Clim. **24**, 3878–3891 (2011b). https://doi.org/10.1175/2011JCLI3988.1

Monerie, P.A., Fontaine, B., Roucou, P.: Expected future changes in the African monsoon between 2030 and 2070 using some CMIP3 and CMIP5 models under a medium-low RCP scenario. J. Geophys. Res. Atmos. **117**, 1–12 (2012). https://doi.org/10.1029/2012JD017510

Nicholson, S.E., Klotter, D., Dezfuli, A.K.: Spatial reconstruction of semi-quantitative precipitation fields over Africa during the nineteenth century from documentary evidence and gauge data. Quat. Res. **78**, 13–23 (2012)

Pu, B., Cook, K.H.: Dynamics of the West African westerly jet. J. Clim. **23**, 6263–6276 (2010)

Rodríguez-Fonseca, B., Janicot, S., Mohino, E., Losada, T., Bader, J., Caminade, C., Chauvin, F., Fontaine, B., García-Serrano, J., Gervois, S., et al.: Interannual and decadal SST-forced responses of the West African monsoon. Atmos. Sci. Lett. **12**, 67–74 (2011)

Rowell, D.P., Folland, C.K., Maskell, K., Owen, J.A., Ward, M.N.: Modelling the influence of global sea surface temperatures on the variability and predictability of seasonal Sahel rainfall. Geophys. Res. Lett. **19**, 905–908 (1992)

Thompson, D.W.J., Wallace, J.M., Kennedy, J.J., Jones, P.D.: An abrupt drop in Northern Hemisphere sea surface temperature around 1970. Nature **467**, 444–447 (2010). https://doi.org/10.1038/nature09394

Ting, M., Kushnir, Y., Seager, R., Li, C.: Forced and Internal Twentieth-Century SST Trends in the North Atlantic*. J. Clim. **22**, 1469–1481 (2009). https://doi.org/10.1175/2008JCLI2561.1

Vellinga, M., Roberts, M., Vidale, P.L., Mizielinski, M.S., Demory, M.-E., Schiemann, R., Strachan, J., Bain, C.: Sahel decadal rainfall variability and the role of model horizontal resolution. Geophys. Res. Lett. **43**, 326–333 (2016)

Villamayor, J., Mohino, E., Khodri, M., Mignot, J., Janicot, S.: Atlantic control of the late nineteenth-century Sahel Humid period. J. Clim. **31**, 8225–8240 (2018b)

Vizy, E.K., Cook, K.H., Crétat, J., Neupane, N.: Projections of a wetter Sahel in the twenty-first century from global and regional models. J. Clim. **26**, 4664–4687 (2013)

Zhang, R., Delworth, T.L.: Impact of Atlantic multidecadal oscillations on India/Sahel rainfall and Atlantic hurricanes. Geophys. Res. Lett. **33**, L17 712, (2006). https://doi.org/10.1029/2006GL026267

Part IV
Concluding Remarks

Chapter 8
Conclusions and Future Work

The main conclusions of this Thesis are presented in this chapter together with potential lines of future work.

8.1 Main Conclusions

Following the order of the objectives pursued, the main conclusions that have been drawn along the Thesis are brought together in a point-by-pint structure to give a synthesized insight of the reached results.

In the first part of this Thesis, a multi-model analysis is done in order to find out whether CMIP5 models reproduce the influence of the GW, AMV and IPO on the precipitation variability in the regions of the Sahel, the Amazonia and the Northeast of Brazil. The first objective in this part was to characterize the main modes of decadal-to-multidecadal variability of SST (GW, AMV and IPO). The results from the analysis of the CMIP5 simulations reveal that:

- The GW signal simulated by the CMIP5 models, on average, reproduces the main features of the observed one. Individually, models show important differences among themselves, with some simulating unrealistically wide oscillations of the global SSTA in the overall warming trend. This discrepancy among models is attributed to the simulated effects of the aerosol radiative forcing that they simulate, which are highly model-dependent. In contrast, the radiative forcing effect of the GHGs on the SST is strongly consistent among models. The simulated model-mean GW pattern of SSTA is similar to the observed one, with stronger warming over the tropics than the extratropics. But models fail in reproducing the observed

J. Villamayor, *Influence of the Sea Surface Temperature Decadal Variability on Tropical Precipitation: West African and South American Monsoon*, Springer Theses, https://doi.org/10.1007/978-3-030-20327-6_8

tropical Pacific cooling associated with the GW, where, instead, they simulate an intense warming.

- CMIP5 models, on average, succeed in reproducing the observed AMV. They reproduce the interhemispheric SST gradient over the Atlantic Ocean, although with underestimated anomalies. Individually, some models simulate unrealistic AMV patterns, with a poorly defined tropical SSTA gradient. However, they reproduce characteristic oscillation periodicities of the AMV index that are distributed in a range around 65 years, as in observations.
- The simulated IPO by CMIP5 models, on average, also reproduces the main features of the observed one, except for the strength of the associated SSTA. CMIP5 models reproduce the characteristic IPO pattern with strong SSTA of the same sign in the tropical and eastern Pacific and opposite ones in the extratropical parts of the basin. As in observations, the IPO indices individually simulated by models present characteristic periodicities in two ranges around 15–25 and 50–70 years.

Once the main modes of decadal-to-multidecadal SST variability are characterized in observations and CMIP5 simulations, a study of the influence of the GW, AMV and IPO on precipitation in the Sahel, Amazonia and Northeast is done aiming to gain a better understanding on these links. The results obtained from the CMIP5 simulations and observations show that:

- In response to the GW, the Sahel JAS rainfall decreases during the historical period, while in the Northeast of Brazil it is enhanced in the DJFMAM season. In both regions, the observed impact of the GW on rainfall is weak but consistent among three precipitation data bases and supported by CMIP5 models. The observed impact of the GW on the Amazon is to increase DJFMAM precipitation, though models show high uncertainty as to this relationship due to the high disagreement among themselves.
- In its positive phases, the AMV induces more rainfall over the Sahel and the Amazon and reduces it in the Northeast of Brazil in their respective rainy seasons. During negative AMV phases the impacts are the opposite. Such a precipitation response is consistent among observations and CMIP5 simulations. But models, in general, simulate weaker precipitation anomalies than the observed ones. The underestimated rainfall response to the AMV is associated with the intensity of the interhemispheric SSTA gradient in the tropical Atlantic simulated by the CMIP5 models. This relationship also accounts for the models spread in reproducing the link between the AMV and precipitation in the three regions.
- The relationship between the IPO and precipitation in the Sahel, the Amazon and the Northeast regions is negative. This link is noticeable using different observational data bases of precipitation and is also successfully reproduced by CMIP5 simulations of most models, though generally underestimated. The differences found among GCMs has been related to the accuracy with which they reproduce the observed intensity of the tropical Pacific SSTA. This is linearly related with the intensity of the rainfall anomalies that models simulate and may partially explain the underestimation of the precipitation anomalies.

Another objective of this Thesis was to understand the atmospheric teleconnections explaining the link between the modes of SST and precipitation. From the analysis of the CMIP5 models and observations, the main conclusions drawn are:

- The GW weakens the WAM low-level circulation associated with enhanced subsidence over West Africa, which results in reduced Sahel JAS precipitation. In northern South America, the atmospheric response to the GW is tightly related to the tropical Pacific SSTA, which induces anomalous Walker circulation. In observations, this mechanisms produces anomalous subsidence over the tropical Pacific and anomalous convection over northern South America (remember that SST trends are negative over the equatorial Pacific). This is consistent with the observed enhancement of the DJFMAM precipitation, specially in the Amazonia. However, since CMIP5 models do not reproduce the observed Pacific SSTA, the simulated mechanism is different, making the simulated rainfall response to the GW unreliable.
- The AMV induces an interhemispheric surface pressure gradient which favors anomalous shifts of the ITCZ and the tropical rainfall over the Atlantic sector. Under positive (negative) AMV phase conditions, an anomalous northward (southward) shift of the ITCZ is favored enhancing (reducing) Sahel and Amazon rainfall and reducing (enhancing) it in the Northeast of Brazil, in their respective rainy seasons. CMIP5 models, on average, successfully reproduce this teleconnection.
- The IPO forces Walker circulation anomalies connecting the tropical Pacific with rainfall in the Sahel, Amazon and Northeast. In positive (negative) IPO phases, the characteristic warm (cold) tongue of SSTA enhances (reduces) deep convection over the tropical Pacific inducing anomalous subsidence (rise) over West Africa in JAS and over northern South America in DJFMAM, resulting in less (more) precipitation in the three regions studied. This teleconnection is robustly reproduced by CMIP5 models.

In order to discuss the eventual role of the external forcing in the AMV and IPO, the results from forced (historical) and unforced (piControl) simulations have been compared. The RCP8.5 future projections have been also analyzed to find whether the Sahel, Amazon and Northeast rainfall response to the GW, AMV and IPO are expected to change in the future. The results from this analysis show that:

- The effects of the aerosol radiative forcing have been shown to be an important source of uncertainty among CMIP5 models as to the simulation of the GW and are crucial to produce changes in the SSTA gradients of the GW pattern. Regarding the AMV, the similarity found between the results obtained with the historical and piControl simulations, in broad terms, suggests that it has an important component of internal variability. Nevertheless, an externally forced component is also identified in the AMV simulated by some CMIP5 models. This effect is attributed to the aerosols and is highly model-dependent. Hence, it is suggested that aerosols may play a role in the AMV with an extent that is hard to assess because their effects are poorly constrained by the GCMs. In contrast, no relevant impacts of the external radiative forcing have been identified in the IPO, suggesting that it is a predominantly internal mode of variability.

- The RCP8.5 future projection of the CMIP5 models, on average, reveal that the effect of the GW on precipitation in the regions studied is likely to change in the future. Abundant precipitation is projected in most of the Sahel, associated with a strengthening of the low-level WAM circulation, and a drying in the westernmost part. This different rainfall response relative to the historical period is attributed to a change in the projected SST warming, with more intense SSTA in the Northern hemisphere and the tropics than to the south. In the Amazonia the projected GW induces more rainfall in the western side of the region and less to the east. Over most of the Northeast of Brazil the projected GW is associated with a drying and with enhanced precipitation along the northern coast. The projected rainfall response in northern South America is related with anomalous Walker circulation induced by the tropical Pacific SSTA. Hence, we have to be skeptical about these results because of the models inability to reproduce the observed tropical Pacific SSTA in the GW pattern of the historical simulation, which may be hampering the future projections skill as well. In the three areas studied, the projected GW induces different rainfall anomalies within the regions. As a consequence, the uncertainty among models in simulating a positive or negative rainfall response to the GW throughout the area of each region is high. Hence, it is advisable not to use these regions as a whole when studying the long-term precipitation tendency projected for the future.
- The main features of the AMV and IPO modes (the characteristic SSTA spatial patterns and oscillatory periodicities), their impacts on rainfall and the associated atmospheric mechanisms are similarly simulated in the RCP8.5 future projections as in the historical and piControl simulations. Therefore, it is concluded that the AMV and IPO and their impacts on precipitation are not expected to change in the future.

The last goal of this part of the Thesis was to assess the contribution of the GW, AMV and IPO to the total rainfall variance at decadal-to-multidecadal time scales in the Sahel, Amazonia and Northeast of Brazil, both in observations and CMIP5 simulations. The most relevant conclusions drawn from the multi-linear regression analysis between the three time series of the SST modes and the low-frequency index of precipitation at each one of the three regions studied are:

- The modulation of the observed decadal-to-multidecadal Sahel rainfall variability is principally led by the AMV. In case of the models, most of them suggest a dominant role of the GW, though the contribution of each one of the three SST modes on the precipitation variability is model-dependent.
- The GW is the main modulator of the decadal-to-multidecadal changes in the observed Amazon precipitation. Most CMIP5 models support the observed dominant role of the GW, although its effect on the Amazon precipitation is uncertain among themselves.

- The contribution of the GW, AMV and IPO to the Northeast rainfall variability at decadal-to-multidecadal time scales is uncertain since both, observations and CMIP5 historical simulations, show very inconsistent results.
- The modulation of the SST modes on rainfall variability at decadal-to-multidecadal time scales simulated by CMIP5 models is, in general, underestimated with respect to observations. The underestimation of the simulated link between the SST modes and rainfall has been attributed to inaccurate distribution of the associated SSTA and to the low sensitivity of the monsoon atmospheric dynamics to SST changes that GCMs simulate.
- In the RCP8.5 future projections, the GW prominently dominates the decadal-to-multidecadal precipitation variability in the three regions studied. Hence, the uncertainties among models as to the projected GW impacts on these regions importantly revert in the precipitation trends projected for the future.

In summary, this part of the Thesis shows that CMIP5 models robustly reproduce the main observed features of the AMV and IPO patterns of SST and, consequently, their influence on precipitation in the Sahel, Amazonia and Northeast regions. The accuracy with which models reproduce these links is related to the simulated SSTA intensity of the characteristic tropical Atlantic gradient and in the tropical Pacific in the AMV and IPO patterns, respectively. The simulated GW is more controversial. It can thereby be suggested that an improvement of the ability of the GCMs to reproduce the SST spatial pattern, the time evolution of the AMV and the IPO and their teleconnection with the atmosphere, will directly convert into a better simulation of the low-frequency variability of rainfall and an improved skill of the long-term forecasting in the Sahel, Amazonia and Northeast regions during their respective rainy seasons. Nevertheless, the low skill in reproducing the contribution of each of the three SST modes on rainfall constraints the way in which models simulate the decadal-to-multidecadal precipitation variability. Furthermore, the high uncertainty among models as to the future impacts of the GW is a challenge for the model developers to have reliable long-term climate projections.

In the second part of this Thesis, the focus in on the particular case of the late-19th century humid period in the Sahel. In response to the initially posed key questions, the main conclusions raised from this analysis are:

- The long rainy period of the late-19th century can be reproduced with an AGCM forced with observational SST data, supporting the evidences of the existence of such a period.
- The abundance of Sahel precipitation in the late-19th century is associated with anomalous deep convection in the middle and high troposphere and enhanced humidity supply related to low-level wind anomalies.
- The key basin inducing this decadal rainy period in the Sahel is the Atlantic. The Atlantic SST accounts for most of the Sahel decadal shifts in rainfall during the late-19th century.

Not only do the results of this part support the existence of a long rainy period in the Sahel prior to the observational records, but they also reveal for the first time that it occurred under similar circumstances to those inducing the widely documented Sahel rainfall decadal variability over the 20th century.

8.2 Future Work

In this section, some possible lines of work to follow from the results obtained in this Thesis are presented:

- According to the results of this Thesis, the AMV has similar effects on the Sahel and on the Amazon rainfall. Also a wet decadal period in the Sahel during the second half of the 19th century has been reproduced in this Thesis, which has been related to the variability of the Atlantic SST at these time scales. Therefore, by analyzing the same simulations performed with the LMDZ model and presented in Chap. 7, we can study whether during the late-19th century there was also an anomalously wet period in the Amazon that coincides with the one of the Sahel.
- It would be interesting to assess to what extent CMIP5 models overestimate the radiative forcing effects of the GHGs on the SST with respect to aerosols, or vice versa. For this purpose, a methodology should be developed to separate the radiative forcing effect on the SST induced by the GHGs from the rest in observations.
- The biases of the GCMs have been primarily related to errors in reproducing the observed climatology. Nevertheless, we could also study the impacts of these biases on the simulated decadal-to-multidecadal SST variability. This can be done by relating some defining features of the SST modes characterized in this Thesis (GW, AMV and IPO) using CMIP5 simulations with the errors of the mean state that they represent with respect to observations.
- Based on the methodology developed and used in this Thesis, the climate variability in other regions of the globe at decadal-to-multidecadal time scales can be also studied.
- The ongoing phase 6 of the Coupled Model Intercomparison Project provide the opportunity to study the influence of the SST on the regions studied in this Thesis with a new generation of the most state-of-the-art GCMs.

Appendix A

This Appendix shows the regression patterns associated with the GW obtained from the 17 CMIP5 models individually.

See Figs. A.1, A.2, A.3, A.4, A.5 and A.6.

J. Villamayor, *Influence of the Sea Surface Temperature Decadal Variability on Tropical Precipitation: West African and South American Monsoon*, Springer Theses, https://doi.org/10.1007/978-3-030-20327-6

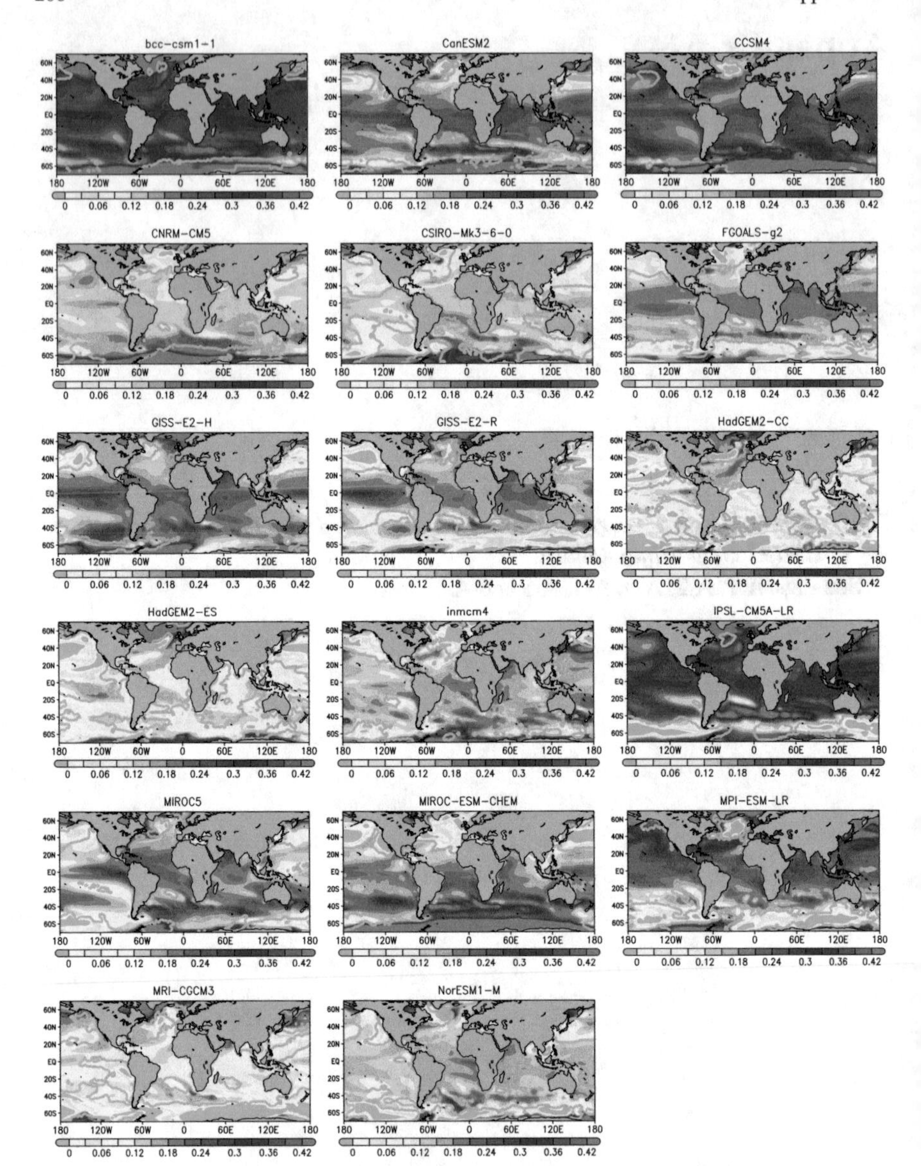

Fig. A.1 Single models regression patterns of SSTA onto the standardized GW index (K per standard deviation) from historical simulations. Grey contours indicate the regions where the correlation is significant at the 5% level from a "random-phase" test

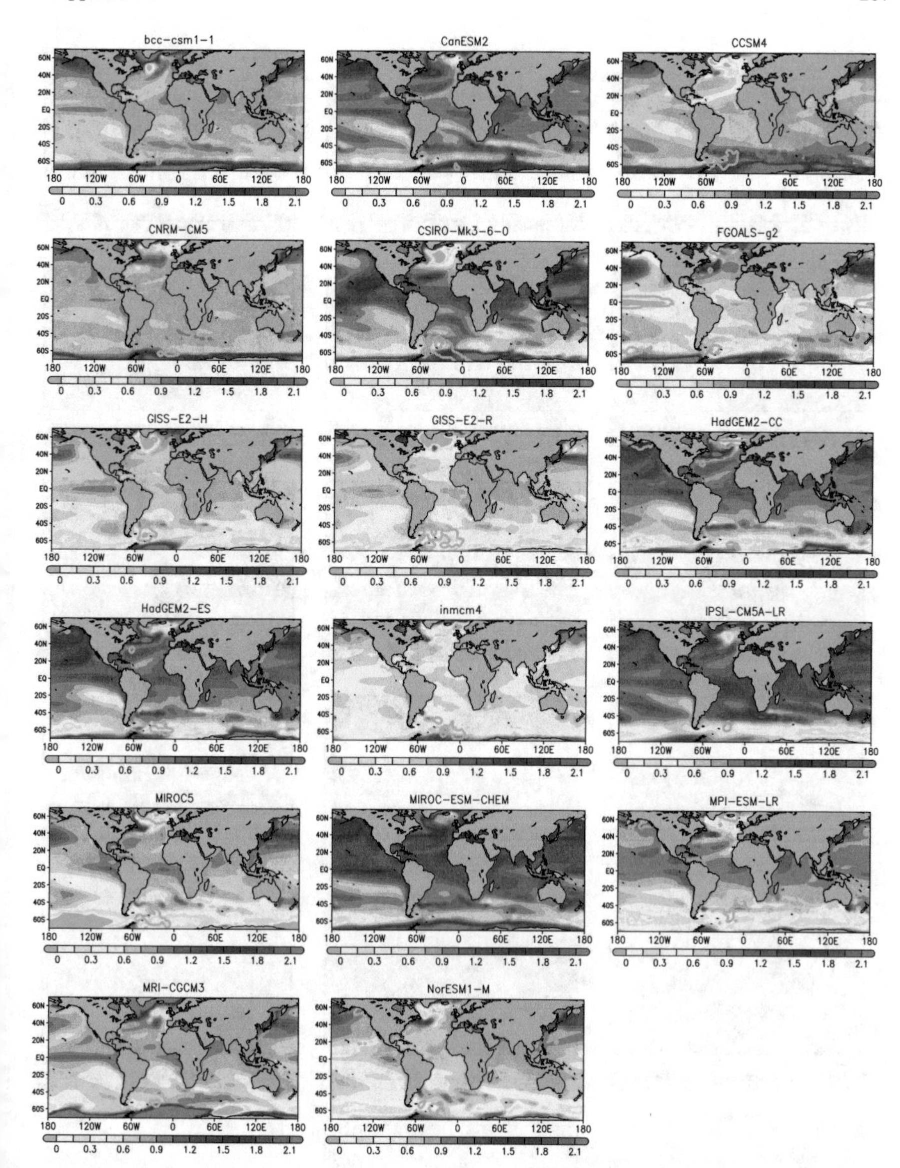

Fig. A.2 Single models regression patterns of SSTA onto the standardized GW index (K per standard deviation) from RCP8.5 future projection. Grey contours indicate the regions where the correlation is significant at the 5% level from a "random-phase" test

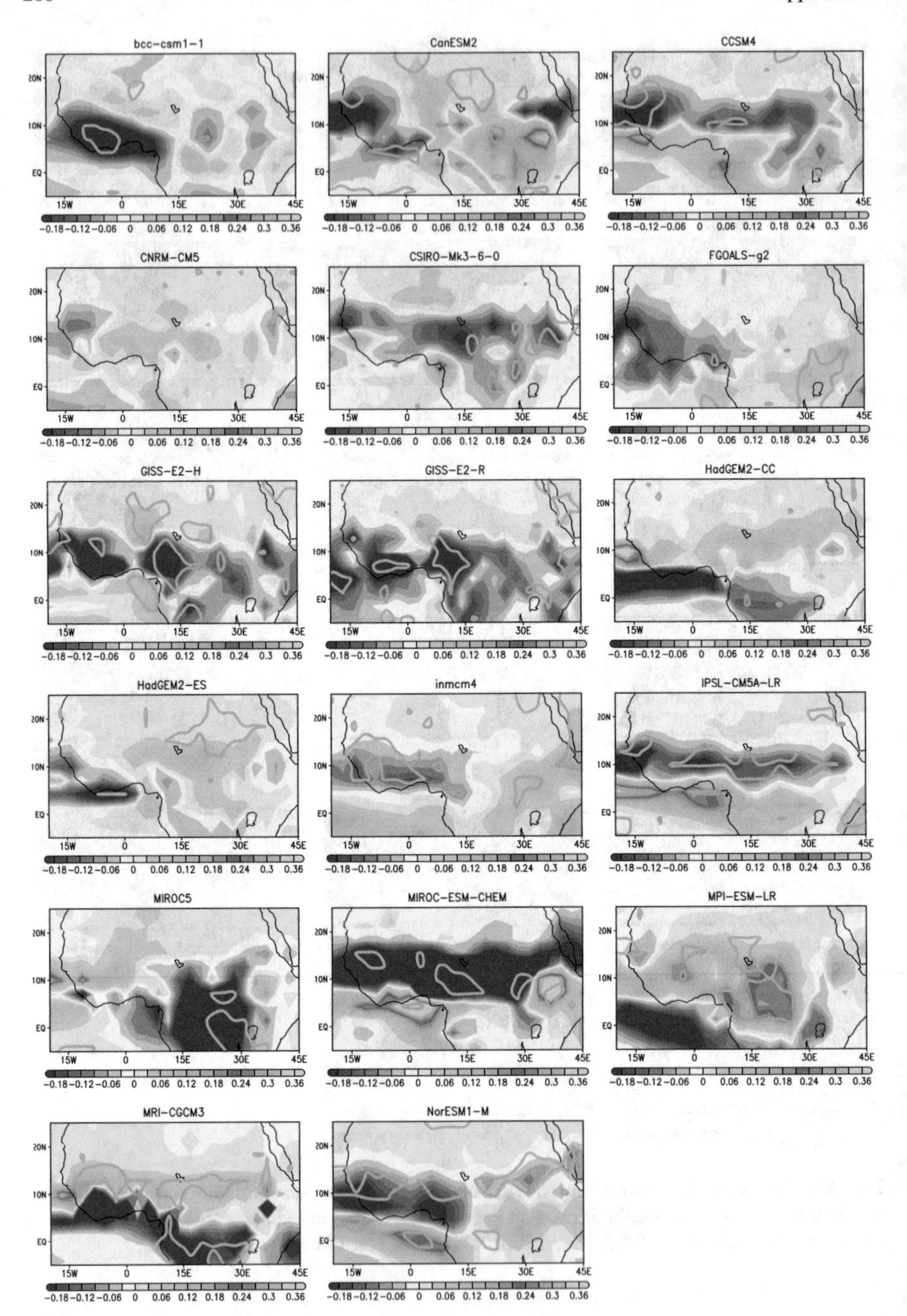

Fig. A.3 Single models regression maps of JAS precipitation anomalies onto the standardized GW index (mm day^{-1} per standard deviation) from historical simulations. Grey contours indicate the regions where the correlation is significant at the 5% level from a "random-phase" test

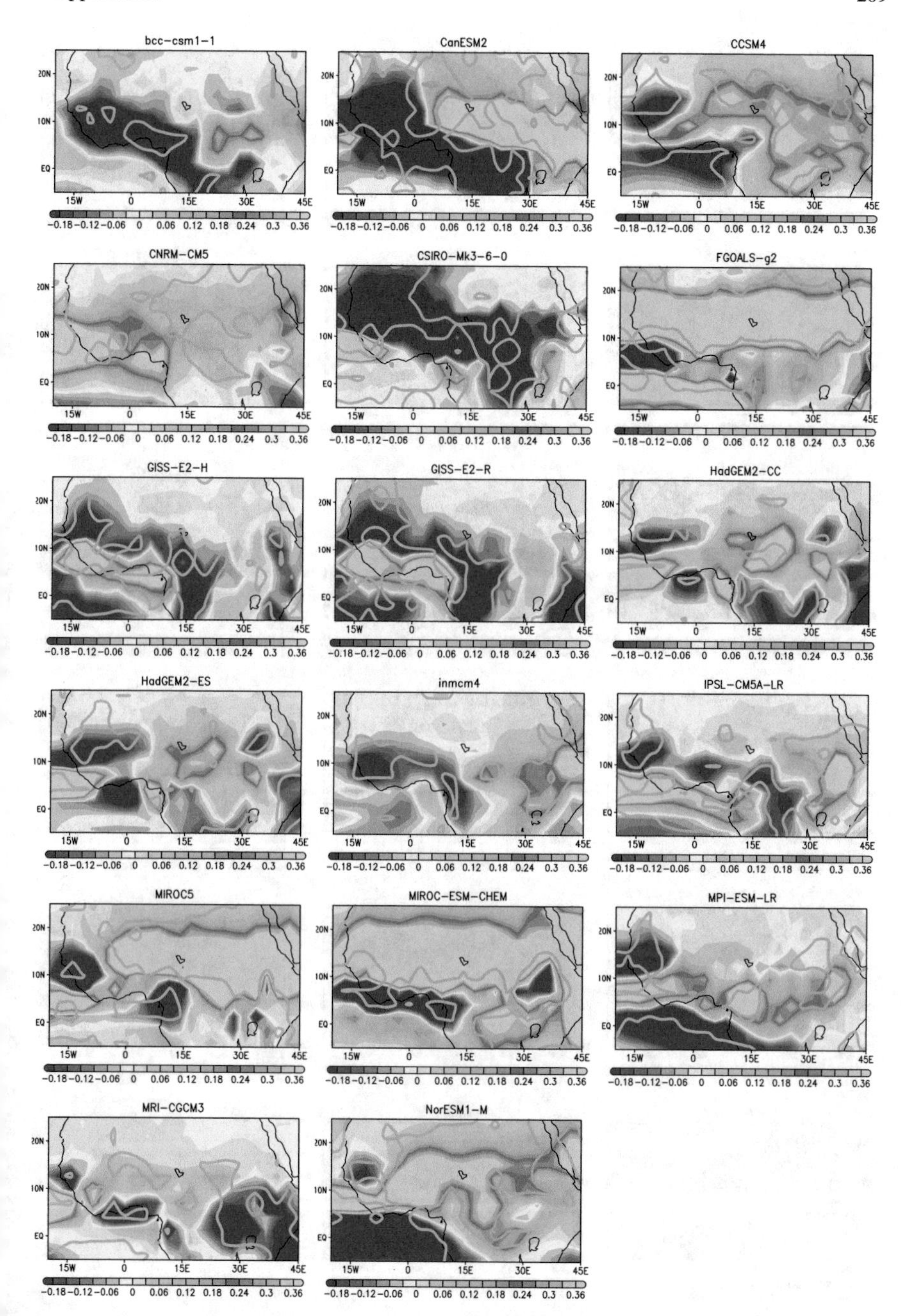

Fig. A.4 Single models regression maps of JAS precipitation anomalies onto the standardized GW index (mm day^{-1} per standard deviation) from RCP8.5 future projection. Grey contours indicate the regions where the correlation is significant at the 5% level from a "random-phase" test

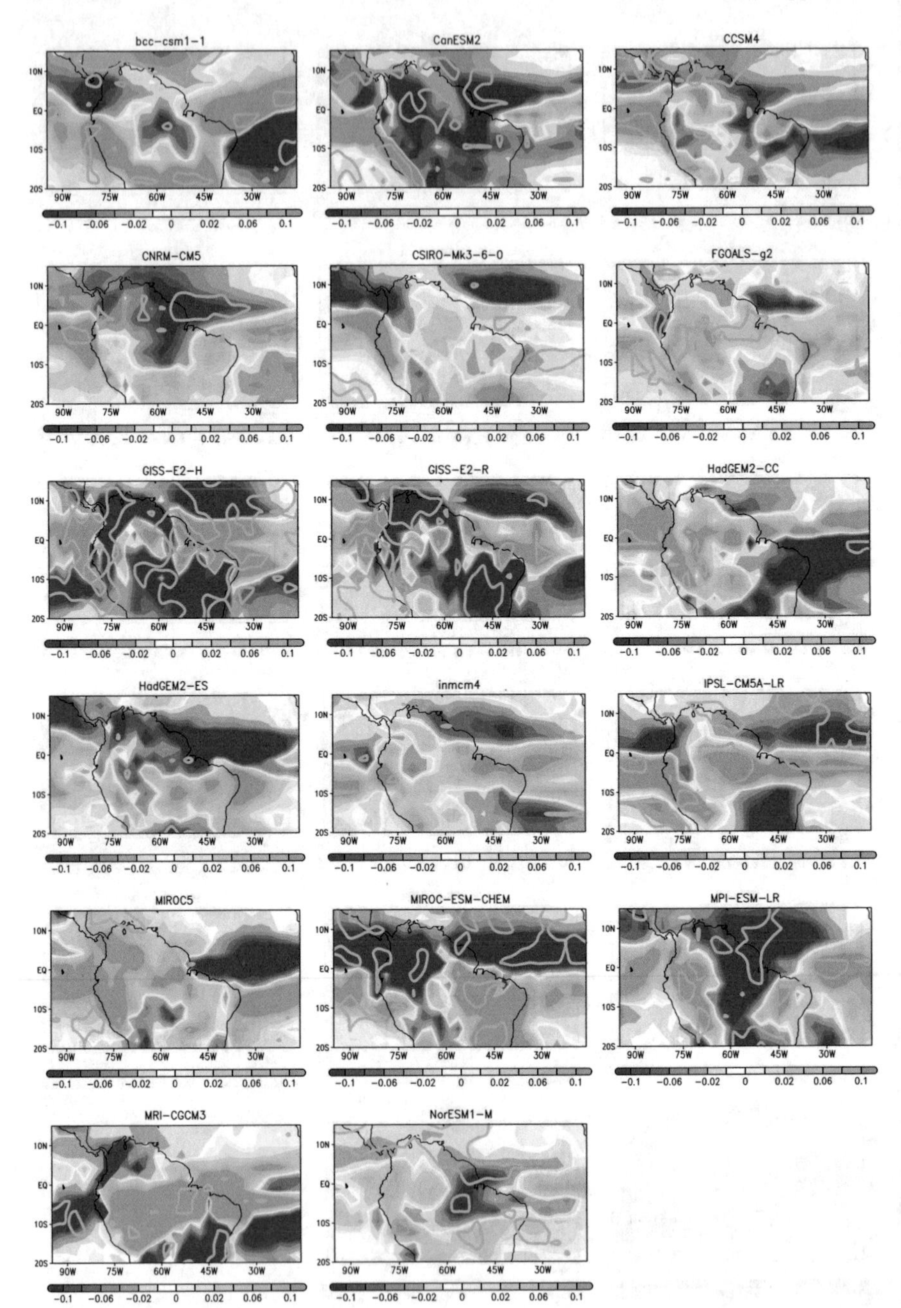

Fig. A.5 Single models regression maps of DJFMAM precipitation anomalies onto the standardized GW index (mm day^{-1} per standard deviation) from historical simulations. Grey contours indicate the regions where the correlation is significant at the 5% level from a "random-phase" test

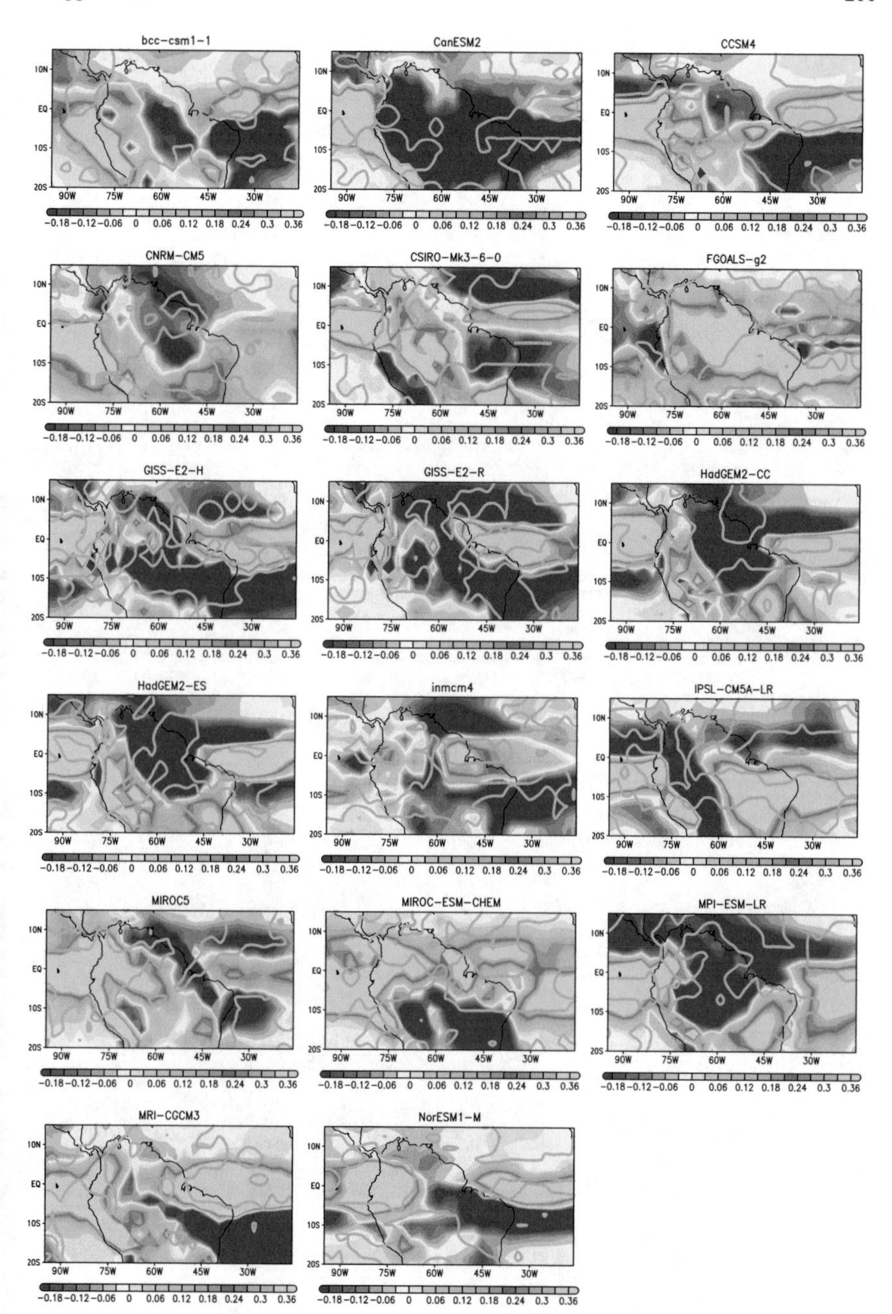

Fig. A.6 Single models regression maps of DJFMAM precipitation anomalies onto the standardized GW index (mm day^{-1} per standard deviation) from RCP8.5 future projection. Grey contours indicate the regions where the correlation is significant at the 5% level from a "random-phase" test

Appendix B

The individual regression patterns of the AMV of all the CMIP5 models used in the Thesis are presented in this Appendix.

See Figs. B.1, B.2, B.3, B.4, B.5, B.6 and B.7.

J. Villamayor, *Influence of the Sea Surface Temperature Decadal Variability on Tropical Precipitation: West African and South American Monsoon*, Springer Theses, https://doi.org/10.1007/978-3-030-20327-6

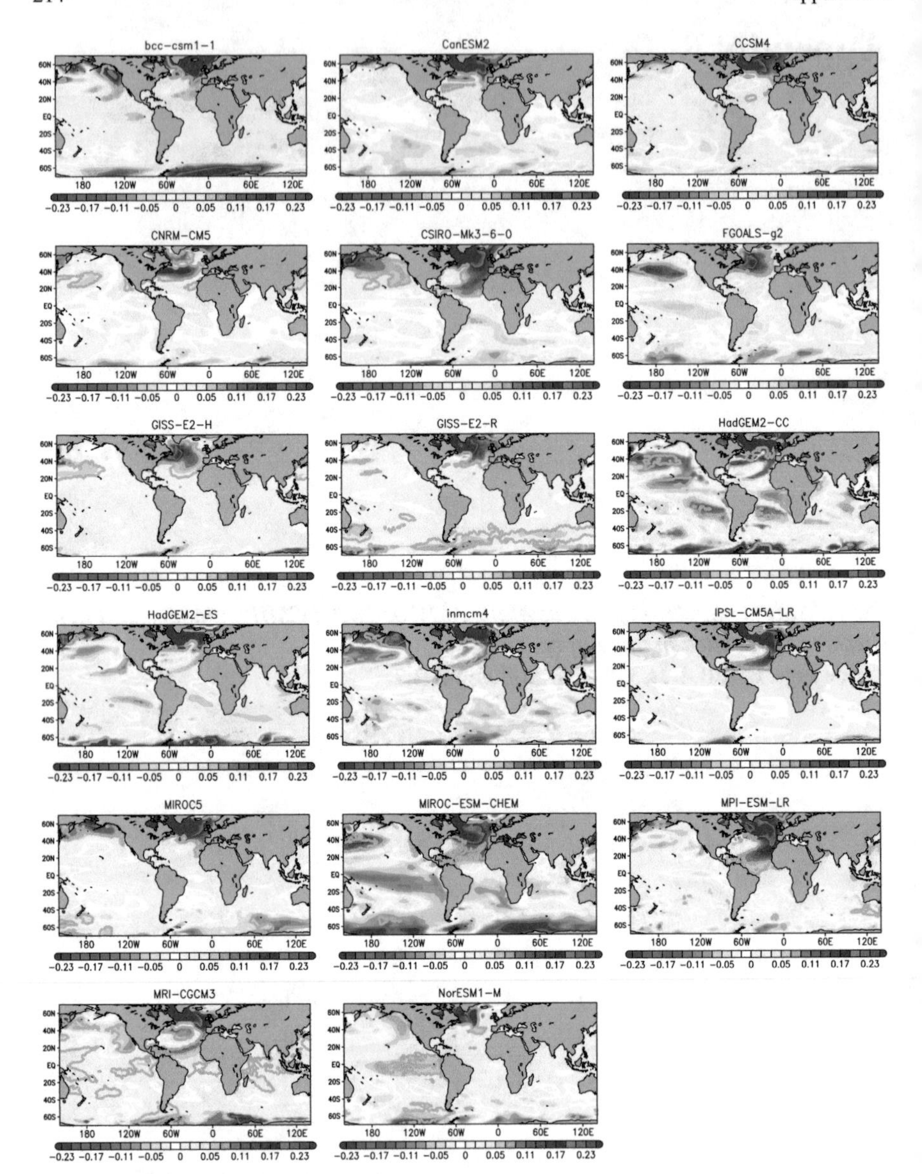

Fig. B.1 Single models regression patterns of SSTA onto the standardized AMV index (K per standard deviation) from historical simulations. Grey contours indicate the regions where the correlation is significant at the 5% level from a "random-phase" test

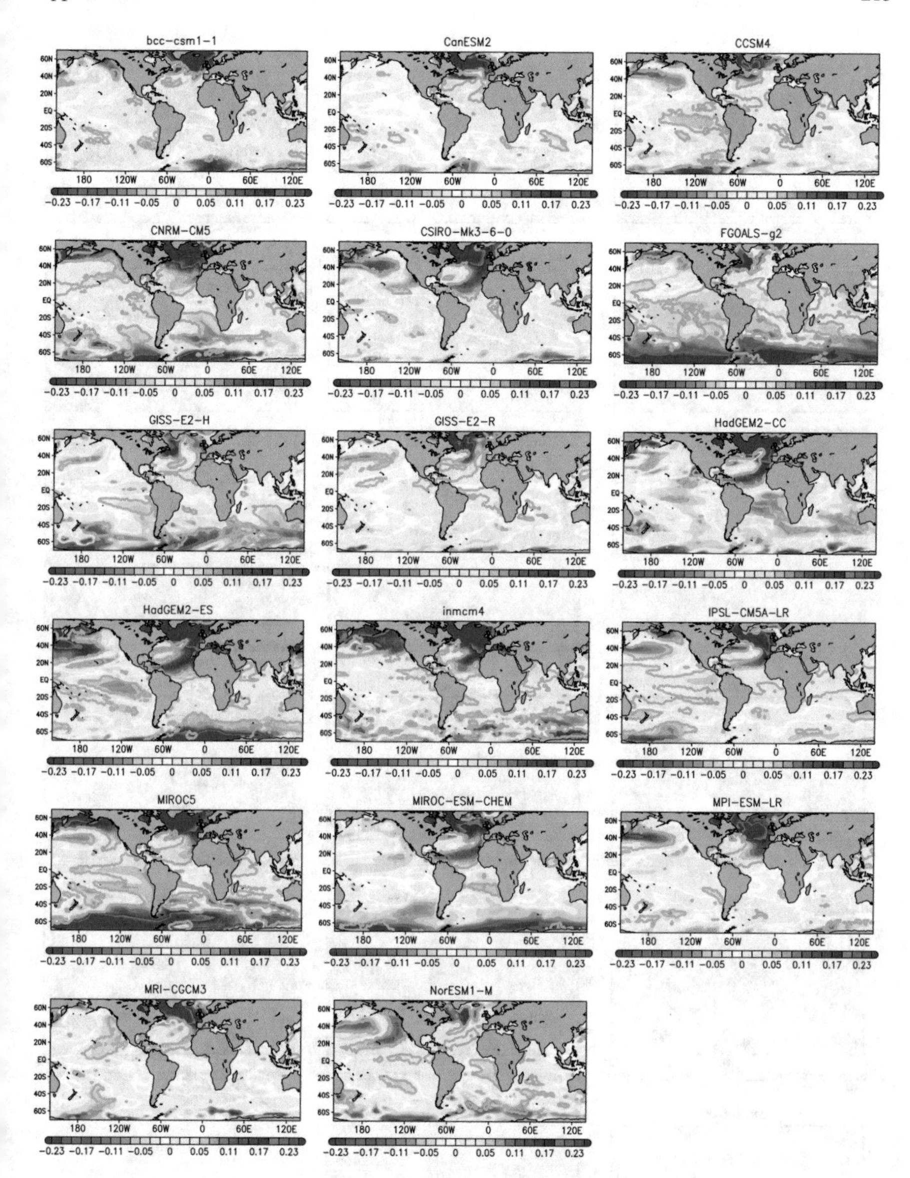

Fig. B.2 Single models regression patterns of SSTA onto the standardized AMV index (K per standard deviation) from piControl simulations. Grey contours indicate the regions where the correlation is significant at the 5% level from a "random-phase" test

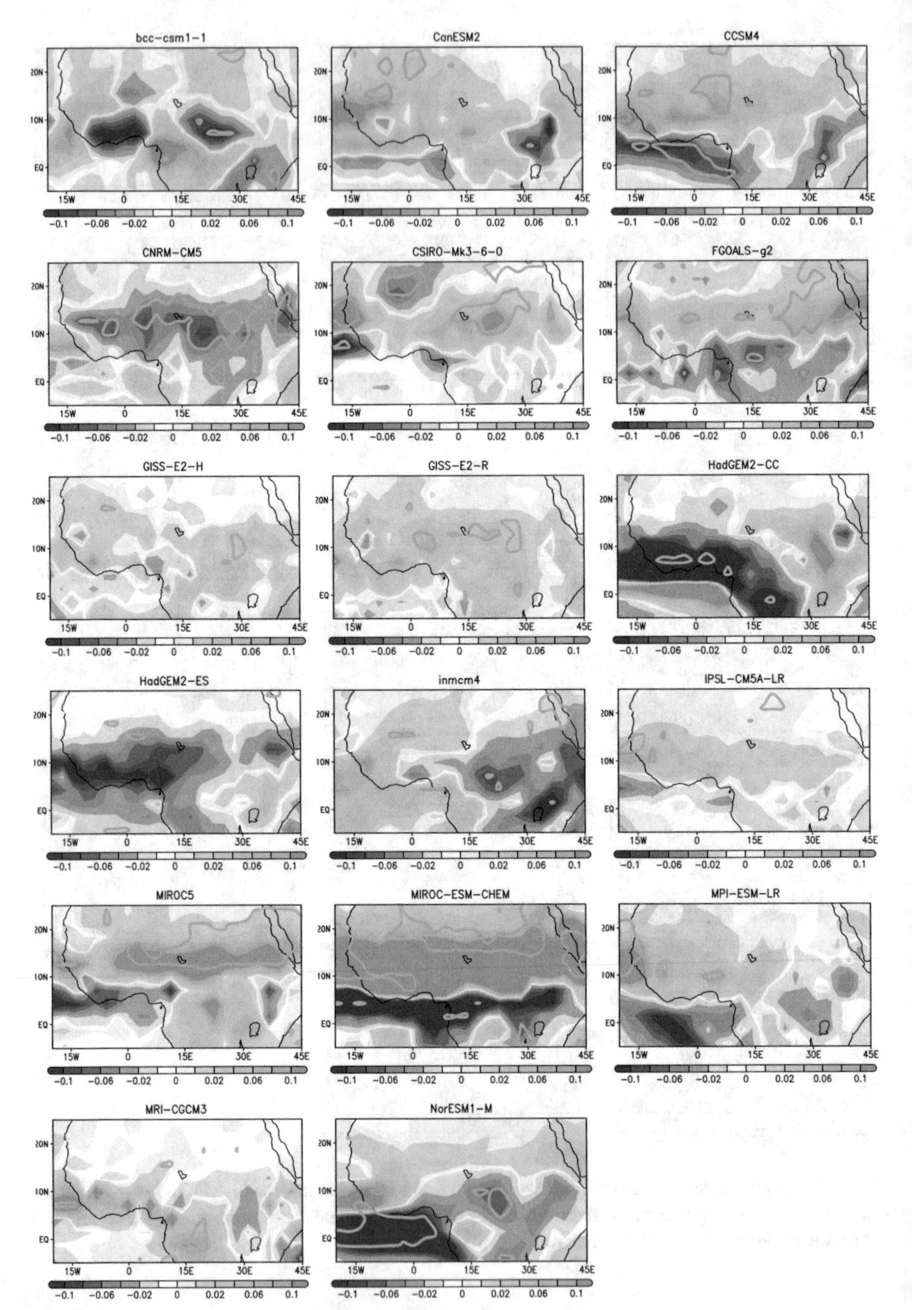

Fig. B.3 Single models regression maps of JAS precipitation anomalies onto the standardized AMV index (mm day^{-1} per standard deviation) from historical simulations. Grey contours indicate the regions where the correlation is significant at the 5% level from a "random-phase" test

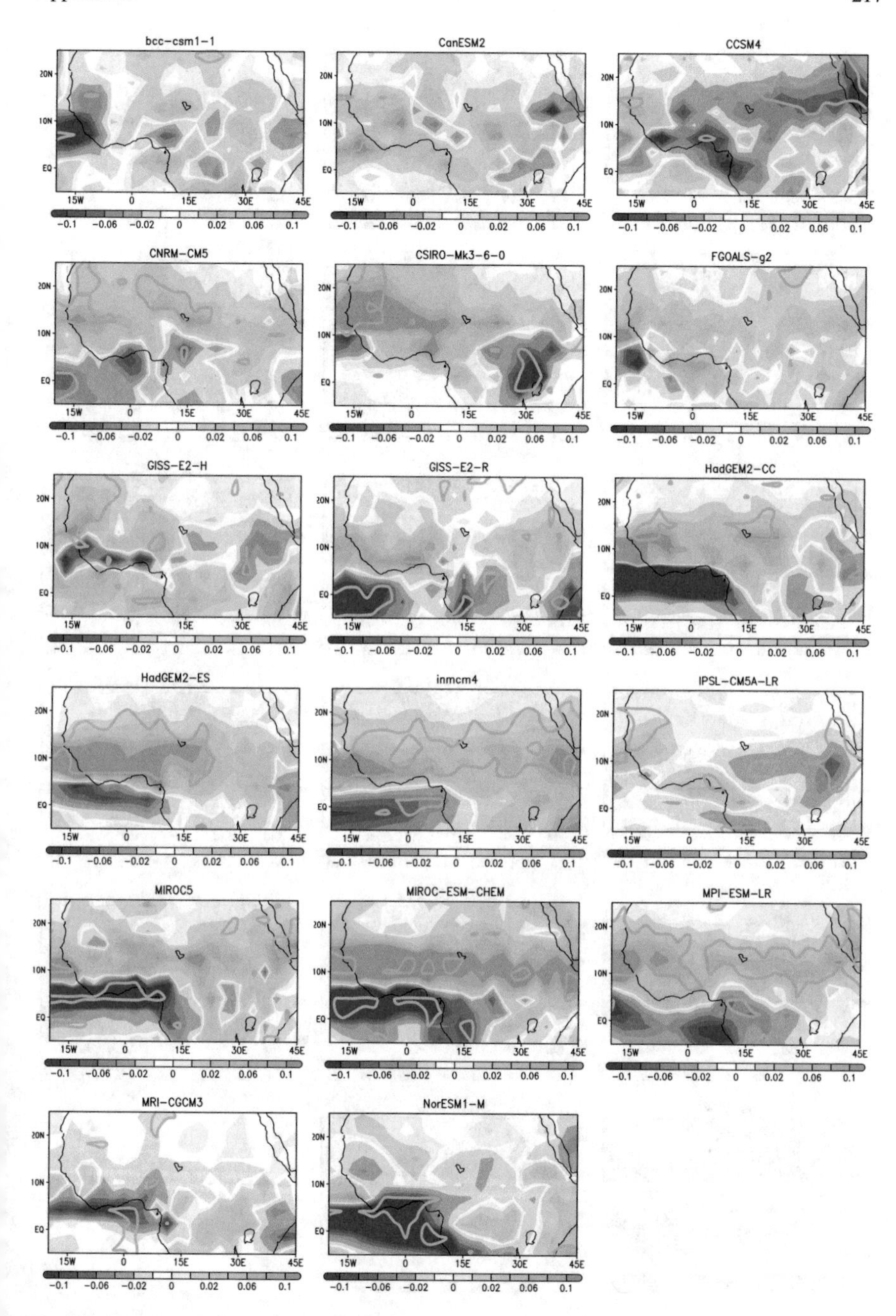

Fig. B.4 Single models regression maps of JAS precipitation anomalies onto the standardized AMV index (mm day^{-1} per standard deviation) from piControl simulations. Grey contours indicate the regions where the correlation is significant at the 5% level from a "random-phase" test

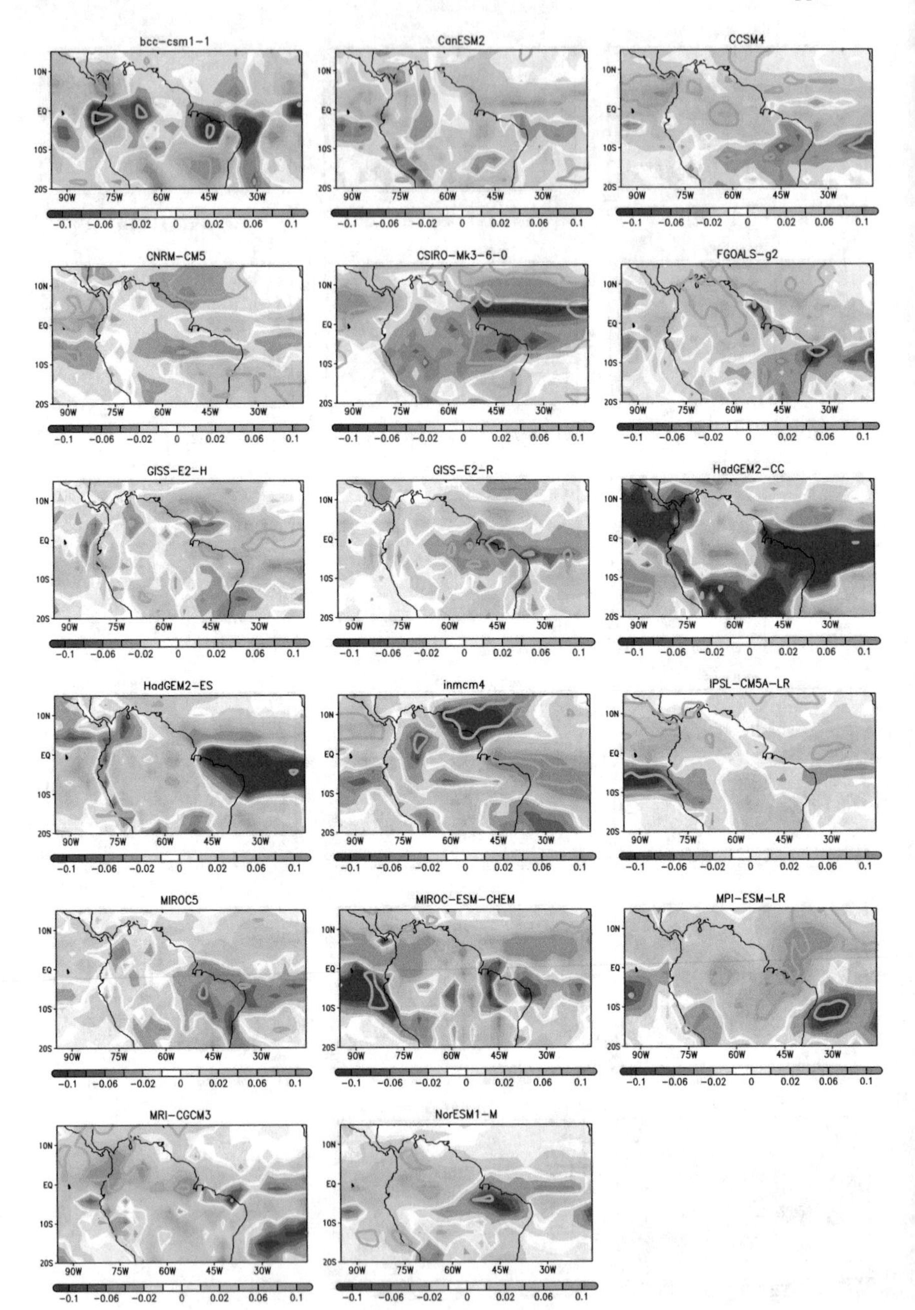

Fig. B.5 Single models regression maps of DJFMAM precipitation anomalies onto the standardized AMV index (mm day^{-1} per standard deviation) from historical simulations. Grey contours indicate the regions where the correlation is significant at the 5% level from a "random-phase" test

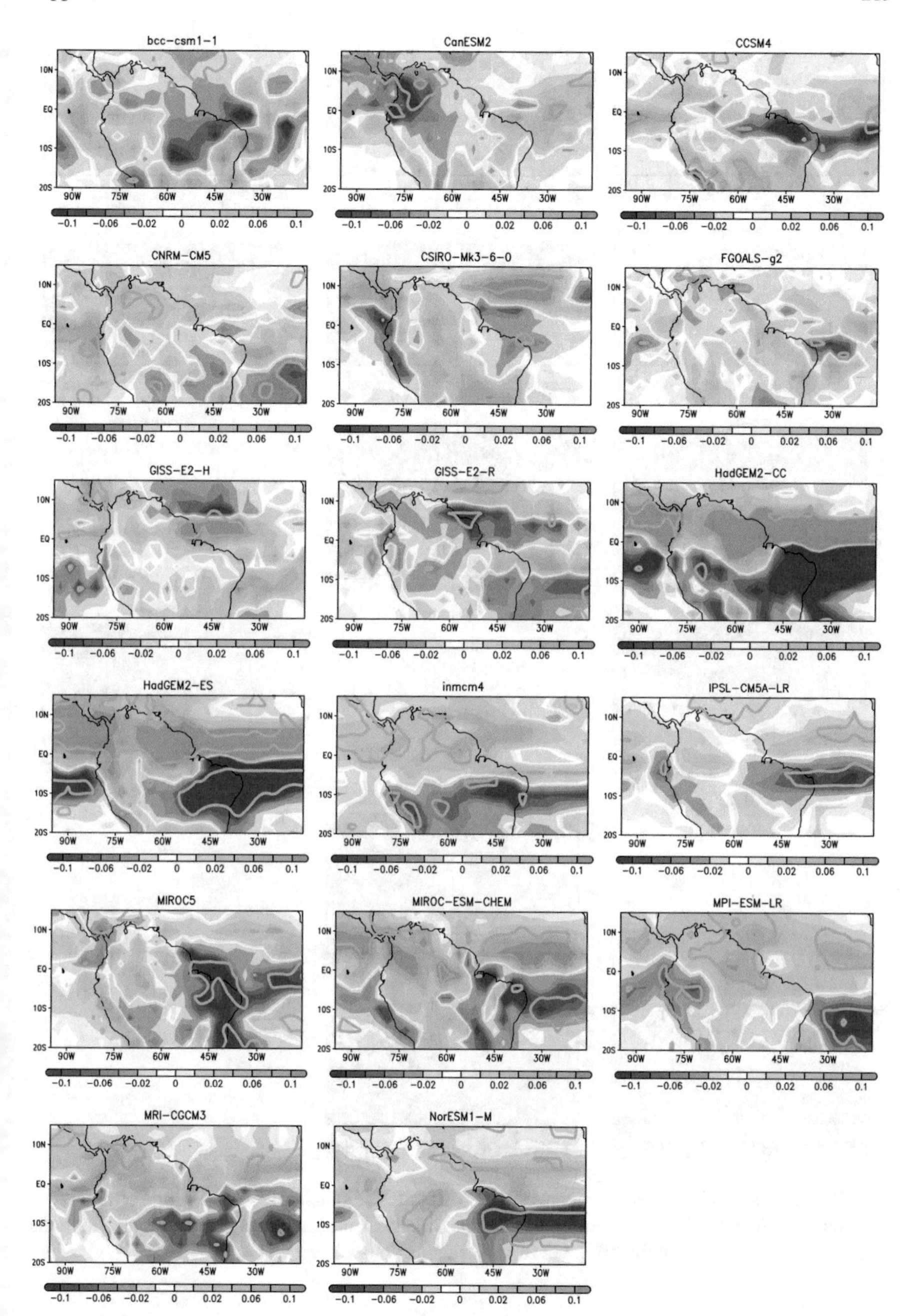

Fig. B.6 Single models regression maps of DJFMAM precipitation anomalies onto the standardized AMV index (mm day^{-1} per standard deviation) from piControl simulations. Grey contours indicate the regions where the correlation is significant at the 5% level from a "random-phase" test

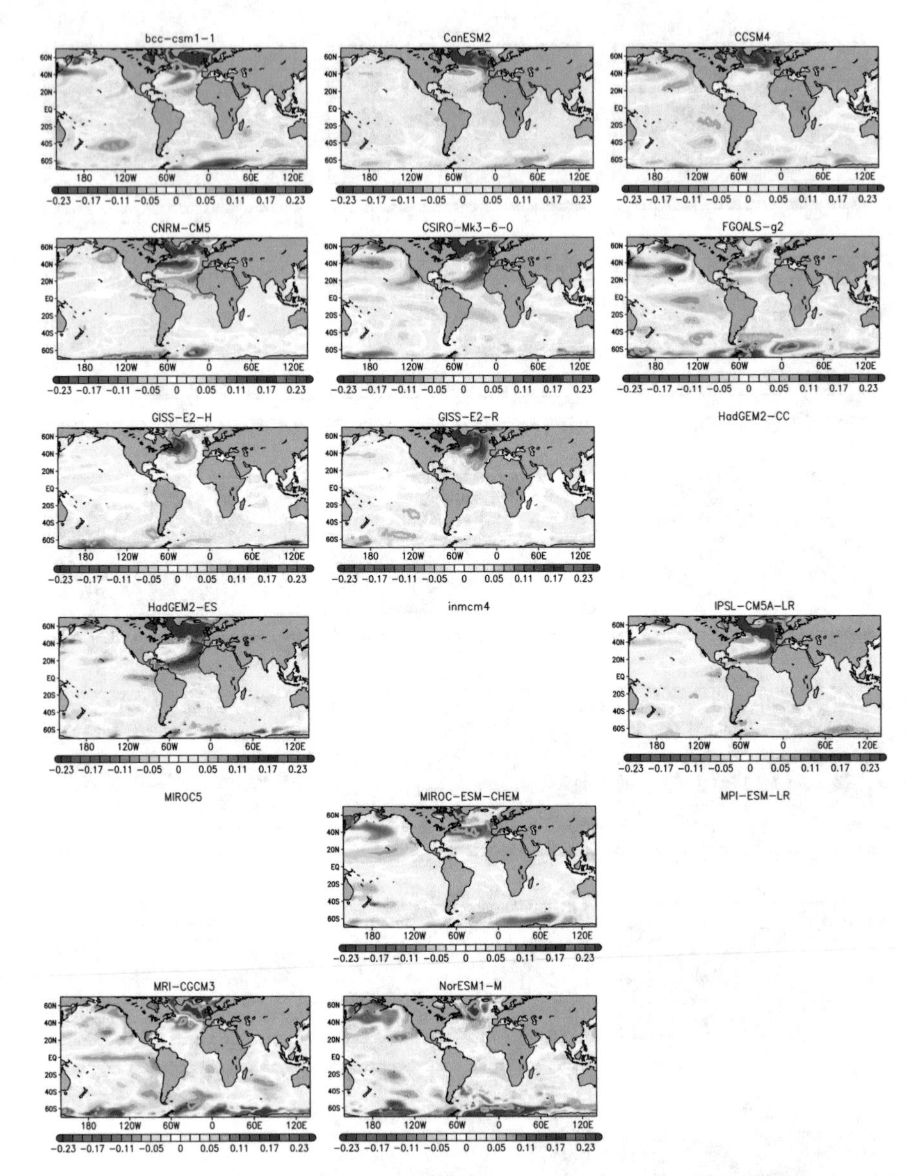

Fig. B.7 Single models regression patterns of SSTA onto the standardized AMV index (mm day^{-1} per standard deviation) from historicalGHG simulations. Contours indicate the regions where the regression is significant at the 5% level from a "random-phase" test

Appendix C

In this Appendix the regression patterns associated with the IPO of the 17 CMIP5 models separately are shown.

See Figs. C.1, C.2, C.3, C.4, C.5 and C.6.

J. Villamayor, *Influence of the Sea Surface Temperature Decadal Variability on Tropical Precipitation: West African and South American Monsoon*, Springer Theses, https://doi.org/10.1007/978-3-030-20327-6

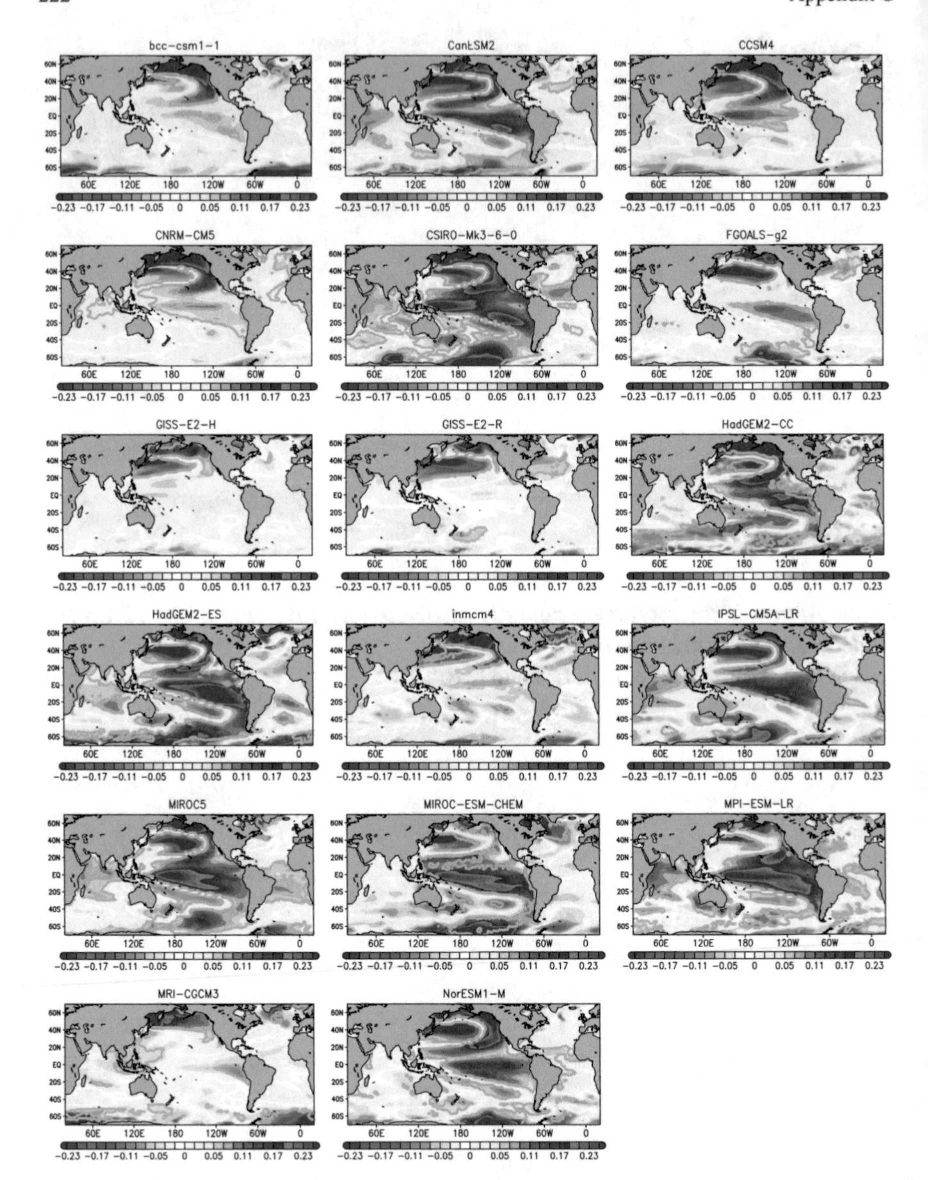

Fig. C.1 Single models regression patterns of SSTA onto the standardized IPO index (K per standard deviation) from historical simulations. Grey contours indicate the regions where the correlation is significant at the 5% level from a "random-phase" test

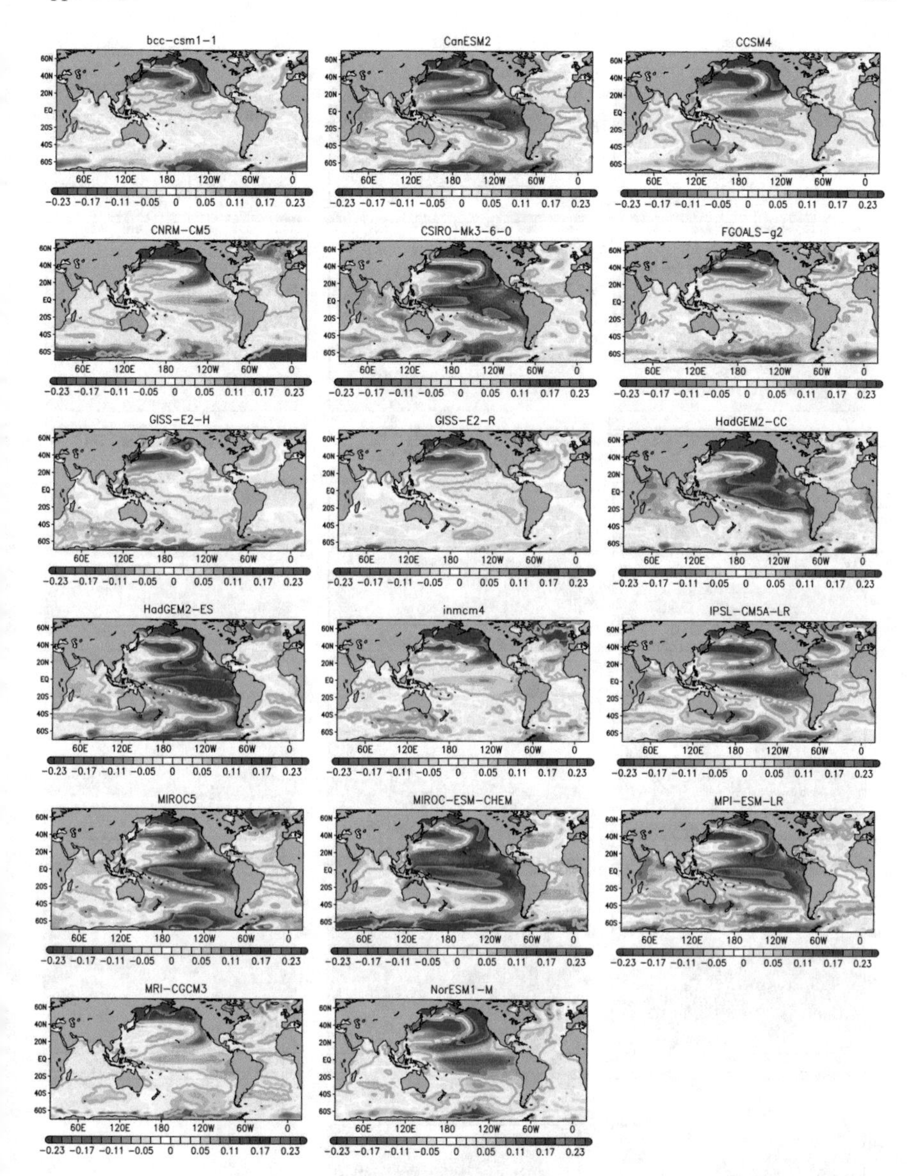

Fig. C.2 Single models regression patterns of SSTA onto the standardized IPO index (K per standard deviation) from piControl simulations. Grey contours indicate the regions where the correlation is significant at the 5% level from a "random-phase" test

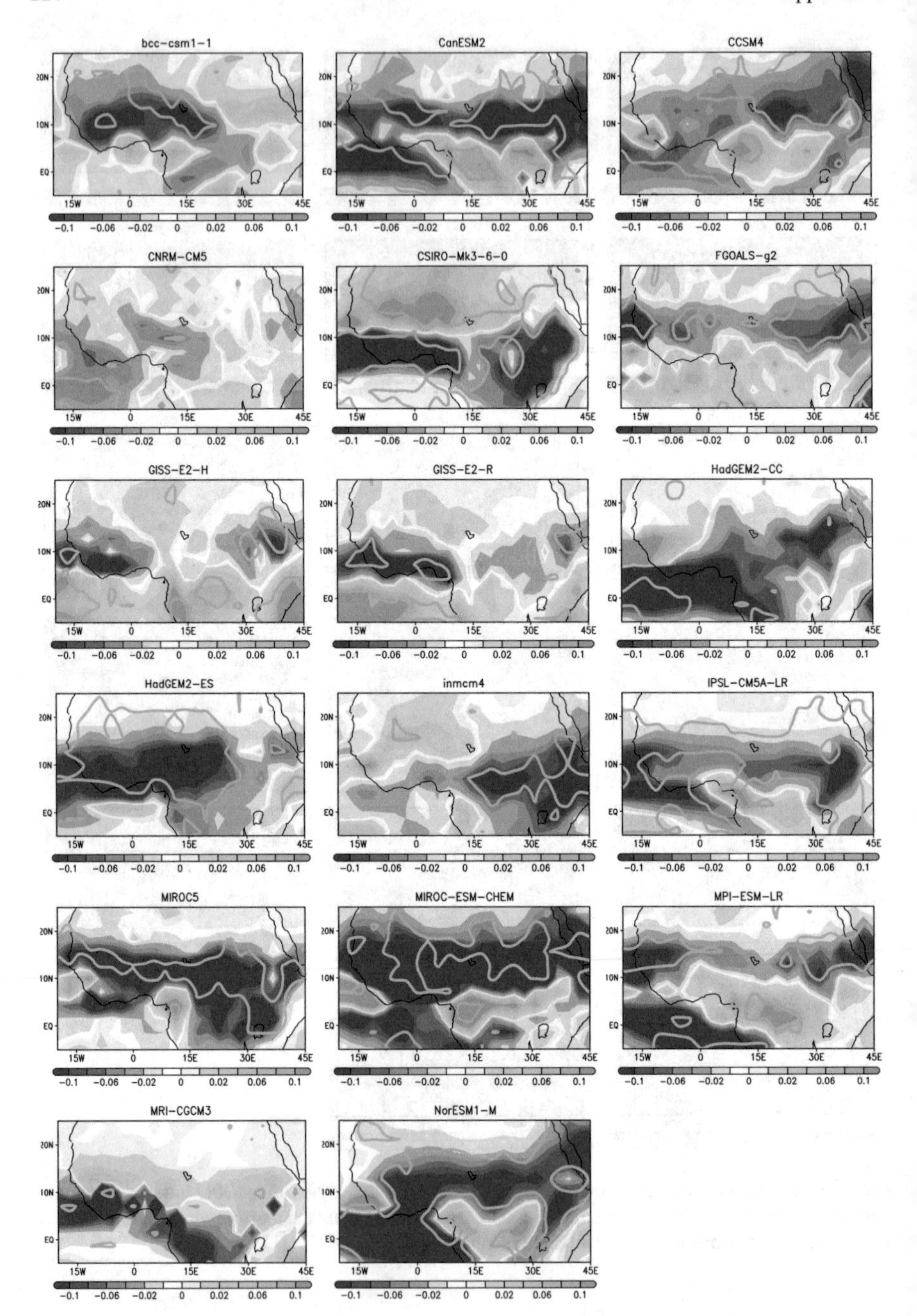

Fig. C.3 Single models regression maps of JAS precipitation anomalies onto the standardized IPO index (mm day^{-1} per standard deviation) from historical simulations. Grey contours indicate the regions where the correlation is significant at the 5% level from a "random-phase" test

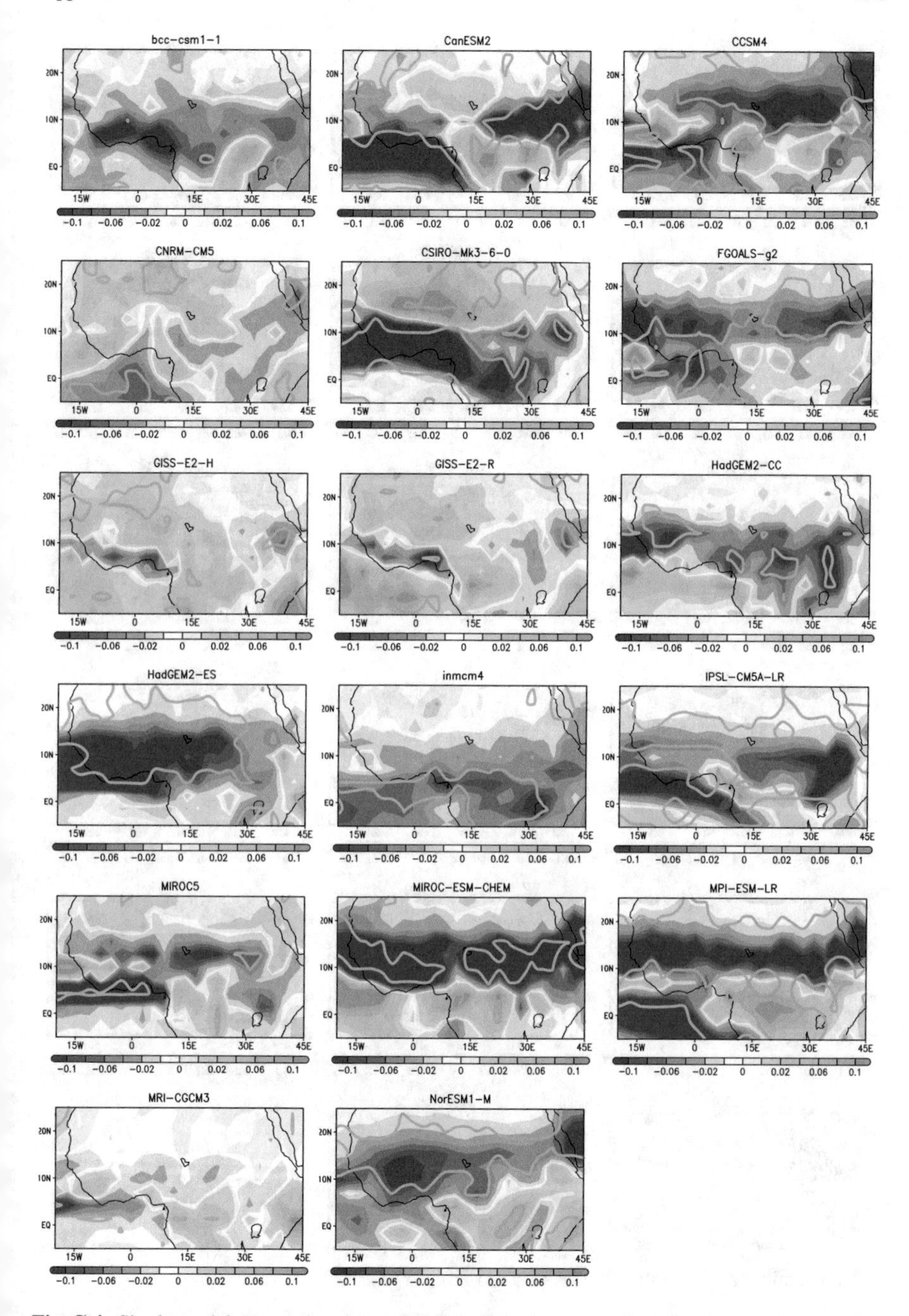

Fig. C.4 Single models regression maps of JAS precipitation anomalies onto the standardized IPO index (mm day^{-1} per standard deviation) from piControl simulations. Grey contours indicate the regions where the correlation is significant at the 5% level from a "random-phase" test

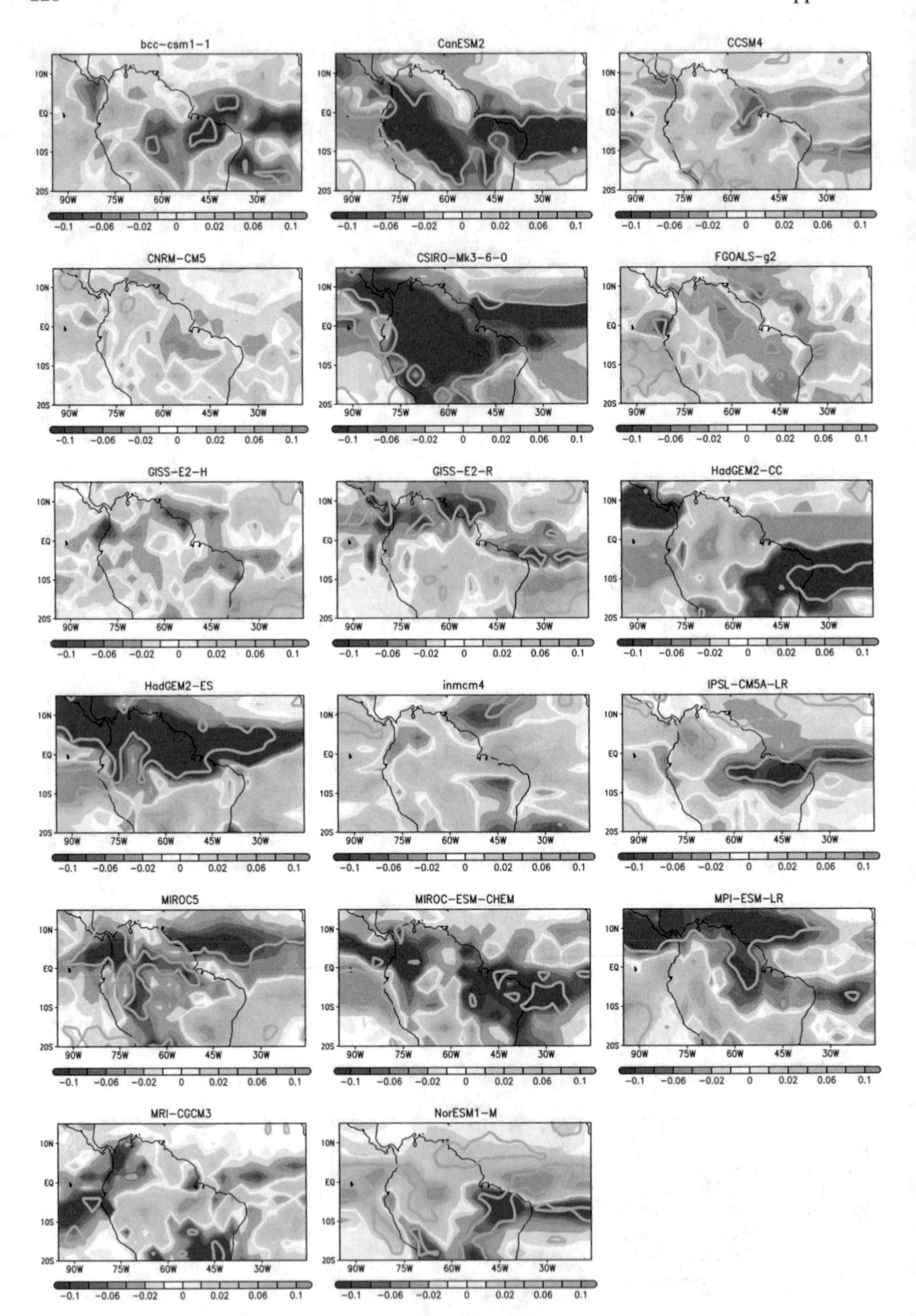

Fig. C.5 Single models regression maps of DJFMAM precipitation anomalies onto the standardized IPO index (mm day^{-1} per standard deviation) from historical simulations. Grey contours indicate the regions where the correlation is significant at the 5% level from a “random-phase” test

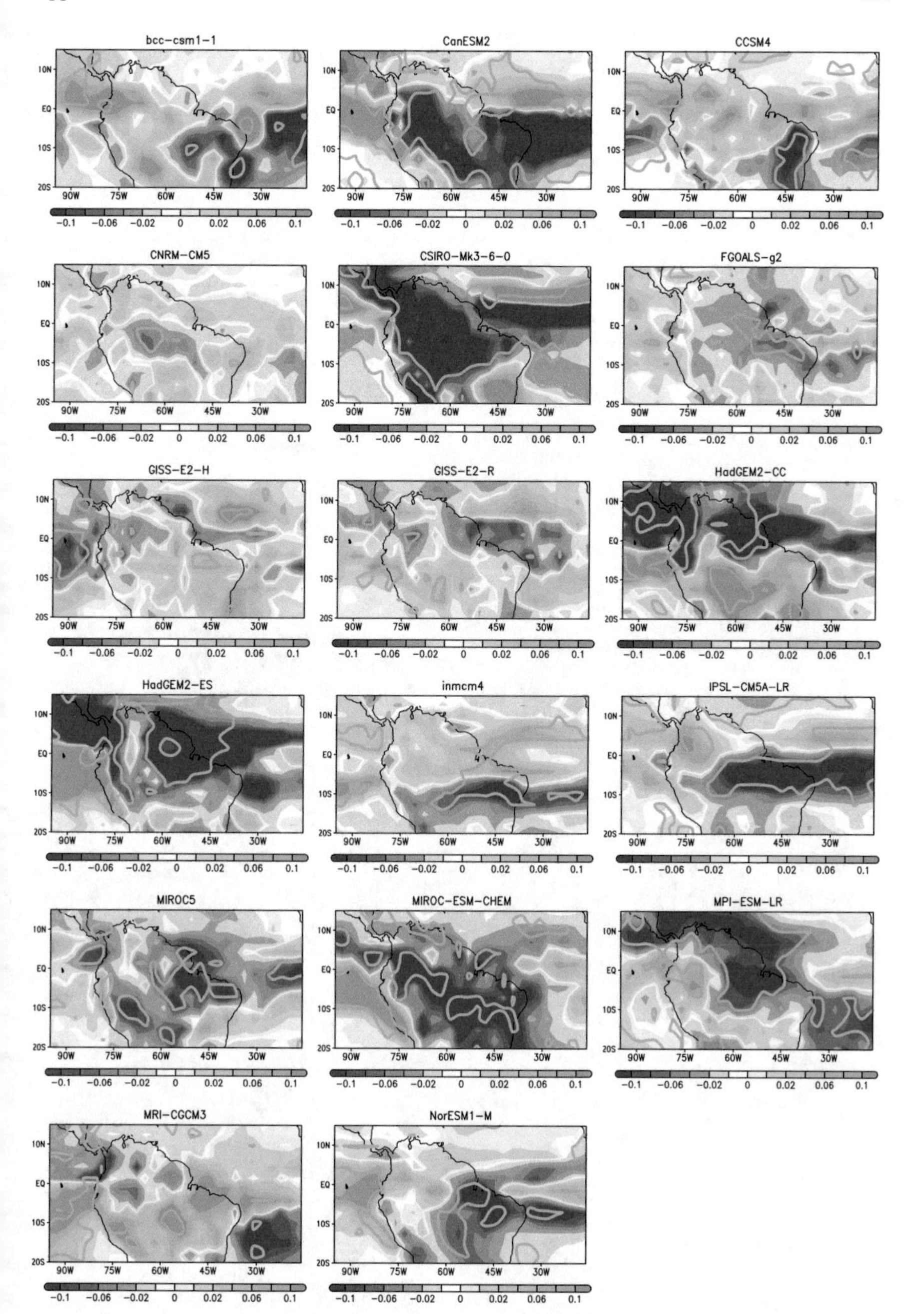

Fig. C.6 Single models regression maps of DJFMAM precipitation anomalies onto the standardized IPO index (mm day^{-1} per standard deviation) from piControl simulations. Grey contours indicate the regions where the correlation is significant at the 5% level from a "random-phase" test

Printed in the United States
By Bookmasters